Results and Problems in Cell Differentiation

Series Editors:
W. Hennig, L. Nover, U. Scheer

23

W0245899

Springer-Verlag Berlin Heidelberg GmbH

S. Kumar (Ed.)

Apoptosis: Biology and Mechanisms

With 13 Figures

 Springer

Dr. Sharad Kumar

Hanson Centre for Cancer Research
Institute of Medical and Veterinary Science
P.O. Box 14, Rundle Mall
Adelaide, SA 5000
Australia

ISSN 0080-1844
ISBN 978-3-662-21623-1

Library of Congress Cataloging-in-Publication Data
Apoptosis : biology and mechanisms / S. Kumar (ed.).
    p.   cm. — (Results and problems in cell differentiation; 23)
    Includes bibliographical references and index.
    ISBN 978-3-662-21623-1   ISBN 978-3-540-69184-6 (eBook)
    DOI 10.1007/978-3-540-69184-6
    1. Apoptosis. I. Kumar, Sharad. II. Series.
    [DNLM: 1. Apoptosis.   W1RE248X v.23   1998]
    QH607.R4   vol. 23
    [QH6714]
    571.8'35 s—dc21
    [571.9'36]

Cover Design: Meta Design, Berlin
Typesetting: Best-set Typesetter Ltd., Hong Kong
SPIN 10571118   39/3137 – 5 4 3 2 1 0 – Printed on acid-free paper

Preface

Apoptosis is currently one of the fastest moving fields in biology with spectacular progress made over the past few years in delineating the molecular mechanisms which underlie this process. It is now indisputable that apoptosis plays an essential role in normal cell physiology and that aberrant apoptosis can manifest itself in a variety of human disorders. Published in two parts (Volumes 23 and 24 of the series entitled *Results and Problems in Cell Differentiation*), this is an attempt to bring together many different aspects of apoptosis. Given that this is such a vast and rapidly expanding field, it is almost impossible to cover everything that is now known about apoptosis in two short books, but I hope these volumes prove to be a guidepost, providing basic essential information on the biology and molecular mechanisms of apoptosis and its implications in some human diseases.

As a significant amount of new information on apoptosis is emerging every week, it is unrealistic to expect that by the time these two books are published, all the articles will deliver up-to-date information. Nevertheless, I believe that the fundamentals of the apoptotic phenomenon are now firmly in place and are discussed at length in various chapters. Readers may find a small degree of overlap between some chapters. This was unavoidable since closely related areas of apoptosis research have been covered by more than one author.

Such an endeavour would not have been possible without the help of many distinguished scientists who contributed the articles assembled in these books. I am very greateful and indebted to all the authors who made considerable efforts to submit their manuscripts as soon as they could. I am also thankful to many other colleagues and members of my laboratory for various suggestions. I have thoroughly enjoyed reading various contributions and have learnt a great deal in the process of compiling these volumes. I hope the readers will find these books a useful resource for both teaching and research purposes.

Adelaide, May 1998   *Sharad Kumar*

Contents

Lipid and Glycolipid Mediators in CD95-Induced Apoptotic Signaling
F. Malisan, M. R. Rippo, R. De Maria and R. Testi

Lymphocyte-Mediated Cytolysis:
Dual Apoptotic Mechanisms with Overlapping Cytoplasmic
and Nuclear Signalling Pathways
J. A. Trapani and D. A. Jans

Granule-Mediated Cytotoxicity
A. J. Darmon, M. J. Pinkoski and R. C. Bleackley

The Cell Cycle and Apoptosis
H. J. M. Brady and G. Gil-Gómez

The p53 Tumor Suppressor Gene:
Structure, Function and Mechanism of Action
C. Choisy-Rossi, Ph. Reisdorf and E. Yonish-Rouach

The Bcl-2 Protein Family
L. O'Connor and A. Strasser

Apoptosis and the Proteasome
L. M. Grimm and B. A. Osborne

A Personal Account of Events Leading to the Definition of the Apoptosis Concept

John F. R. Kerr

1
Introduction

In this chapter I will describe the sequence of events that culminated in the proposal of the apoptosis concept (Kerr et al. 1972). Since it is now many years since these events took place, it is possible that memory has become coloured by subsequent insights. The main steps in the story were, however, recorded in publications written at the time.

2
Delineation of Two Types of Cell Death with Distinctive Lysosomal Changes

I first developed a special interest in cell death when I went to London in 1962 as a young Queensland medical graduate specializing in pathology to undertake a PhD under the supervision of Professor Sir Roy Cameron in the Department of Morbid Anatomy, University College Hospital Medical School. Cameron was born and educated in Australia and, perhaps because of this, he actively fostered academic links with British Commonwealth countries; a long procession of young pathologists from the Commonwealth obtained their research training in his Department. He suggested that I study the effects on liver tissue of interrupting its portal venous blood supply, repeating experiments conducted many years previously by Peyton Rous (Rous and Larimore 1920), the same Peyton Rous who had subsequently become famous for his discovery of virus induction of tumours. He and Larimore had shown that obstruction of portal vein branches supplying several lobes of the liver resulted in rapid and marked shrinkage of these lobes with simultaneous compensatory enlargement of the rest of the organ. I therefore ligated the portal vein branches supplying the left and median lobes of the liver in rats and studied the resulting changes microscopically (Kerr 1965).

The shrinkage of the ischaemic tissue was found to be due to two distinct processes. Firstly, within a few hours of operation, confluent necrosis devel-

29 Hipwood Road, Hamilton, Brisbane, Queensland 4007, Australia

oped in circumscribed areas around the terminal hepatic venules (that is, in areas furthest from the blood supply) and these groups of necrotic cells were removed by mononuclear phagocytes, the involved areas undergoing collapse. Secondly, in the periportal parenchyma, which remained viable as a result of still receiving a blood supply from branches of the hepatic artery, scattered individual liver cells were progressively deleted by a process that was morphologically quite different from necrosis. The affected cells were converted into small round or ovoid cytoplasmic masses, which often contained one or more specks of condensed nuclear chromatin. These masses were phagocytosed by Kupffer cells or by intact parenchymal epithelial cells. The process was prominent during the period of rapid shrinkage of the ischaemic lobes, and thereafter decreased as the lobes approached a new equilibrium with their residual blood supply. Importantly, the same process was found to occur at a very low rate in the livers of healthy rats; a statement to this effect is buried rather inconspicuously on page 422 of the 1965 paper.

At the time I was starting my experiments, there was a good deal of interest in the possible involvement of lysosomes in the production of cell death following various types of injury (de Reuck and Cameron 1963). These organelles had been defined only a few years previously by de Duve and his colleagues and it had been suggested that injurious agents might cause release of their digestive enzymes, thus killing the cell, an idea encapsulated in the term "suicide bag". Professor J. F. Smith, Cameron's deputy, introduced me to recently developed histochemical methods for demonstrating the distribution of lysosomal enzymes in tissues and I applied these to the rat livers (Kerr 1965). In normal animals, strings of discretely stained lysosomes were observed in the paracanalicular cytoplasm of the parenchymal cells, these organelles being normally concentrated in paracanalicular locations. In the areas of confluent necrosis, on the other hand, the affected parenchymal cells showed diffuse paracanalicular staining, which did indeed suggest release of lysosomal enzymes. The diffuse staining was, however, not an early change and it seemed likely that lysosome rupture was merely part of the general cellular degeneration that accompanies necrosis rather than being the initiating event in the production of cell death. But the most interesting finding was that the lysosomes in the small round and ovoid cytoplasmic masses were stained discretely by several different histochemical methods, indicating that they were still intact. Moreover, histochemical staining by appropriate methods suggested that ribosomes and mitochondria were also preserved in the rounded cytoplasmic masses. It was proposed that the cellular shrinkage that leads to their formation might be effected by autophagocytosis, with cytoplasmic components being progressively digested within the cell's own membrane-bounded lysosomes. This hypothesis was later refuted by electron microscopy (Kerr 1969).

At this time it seemed clear that the small cytoplasmic masses represented a distinctive type of cell death, which differed from classical necrosis in its histological appearance, in not being degenerative in nature, in affecting only scattered single cells and in not being accompanied by inflammation. The name shrinkage necrosis was suggested for it (Kerr 1965).

At the beginning of 1965 I returned to Brisbane and took up a position in the University of Queensland Pathology Department. My first project was to apply the histochemical methods for lysosomal enzymes that I had used in London to the livers of rats given the pyrrolizidine alkaloid heliotrine, an hepatotoxic agent that produces zonal necrosis rather like that seen in the liver lobes deprived of their portal venous blood supply (Kerr 1967). Cells undergoing shrinkage necrosis were numerous in the viable parenchyma between the areas of necrosis. The staining patterns for lysosomal enzymes in the two types of cell death were found to be the same as those seen previously in ischaemic injury (Kerr 1967).

3
Definition of the Sequence of Ultrastructural Events Involved in Shrinkage Necrosis Occurring in the Liver

In 1967, the Queensland University Pathology Department acquired its first electron microscope and I embarked on a systematic study of the ultrastructural events involved in the evolution of shrinkage necrosis in the rat liver with the help first of David Collins and later of Brian Harmon (Kerr 1969, 1970, 1971, 1973). The stereotyped morphological sequence that emerged at that time was subsequently confirmed in many other tissues. The appearances were noted to be consistent with several recently published electron microscopic studies of so-called acidophilic or Councilman-like bodies found in the liver in naturally occurring diseases (Biava and Mukhlova-Montiel 1965; Klion and Schaffner 1966; Moppert et al. 1967). These latter bodies clearly represented the same process as I had referred to as shrinkage necrosis in 1965.

In my electron microscopic studies of rat liver I found that rounded bodies that still lay free in the extracellular space comprised membrane-bounded fragments of condensed parenchymal cell cytoplasm in which the closely packed organelles were well preserved. Some of these bodies contained masses of condensed chromatin, which only occasionally appeared to be surrounded by membranes. This lack of bounding membranes was subsequently shown to be an artefact resulting from the electron microscopic preparative techniques used in those days. It is now known that nuclear fragmentation in apoptosis is associated with preservation of the nuclear envelope (Kerr et al. 1995). The earliest nuclear changes, with condensation and margination of chromatin prior to fragmentation, were described in these early liver studies, but were seen only infrequently. The fact that the condensed cytoplasmic masses often occurred in clusters, taken in conjunction with the extremely small size of some of the masses, indicated that they arose by a process of budding-off of protuberances that developed on the surface of condensing cells. The actual process of budding was, however, rarely observed. This was correctly interpreted as indicating that it occurs very quickly (Kerr 1969), a conclusion that appeared to be supported by the time-lapse microcinematographic observations of cell death taking place in vitro that had been made by Marcel Bessis (1964). Bessis's classical contributions to the understanding of cell death have recently been reviewed by Majno and Joris (1995). Phagocytosis of the condensed, membrane-bounded liver cell fragments by Kupffer and parenchymal epithelial cells was confirmed and their progressive degradation within phagolysosomes was followed by electron microscopy.

In a paper submitted in November 1970 (Kerr 1971) it was suggested that, whilst severe damage to a tissue causes classical necrosis, a moderately noxious environment induces scattered cells to undergo shrinkage necrosis. Secondly it was stated that shrinkage necrosis constitutes one type of cell death occurring in normal tissues. Thirdly it was concluded that the prolific cellular budding that occurs in shrinkage necrosis is likely to be the result of inherent activity of the cells themselves. However, it was noted that the involvement of only scattered, individual cells by the process would make its biochemical analysis difficult.

4
The Occurrence of Shrinkage Necrosis in Tumours

About the middle of 1970, I attended a seminar on tumours. I remember the speaker discussing the slow growth rate of basal cell carcinomas of human skin, which appeared paradoxical in view of the high rate of mitosis observed within them. I was reminded of the fact that Jeffrey Searle, who was then training as a pathologist at the Royal Brisbane Hospital, had recently pointed out to me that cells with the histological features of shrinkage necrosis were often numerous in these tumours. We decided to undertake a light and electron microscopic study of surgically excised basal cell carcinomas, which are particularly common in Queensland.

In a paper submitted in July 1971 (Kerr and Searle 1972a) we confirmed that shrinkage necrosis could be found in all basal cell carcinomas and that it was often extensive. The ultrastructural appearances were similar to those seen in the liver. Many of the condensed cell fragments were taken up and digested by tumour cells.

When we looked at the literature, we found that it had recently become apparent that there is a marked discrepancy between the rate of growth of a variety of malignant tumours and the rate of division of their constituent cells. On the basis of these findings, a number of investigators had concluded that spontaneous loss of cells is often a significant determinant of tumour growth rate. However, the mechanism of the loss was not understood.

We found that the extent of shrinkage necrosis in parts of some basal cell carcinomas was comparable to that seen in the liver lobes deprived of their portal blood supply; the latter decreased to about one-sixth of their original weight within 8 days of operation. Cellular deletion by shrinkage necrosis in basal cell carcinomas was thus likely to be very significant kinetically. Further, we found that extensive shrinkage necrosis could also be detected by light microscopy in other types of malignant tumours. We suggested that this process makes a major contribution to cell loss occurring in untreated tumours in general (Kerr and Searle 1972a). We also speculated about its possible causes. Whilst the fact that it was often prominent near the centres of large tumour nodules suggested that it was caused by mild ischaemia, we sometimes saw it in thin tumour trabeculae that were no more than two cells thick. Significant ischaemia was unlikely to be present in such trabeculae. By this time we had seen shrinkage necrosis in many of the tissues of healthy animals and had concluded that it is involved in normal cellular turnover in these tissues (Kerr and Searle 1972a). We quoted the seminal statement by Laird (1969) that death of both normal and neoplastic cells may be a pre-ordained, genetically determined phenomenon.

In a second study (Kerr and Searle 1972b) we confirmed by electron microscopic histochemistry that fragments of cells that had undergone shrinkage necrosis in basal cell carcinomas and that had been phagocytosed by viable tumour cells were degraded within lysosomes. We also reported preliminary observations indicating that the extent of shrinkage necrosis in squamous cell carcinomas of human skin increases in response to radiotherapy.

5
Proposal of the Apoptosis Concept

In late 1970, Professor A. R. (subsequently Sir Alastair) Currie, who at that time was Head of the Department of Pathology in the University of Aberdeen, Scotland, came to Brisbane for a month as a guest professor in the University of Queensland. I showed him electron micrographs of shrinkage necrosis occurring in the liver and discussed the plans to study basal cell carcinomas. He was intensely interested.

Currie had a major interest in endocrine pathology and had long mused upon the reversible increase and decrease in the size of endocrine-dependent tissues that follows changes in blood levels of trophic hormones; he had studied the regression of rat breast tumours induced by 9,10-dimethyl-1,2-benzanthracene (DMBA) that often follows surgical removal of the ovaries; and with Andrew Wyllie, who had recently commenced PhD studies under his guidance, he had observed scattered, single, dead cells by light microscopy in the adrenal cortices of animals in which adrenocorticotrophic hormone (ACTH) secretion had been interrupted (Wyllie 1994–1995). He was excited by the possibility that the ultrastructurally distinctive type of cell death I had shown him might be regulated by trophic hormones in endocrine-dependent tissues. I was due to take study leave from the University of Queensland the following year. He suggested that I spend it in his Department in Aberdeen.

As a side issue during his treatment of rats with DMBA to induce breast tumours, Currie had noticed single cell death in the adrenal cortices of some animals; in others there was confluent adrenal cortical necrosis. At his suggestion, I studied these lesions before going to Aberdeen. The single cell death was found to display the ultrastructural features of shrinkage necrosis (Kerr 1972).

Following my arrival in Aberdeen in September 1971, electron microscopy was performed on tissues obtained from several series of animal experiments that had been under way in the Pathology Department there for some time. In each case, single cell death with the ultrastructural features of shrinkage necrosis was found to occur.

Andrew Wyllie had shown that decrease in the size of the adrenal cortex that follows experimental interruption of ACTH secretion in both adult and fetal rats is accompanied by death of many scattered epithelial cells in the inner part of the cortex, and that similar cell death also takes place in normal neonatal rats, where adrenal cortical shrinkage is associated with a physiological decrease in ACTH secretion. In all cases, he had found that the cell death was prevented by coincident injection of ACTH. Two papers describing the occurrence and the ultrastructural features of the adrenal cell death (by then referred to as apoptosis) were returned by the editors of endocrinological journals with scathing referees' comments (Wyllie 1994–1995); they were subsequently published in the *Journal of Pathology* (Wyllie et al. 1973a,b).

Histological sections and blocks of tissue that had been processed for electron microscopy were available from the experiments carried out by Currie and his colleagues on DMBA-induced rat breast tumours. These were restudied. Large numbers of tumour cells were observed to undergo shrinkage necrosis when regression followed removal of the ovaries (Kerr et al. 1972).

At the time of my visit to Aberdeen, Allison Crawford, a developmental biologist, was working in the Pathology Department on the teratogenic effects of 7-hydroxymethyl-12-methylbenz(a)anthracene, one of the principal metabolites of DMBA. Injection of this substance in Sprague–Dawley rats on day 11–14 of pregnancy resulted in the occurrence of encephalocoele and spina bifida in the mature fetuses. She had found that these developmental defects were explicable on the basis of tissue deletion resulting from massive but localized single cell death occurring within 24 h of injection of the teratogen. Electron microscopy performed on her experimental fetuses showed that the ultrastructural features of this death were those of shrinkage necrosis (Crawford et al. 1972). Of perhaps even greater importance to the evolution of the apoptosis concept, however, was the fact that she introduced us to the literature on the occurrence of cell death during normal embryonic and fetal development; precisely controlled, localized cell death had long been known to be involved in a variety of developmental processes (Glücksmann 1951; Saunders 1966). At that time, however, most pathologists, and indeed many general biologists, were unaware of the phenomenon. Examination of published electron micrographs of normal developmental cell death (Saunders and Fallon 1966; Farbman 1968; Webster and Gross 1970) indicated to us that it is morphologically essentially the same as shrinkage necrosis occurring in postnatal life.

The stage was now set for the formulation of the apoptosis concept (Kerr et al. 1972). Here we had a distinctive type of cell death with features suggesting an active process of cellular self-destruction, not degeneration as in classical necrosis. It occurred in normal adult tissues, where it was probably implicated in cellular turnover under steady state conditions, and it was involved in both physiological involution and pathological atrophy of tissues; in endocrine-dependent tissues it was subject to hormonal control. It subserved a variety of essential morphogenetic functions during normal development. It occurred spontaneously in malignant tumours, often substantially retarding their growth, and it was enhanced in tumours responding to at least some types of therapy. In many situations, it clearly played an opposite role to mitosis in regulating tissue size. That it could be induced by certain injurious agents that were also capable of causing necrosis, however, posed an enigma which was not resolved.

The term apoptosis was suggested by Professor James Cormack of the Department of Greek, University of Aberdeen. It was used in classical Greek to describe the falling of leaves from trees. It seemed to encapsulate many of the ideas inherent in the apoptosis concept. Cormack advised us that the second "p" should not be pronounced. Modern Greek speakers disagree (Georgatsos 1995).

While we were finishing the paper for the *British Journal of Cancer*, Alastair Currie accepted an invitation to head the Department of Pathology in the University of Edinburgh Medical School and moved there in April 1972. Andrew Wyllie followed him in October (Wyllie 1994–1995). I returned to Brisbane. Progress in the field, which was still very modest, was reviewed by the three of us when I spent another period of study leave in Scotland in 1979 (Wyllie et al. 1980). Sadly, Alastair Currie died in January 1994.

6
Why was Apoptosis Neglected for so Long?

It is now apparent that we were not the first to recognize the distinctive morphology of apoptosis or to appreciate its kinetic significance. Majno and Joris (1995) have drawn attention to the fact that German anatomists described the light microscopic features of what we called apoptosis around the turn of last century and clearly understood the role of the process in regulating tissue size. Thus, in 1885, Walther Flemming, who also coined the terms chromatin and mitosis, published beautiful camera lucida drawings of typical apoptosis occurring during normal regression of ovarian follicles. He called it chromatolysis. In 1914, another German anatomist, Ludwig Gräpper, pointed out that a mechanism must exist to counterbalance mitosis and proposed Flemming's chromatolysis as being the answer. Some of the figures from these early German papers have been reproduced by Majno and Joris (1995); they are well worth looking at.

Remarkably, these profound insights had no general impact and, except in the field of developmental biology, normal cell death, in sharp contrast to mitosis, was ignored for many years. It is uncertain why this was so. Perhaps there is a natural human tendency to look at the positive rather than the negative side of life processes. Apoptosis is a cryptic phenomenon morphologically, and the small size of many apoptotic bodies, the speed with which they are disposed of by adjacent cells and the absence of exudative inflammation would all have militated against detection of the process in most normal adult tissues, where its rate of occurrence is low. It is only in a few locations, such as the centres of reactive lymphoid follicles, that apoptosis in healthy adult animals is so extensive that it could not be ignored by early histologists; the apoptotic cells in this situation were traditionally referred to as tingible bodies. Likewise, the locally massive extent of apoptosis in parts of the embryo undoubtedly contributed to the early recognition of the importance of cell death during normal development. Lastly, some of the light microscopic changes seen within dying cells may have inhibited recognition of the distinctive nature of apoptosis; the fact that chromatin condensation can occur in necrosis as well as being a characteristic feature of apoptosis probably led to confusion. It was only with the advent of electron microscopy that it was widely realized that the details of the chromatin condensation are different in the two types of cell death and that the coincident cytoplasmic changes are radically dissimilar.

7
Recent Accounts of the History of Apoptosis

Two different perspectives of the history of apoptosis have recently been published in the journal *Cell Death and Differentiation*. Richard Lockshin (1997) has given a qualitative account, whereas Garfield and Melino (1997) have traced the development of the field by applying bibliometric techniques to quantitative data on rates of publication and of citation of publications available through the Institute for Scientific Information in Philadelphia. The two papers complement one another nicely.

References

Bessis M (1964) Studies on cell agony and death: an attempt at classification. In: de Reuck AVS, Knight J (eds) Ciba Foundation symposium on cellular injury. Churchill, London, pp 287–316

Biava C, Mukhlova-Montiel M (1965) Electron microscopic observations on Councilman-like acidophilic bodies and other forms of acidophilic changes in human liver cells. Am J Pathol 46:775–802

Crawford AM, Kerr JFR, Currie AR (1972) The relationship of acute mesodermal cell death to the teratogenic effects of 7-OHM-12-MBA in the foetal rat. Br J Cancer 26:498–503

de Reuck AVS, Cameron MP (eds) (1963) Ciba Foundation symposium on lysosomes. Churchill, London

Farbman AI (1968) Electron microscope study of palate fusion in mouse embryos. Dev Biol 18:93–116

Garfield E, Melino G (1997) The growth of the cell death field: an analysis from the ISI-Science citation index. Cell Death Differ 4:352–361

Georgatsos JG (1995) The s(p)elling of apo(p)tosis. Nature 375:100

Glücksmann A (1951) Cell deaths in normal vertebrate ontogeny. Biol Rev 26:59–86

Kerr JFR (1965) A histochemical study of hypertrophy and ischaemic injury of rat liver with special reference to changes in lysosomes. J Pathol Bacteriol 90:419–435

Kerr JFR (1967) Lysosome changes in acute liver injury due to heliotrine. J Pathol Bacteriol 93:167–174

Kerr JFR (1969) An electron-microscope study of liver cell necrosis due to heliotrine. J Pathol 97:557–562

Kerr JFR (1970) An electron microscopic study of liver cell necrosis due to albitocin. Pathology 2:251–259

Kerr JFR (1971) Shrinkage necrosis: a distinct mode of cellular death. J Pathol 105:13–20

Kerr JFR (1972) Shrinkage necrosis of adrenal cortical cells. J Pathol 107:217–219

Kerr JFR (1973) Some lysosome functions in liver cells reacting to sublethal injury. In: Dingle JT (ed) Lysosomes in biology and pathology 3. Frontiers of biology, vol 29. North-Holland, Amsterdam, pp 365–394

Kerr JFR, Searle J (1972a) A suggested explanation for the paradoxically slow growth rate of basal-cell carcinomas that contain numerous mitotic figures. J Pathol 107:41–44

Kerr JFR, Searle J (1972b) The digestion of cellular fragments within phagolysosomes in carcinoma cells. J Pathol 108:55–58

Kerr JFR, Wyllie AH, Currie AR (1972) Apoptosis: a basic biological phenomenon with wide-ranging implications in tissue kinetics. Br J Cancer 26:239–257

Kerr JFR, Gobé GC, Winterford CM, Harmon BV (1995) Anatomical methods in cell death. In: Schwartz LM, Osborne BA (eds) Cell death. Methods in cell biology, vol 46. Academic Press, San Diego, pp 1–27

Klion FM, Schaffner F (1966) The ultrastructure of acidophilic "Councilman-like" bodies in the liver. Am J Pathol 48:755–767

Laird AK (1969) Dynamics of growth in tumors and in normal organisms. In: Perry S (ed) Human tumor cell kinetics. National Cancer Institute monograph, no 30. National Cancer Institute, Bethesda, pp 15–28

Lockshin RA (1997) The early modern period in cell death. Cell Death Differ 4:347–351

Majno G, Joris I (1995) Apoptosis, oncosis and necrosis. An overview of cell death. Am J Pathol 146:3–15

Moppert J, v Ekesparre D, Bianchi L (1967) Zur Morphogenese der eosinophilen Einzelzellnekrose im Leberparenchym des Menschen. Eine licht- und elektronenoptisch korrelierte Untersuchung. Virchows Arch Pathol Anat 342:210–220

Rous P, Larimore LD (1920) Relation of the portal blood to liver maintenance: a demonstration of liver atrophy conditional on compensation. J Exp Med 31:609–632

Saunders JW (1966) Death in embryonic systems. Science 154:604–612

Saunders JW, Fallon JF (1966) Cell death in morphogenesis. In: Locke M (ed) Major problems in developmental biology. Academic Press, New York, pp 289–314

Webster DA, Gross J (1970) Studies on possible mechanisms of programmed cell death in the chick embryo. Dev Biol 22:157–184

Wyllie A (1994–1995) The patient and the cell. An appreciation of Sir Alastair Currie. In: Cancer Research Campaign scientific yearbook. Cancer Research Campaign, London, pp 6–9

Wyllie AH, Kerr JFR, Macaskill IAM, Currie AR (1973a) Adrenocortical cell deletion: the role of ACTH. J Pathol 111:85–94

Wyllie AH, Kerr JFR, Currie AR (1973b) Cell death in the normal neonatal rat adrenal cortex. J Pathol 111:255–261

Wyllie AH, Kerr JFR, Currie AR (1980) Cell death: the significance of apoptosis. Int Rev Cytol 68:251–306

Molecular Mechanisms of Apoptosis: An Overview

Anne M. Verhagen and David L. Vaux

1
Introduction

Apoptosis (physiological cell death) has a central role in the normal development and homeostastis of all multicellular organisms. It is also used by the
body's defence system to eliminate dangerous cells, such as those that are
mutated or are harbouring viruses. Deregulation of this process, resulting in
either too much or too little cell death, can cause both developmental defects
and a wide variety of disease states. Much of our understanding of apoptosis
has come from genetic studies in the nematode *Caenorhabditis elegans*. Although the mechanisms of apoptosis are highly conserved, regulation of
apoptosis in higher species is more complicated and involves several large
families of proteins. In this chapter we will provide an overview of the molecular mechanisms of physiological cell death.

2
The Genetic Framework of Apoptosis Revealed
by *C. elegans*

Although the first cell death gene to be identified, *bcl-2*, was a mammalian
gene, much of the genetic framework of the process of apoptosis was learnt
from the study of the nematode *C. elegans*, in which cell death mutants can
easily be generated. Here, we have continued to use the genetic pathway of
programmed cell death in the worm as a basis for understanding apoptosis in
mammalian systems, as they have been shown to be implemented by the same
highly conserved process.

During normal development of the *C. elegans* hermaphrodite, exactly 131 of
the 1090 somatic cells formed undergo apoptosis. Two genes, *ced-3* and *ced-4*,
are essential for all developmental programmed cell death in *C. elegans* as the
cells which are normally destined to die survive in worms with loss of function

Walter and Eliza Hall Institute of Medical Research, Post Office Royal Melbourne Hospital,
Parkville, Victoria 3050, Australia

mutations in either of these genes (Ellis and Horvitz 1986). Ced-3 encodes a cysteine protease or "caspase" for which there are multiple mammalian homologues (Yuan et al. 1993; reviewed by Henkart 1996). The caspases are key death effector molecules and their activation, which can be induced in a multitude of ways, is required for cell death.

The CED-4 protein appears to function as an adaptor protein that activates the CED-3 caspase precursor. These two proteins bind to each other via a shared caspase recruitment domain (CARD) in the N terminus of CED-4 and the pro-domain of CED-3 (Hofmann et al. 1997; Irmler et al. 1997; Chinnaiyan et al. 1997).

Negative regulation of apoptosis in C. elegans is controlled by the product of the gene ced-9 (Hengartner et al. 1992; Shaham and Horvitz 1996). Genetically, CED-9 acts upstream of CED-3 and CED-4, as loss of function mutations in the ced-9 gene result in excessive cell death during development and embryonic lethality, and these excessive deaths can be suppressed by mutation of either ced-4 or ced-3. Conversely, a gain of function mutation in the ced-9 gene results in no cell death during development.

As CED-9 can bind to CED-4, its protective properties are most likely due to inhibition of CED-4 activity, preventing it from activating the caspase CED-3 (Chinnaiyan et al. 1997; Spector et al. 1997; Wu et al. 1997). CED-9 may act by sequestering CED-4 from the cytosol and relocating it to the intracellular membranes (Wu et al. 1997). CED-9 does not appear to function by competition with CED-3 for CED-4 binding since simultaneous interaction of both molecules with CED-4 has been demonstrated (Chinnaiyan et al. 1997).

There are nine known mammalian homologues of CED-9. As one of these, bcl-2, can inhibit programmed cell death in C. elegans (Vaux et al. 1992), and rescue a ced-9 mutant worm (Hengartner and Horvitz 1994), it is likely that mammalian bcl-2 family proteins act just like CED-9, namely to inhibit apoptosis by preventing caspase activation caused by mammalian adaptor proteins that function analogously to CED-4. To date, one mammalian homologue for the ced-4 gene (Apaf-1) has been described which seems to be required for the activation of caspase-3 in a cytochrome c-dependent manner (Zou et al. 1997).

Genetic studies in C. elegans have also shown that multiple genes are needed for efficient engulfment of the dead cell corpse and degradation of the dead cell's DNA (Ellis et al. 1991). These genes, which presumably encode "eat-me" signals on the dead cell and receptors on the engulfing cell, allow efficient recognition and removal of apoptotic bodies. One gene, nuc-1, encodes a nuclease that may function analogously to the endonuclease that causes DNA degradation ("laddering") in apoptotic mammalian cells (Hedgecock et al. 1983).

3
Apoptosis in Mammalian Cells

3.1
The Caspases

As in *C. elegans,* the key effector proteins of apoptosis in mammalian cells appear to be caspases, cysteine proteases that have an aspartate specificity.

The first identified caspase was IL-1β converting enzyme (ICE) which, as the name suggests, was not identified by virtue of its death-promoting activity but by its ability to cleave the inactive precursor of IL-1β into an active cytokine (Cerretti et al. 1992). Indeed, the apoptotic properties of ICE (now known as caspase-1) were not recognized until after the characterization of the *C. elegans* gene *ced-3,* which bore striking homology to ICE (Yuan et al. 1993).

Ten mammalian caspases have been identified to date (reviewed by Henkart 1996). Caspases contain a catalytic cysteine residue located within a conserved QACRG motif that is involved in interaction with and enzymatic cleavage of substrates. All caspases cleave after aspartate residues but vary in their specificity for preceding amino acids as well as the overall tertiary structure of the substrate polypeptide. For example, while caspase-1 preferentially cleaves after YVAD sequences, another mammalian caspase CPP32 or caspase-3, similarly to the CED-3 protein, cleaves preferentially after DEVD.

All caspases are produced as inactive precursor proteins that must be processed by proteolytic cleavage into the active enzyme. The processing of caspases is best defined for caspase-1 (Walker et al. 1994; Wilson et al. 1994). From the 45-kDa caspase-1 precursor, the pro-domain is removed and the remaining protein processed into a p10 and p20 subunit. Two of each of these subunits are assembled into the active heterotetramer. Four cleavage events are required to yield active ICE, and each occurs after an aspartate residue.

The fact that caspase-1 is required for generating the pro-inflammatory cytokine IL-1β illustrates that, contrary to earlier beliefs, apoptosis does not always occur quietly and in the absence of inflammation (Cerretti et al. 1992). It is possible that some caspases will be used to cause apoptosis in normal circumstances such as development, when inflammation is not desirable. On the other hand, others, such as ICE, may be used in cases such as defence against viruses, when inflammation and involvement of the immune system is helpful.

Gene deletion (knockout) experiments in mice have shown that mice lacking the gene for caspase-1 develop normally but cannot produce active IL-1β (Kuida et al. 1995; Li et al. 1995). Mice lacking the gene for caspase-3 are born with giant brains, due to failure of excess neuronal cells to die normally during development (Kuida et al. 1996).

It is likely that caspases are not the only important proteases in apoptosis. Studies with protease inhibitors have implicated other proteases including other cysteine proteases, aspartic proteases, serine proteases and the proteosome (Shi et al. 1992; Deiss et al. 1996; Grimm et al. 1996; Martin et al. 1996). These enzymes may function upstream of the caspases, possibly to activate them, or downstream of the caspases to cause postmortem effects.

3.2
Caspase Substrates

Gene deletion studies in *C. elegans* and in mice have shown that certain caspases are required for apoptosis. These caspases must cleave polypeptide substrates within the cell to cause the morphological changes recognizable as apoptosis. The nature of these substrates is an area of intense investigation (see Chapters in Volume 24 of this series).

Some caspases are known to be able to cleave and thereby activate the precursors of other caspases. For example, caspases-2, -6, -8 and -10 can cleave and activate pro-caspase-3 (Fernandesalnemri et al. 1996; Liu et al. 1996; Srinivasula et al. 1996; Xue et al. 1996; Muzio et al. 1997). Some active caspases can activate their own precursors, for example caspase-1 can cleave pro-caspase-1 (Molineaux et al. 1993). In this way it is possible that the caspases act in a hierarchical fashion resulting in an avalanche of protease activity.

A number of other substrates of the caspases have been identified that are cleaved during apoptosis. So far, the only one that is thought to have its main role in cell death is DNA fragmentation factor (DFF; Liu et al. 1997). DFF is a heterodimer that is activated following its cleavage by caspases, and is required for subsequent activation of the endonucleases that cause the DNA fragmentation that usually accompanies apoptosis.

The growing list of other proteins that can be cleaved by caspases includes poly-ADP-ribose polymerase (PARP), DNA protein kinase, huntingtin, fodrin, lamins A and B, protein kinase A, protein kinase C, U1 ribonuclear protein, actin and retinoblastoma protein to name but a few (Emoto et al. 1995; Lazebnik et al. 1994, 1995; Casciolarosen et al. 1996; Goldberg et al. 1996). There is currently controversy over whether one or a few of these substrates must be cleaved for apoptosis to occur, or whether many proteins are cleaved during apoptosis but none has special significance.

Even when it is known that these proteins play no essential role in apoptosis, cleavage of particular proteins can be used as a marker of caspase activation and apoptosis. For example, PARP is not essential for apoptosis as PARP knockout (KO) mice are normal, but detection of cleaved PARP has been used as an indicator that caspases such as CPP32 have been activated (Wang et al. 1995).

3.3
Adaptor Proteins: The Caspase Activators

Although the only mammalian homologue of CED-4, Apaf-1, has not yet been shown to act as an adaptor molecule, a number of adaptor molecules have been found that function as caspase activators. Like CED-4, these proteins are thought to bind to the pro-domain of the caspase precursors via homotypic interaction motifs, ultimately leading to cleavage and activation of the caspase.

The first mammalian cell death adaptor molecule to be identified was MORT-1/FADD (Boldin et al. 1995; Chinnaiyan et al. 1995; Kischkel et al. 1995). FADD possesses a motif known as the death effector domain (DED) which interacts with DEDs in the pro-domain of caspase-8 (Boldin et al. 1996; Muzio et al. 1996). Subsequent proteolytic processing either by autoprocessing or by another protease releases the active form of caspase-8 from the receptor complex (Medema et al. 1997).

The other end of FADD bears another interaction motif termed a "death domain" (DD) which allows it to bind to signalling molecules and receptors sending the death signal. FADD therefore acts as an adaptor molecule by coupling the death signalling molecules to the caspases via its DD and DED domains.

Another adaptor molecule is RAIDD, which, like FADD, has a death domain (DD) at one end (Duan and Dixit 1997; Hofmann et al. 1997). The other end of RAIDD does not bear a DED domain, but has a "caspase recruitment domain" (CARD). The CARD motif in RAIDD allows it to bind the CARD motif in the pro-domain of caspase-2 (Nedd-2) in a manner analogous to the binding of the DED domain of FADD to the DED domains in caspase-8. CARD domains have also been identified in the pro-domains of ICE and *ced-3*.

3.4
Generation of the Death Signal

The signal for cell death can be generated in a vast number of ways (Fig. 1). In normal circumstances it is often generated by the addition or removal of cytokines such as tumour necrosis factor (TNF) or growth factors. Death signalling pathways activated by members of the TNF receptor family of receptors, which includes CD95 (Fas/Apo-1), are understood in the greatest detail (Fig. 1). The cytoplasmic domain of CD95 possesses a DD. Ligand binding induces the DD in the cytoplasmic domain of CD95 to associate with the DD in FADD (Kischkel et al. 1995). Ligand binding is thus able to stimulate the formation of a death signalling complex that recruits both FADD and caspase-8, which then becomes activated (Boldin et al. 1996; Muzio et al. 1996; Medema et al. 1997).

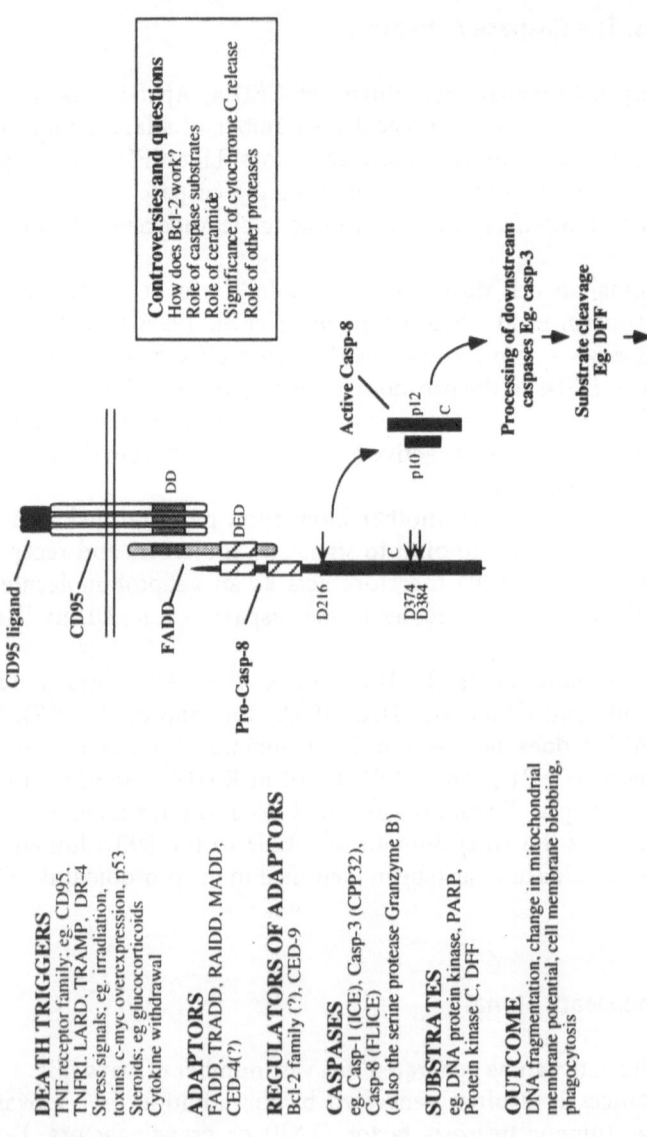

PATHWAY EXAMPLE

Controversies and questions
How does Bcl-2 work?
Role of caspase substrates
Role of ceramide
Significance of cytochrome C release
Role of other proteases

CD95 ligand

CD95

DD

FADD

DED

Pro-Casp-8

Active Casp-8

p10   p12

D216

D374
D384

C   C

Processing of downstream
caspases Eg. casp-3

Substrate cleavage
Eg. DFF

endonuclease activation

DNA fragmentation

DEATH TRIGGERS
TNF receptor family; eg. CD95,
TNFR1, LARD, TRAMP, DR-4
Stress signals; eg. irradiation,
toxins, c-myc overexpression, p53
Steroids; eg glucocorticoids
Cytokine withdrawal

ADAPTORS
FADD, TRADD, RAIDD, MADD.
CED-4(?)

REGULATORS OF ADAPTORS
Bcl-2 family (?), CED-9

CASPASES
eg. Casp-1 (ICE), Casp-3 (CPP32),
Casp-8 (FLICE)
(Also the serine protease Granzyme B)

SUBSTRATES
eg. DNA protein kinase, PARP,
Protein kinase C, DFF

OUTCOME
DNA fragmentation, change in mitochondrial
membrane potential, cell membrane blebbing,
phagocytosis

Fig. 1. Molecular mechanisms of apoptosis, this figure outlines the different effector molecules involved in implement-
ing and regulating apoptosis. An example of an apoptotic pathway, CD95-mediated apoptosis, is shown (*right*) and some
controversial areas are highlighted within the *box*

Many cells are dependent on growth factors, and undergo apoptosis when they are removed. The molecular mechanism involved in the regulation of cell survival by growth factors is not known in any detail, but it is known that apoptosis occurs via caspases, and in most cases cell death can be delayed or inhibited by *bcl-2* (Vaux et al. 1988). Presumably, the pathways controlled by growth factor receptors that regulate cell survival involve conventional signal transduction proteins such as kinases and transcription factors among others. It has been proposed that the second messenger ceramide plays a key role in transduction of cell death signals, but its role remains controversial (Cifone et al. 1994; Zumbansen and Stoffel 1997).

Some death signals originate within the cell. Disturbances to the cell cycle, such as those caused by overexpression of c-myc in resting cells, or radiation-induced damage to DNA, can induce apoptosis via induction of p53, which is inhibitable by *bcl-2* (Yonish et al. 1991; Chiou et al. 1994). Viruses can elicit a defence apoptotic response in infected cells as an altruistic attempt to stop viral spread (Lowin et al. 1994). Presumably, the cell detects changes to cellular metabolism caused by the virus, and kills itself. Thus a cell may kill itself as a stress response, and this may explain the propensity of cells to undergo apoptosis when exposed to a great variety of drugs and toxins with widely varying pharmacological actions (Vaux and Hacker 1995).

3.5
The Bcl-2 Family of Protective and Pro-apoptotic Proteins

The Bcl-2 family are a growing group of proteins which share homology with the product of the *C. elegans* gene *ced-9* (see Chapter by O'Connor and Strasser, this Volume). Like *bcl-2*, some members of this family (*bcl-x, A1, mcl-1, bcl-w*) inhibit apoptosis, whereas others (*bax, bad, bmi, hik, bak*) have the opposite effect. Presumably, *bcl-2* acts like CED-9, namely to inhibit the function of adaptor proteins that activate the caspases.

Death inhibitory Bcl-2 family members protect cells from death resulting from growth factor withdrawal, irradiation and exposure to glucocorticoids but generally offer little protection against apoptosis induced via death receptors such as Fas (Vaux et al. 1988; Alnemri et al. 1992; Strasser et al. 1994, 1995). A possible explanation for this may be that Bcl-2 family proteins act on adaptor proteins that are homologues of CED-4, but do not effectively inhibit adaptors such as FADD.

The precise mechanism of action of Bcl-2-like proteins is not known with certainty, but a large number of possible activities have been proposed. It has been suggested that *bcl-2* forms pores in membranes, acts as a scavenger of free radicals, binds cytochrome c to prevent it leaving the mitochondria, is a G protein, regulates microtubule function, or regulates NFϰB activity, to name just a few (Haldar et al. 1989; Hockenbery et al. 1993; Kluck et al. 1997; Minn et al. 1997; Yang et al. 1997). It is worth keeping in mind that *bcl-2* must function like CED-9 as it can rescue *ced-9* mutant worms (Vaux et al. 1992), and CED-9 expression has no effect in *C. elegans* lacking the gene for the caspase CED-3 (Hengartner et al. 1992; Shaham and Horvitz 1996), so the ultimate function of bcl-2 is likely to regulate caspase activity.

The importance of the Bcl-2 family in both embryonic development and postnatally has been illustrated by studies of knockout mice. Excessive apoptosis is thought to account for abnormalities in Bcl-2 knockout mice including growth retardation in many tissues, excessive cell death in lymphoid tissues and polycystic disease of the kidneys resulting in renal failure and early death (Nakayama et al. 1993; Veis et al. 1993; Kamada et al. 1995). Bcl-X knockout mice die during embryogenesis displaying excessive cell death in both neural and haematopoietic tissues (Motoyama et al. 1995).

3.6
Viruses and Apoptosis

Because cells use cell suicide as a defence against viral replication, viruses have been under selective pressure to develop anti-apoptosis strategies. Studying the anti-apoptosis genes carried by viruses has provided many new insights into the mechanisms of cell death. Several viruses (e.g. Epstein-Barr virus, African swine fever virus, Kaposi's sarcoma-associated herpesvirus) carry homologues of bcl-2 (Henderson et al. 1993; Neilan et al. 1993; Sarid et al. 1997). Some viruses carry caspase inhibitors, such as p35 (carried by some baculoviruses) and crmA (carried by cowpox virus) (Clem et al. 1991; Gagliardini et al. 1994). Other viruses have been found to encode apoptosis inhibitory proteins with DEDs. These proteins, termed FLIPs, function by inhibiting activity of adaptor proteins such as FADD (Thome et al. 1997).

The inhibitor of apoptosis (IAP) proteins were initially identified in baculoviruses, but were also able to inhibit apoptosis in mammalian cells (Crook et al. 1993; Rothe et al. 1995; Hawkins et al. 1996; Uren et al. 1996). These proteins all bear characteristic motifs termed BIRs (baculoviral IAP repeats). The finding that one of the genes frequently mutated in the inherited disease spinal muscular atrophy also bore BIRs raised the possibility that IAP proteins may be important for maintaining neuronal survival in humans (Roy et al. 1995). It is not known how IAP proteins function, but two of them were found to associate with the cytoplasmic domain of the TNF receptor complex, so it is possible that they play a role in the transmission of apoptotic signals (Rothe et al. 1995).

3.7
Cytotoxic T Lymphocyte Killing

The cells which are most important for defence against viral infection are cytotoxic T cells (CTL), lymphocytes which can induce apoptosis in target cells (see Chapters by Trapani and Jans, and Darmon et al. in this Volume). CTL can kill virally infected cells in three ways, via expression of ligands for TNF receptor family members on the surface of the target cell, by a granule exocytosis mechanism that is dependent on perforin, or by the secretion of soluble cytokines such as interferons (Lowin et al. 1994; reviewed by Boehm et al. 1997). Studies comparing mice lacking genes for CD95, TNF receptors or perforin suggest that granule exocytosis is the most important means of CTL killing function (Erickson et al. 1994; Lowin et al. 1994).

The granules of CTL and natural killer (NK) cells contain a variety of serine proteases termed granzymes. Granzyme B is thought to be the most important for CTL killing, and resembles the caspases as its substrate specificity is for aspartate residues (Martin et al. 1996). In vitro, granzyme B can cleave a number of caspase precursor polypeptides, leading to their activation, and it is also capable of cleaving some of the same substrates as certain caspases. These observations have led to speculation that granzyme B in CTL causes target cell death by activating the caspases or by replacing them.

3.8
Apoptosis in Disease

Although apoptosis ultimately serves to benefit the organism, malfunction in the regulation of this process can result in various pathological states. Failure of apoptosis can cause autoimmune disease. For example, mice deficient in expression of the TNFR family member CD95 (the naturally occurring *lpr* mutants and CD95 KO mice) develop lymphadenopathy and splenomegally and have high levels of circulating autoantibodies (Watanabe et al. 1992; Adachi et al. 1996). Autoimmune disease in these mice resembles that occurring in children deficient in CD95 expression (Fisher et al. 1995; Rieux et al. 1995). Overexpression of the cell death inhibitor *bcl-2* in the lymphocytes of transgenic mice also causes an autoimmune disease that resembles systemic lupus erythematosus (McDonnell et al. 1989; Strasser et al. 1991).

Failure of apoptosis can lead to cancer. The common lymphoid malignancy follicular lymphoma is associated with a translocation activating the *bcl-2* gene (Tsujimoto et al. 1985). Transgenic mice expressing *bcl-2* in their lymphoid cells also develop lymphomas (Strasser et al. 1993). The product of the tumour suppressor gene p53 functions to prevent cancer in two ways, firstly by arresting the cell cycle to allow DNA repair, and secondly by generating a signal for apoptosis. Individuals inheriting a mutated allele of p53 (Li-Fraumeni syndrome) are highly susceptible to cancer, as are mice lacking the p53 gene (Malkin et al. 1990; Srivastava et al. 1990; Donehower et al. 1992).

Inappropriate activation of the cell death pathways can also cause or exacerbate disease. For example, it is now thought that death of many heart muscle cells following a heart attack, and death of many neurons following a stroke, is due to apoptosis as a stress response (reviews by Bromme and Holtz 1996; Choi

1996). Rejection of grafts mediated by cytokines and CTL is due to apoptosis in the donor tissue (Heusel et al. 1994). A role for apoptosis has also been proposed in neurodegenerative diseases (Gschwind and Huber 1995; Roy et al. 1995).

It is hoped that by understanding the mechanisms of apoptosis new therapies can be developed to help combat these diseases.

References

Adachi M, Suematsu S, Suda T, Watanabe D, Fukuyama H, Ogasawara J, Tanaka T, Yoshida N, Nagata S (1996) Enhanced and accelerated lymphoproliferation in Fas-null mice. Proc Natl Acad Sci USA 93:2131–2136

Alnemri ES, Fernandes TF, Haldar S, Croce CM, Litwack G (1992) Involvement of bcl-2 in glucocorticoid-induced apoptosis of human pre-B-leukemias. Cancer Res 52:491–495

Boehm U, Klamp T, Groot M, Howard JC (1997) Cellular responses to interferon gamma (Review). Annu Rev Immunol 15:749–795

Boldin MP, Mett IL, Varfolomeev EE, Chumakov I, Shemeravni Y, Camonis JH, Wallach D (1995) Self-association of the death domains of the p55 tumor necrosis factor (TNF) receptor and Fas/Apo-1 prompts signaling for TNF and Fas/Apo-1 effects. J Biol Chem 270:387–391

Boldin MP, Goncharov TM, Goltsev YV, Wallach D (1996) Involvement of MACH, a novel MORT1/FADD-interacting protease, in Fas/Apo-1- and TNF receptor-induced cell death. Cell 85:803–815

Bromme HJ, Holtz J (1996) Apoptosis in the heart – when and why? Mol Cell Biochem 164:261–275

Casciolarosen L, Nicholson DW, Chong T, Rowan KR, Thornberry NA, Miller DK, Rosen A (1996) Apopain/CPP32 cleaves proteins that are essential for cellular repair – a fundamental principle of apoptotic death. J Exp Med 183:1957–1964

Cerretti DP, Kozlosky CJ, Mosley B, Nelson N, Van NK, Greenstreet TA, March CJ, Kronheim SR, Druck T, Cannizzaro LA, et al. (1992) Molecular cloning of the interleukin-1 beta converting enzyme. Science 256:97–100

Chinnaiyan AM, O'Rourke K, Tewari M, Dixit VM (1995) FADD, a novel death domain-containing protein, interacts with the death domain of Fas and initiates apoptosis. Cell 81:505–512

Chinnaiyan AM, O'Rourke K, Lane BR, Dixit VM (1997) Interaction of ced-4 with ced-3 and ced-9 – a molecular framework for cell death. Science 275:1122–1126

Chiou SK, Rao L, White E (1994) Bcl-2 blocks p53 dependent apoptosis. Mol Cell Biol 14:2556–2563

Choi DW (1996) Ischemia-induced neuronal apoptosis. Curr Opin Neurobiol 6:667–672

Cifone MG, De MR, Roncaioli P, Rippo MR, Azuma M, Lanier LL, Santoni A, Testi R (1994) Apoptotic signaling through CD95 (Fas/Apo-1) activates an acidic sphingomyelinase. J Exp Med 180:1547–1552

Clem RJ, Fechheimer M, Miller LK (1991) Prevention of apoptosis by a baculovirus gene during infection of insect cells. Science 254:1388–1390

Crook NE, Clem RJ, Miller LK (1993) An apoptosis inhibiting baculovirus gene with a zinc finger like motif. J Virol 67:2168–2174

Deiss LP, Galinka H, Berissi H, Cohen O, Kimchi A (1996) Cathepsin D protease mediates programmed cell death induced by interferon-gamma, Fas/Apo-1 and TNF-alpha. EMBO J 15:3861–3870

Donehower LA, Harvey M, Slagle BL, McArthur MJ, Montgomery CJ, Butel JS, Bradley A (1992) Mice deficient in p53 are developmentally normal but susceptible to spontaneous tumours. Nature 356:215–221

Duan H, Dixit VM (1997) RAIDD is a new death adaptor molecule. Nature 385:86–89

Ellis HM, Horvitz HR (1986) Genetic control of programmed cell death in the nematode *C. elegans*. Cell 44:817–829

Ellis RE, Jacobson DM, Horvitz HR (1991) Genes required for the engulfment of cell corpses during programmed cell death in *Caenorhabditis elegans*. Genetics 129:79–94

Emoto Y, Manome Y, Meinhardt G, Kisaki H, Kharbanda S, Robertson M, Ghayur T, Wong WW, Kamen R, Weichselbaum R, Kufe D (1995) Proteolytic activation of protein kinase C delta by an ICE-like protease in apoptotic cells. EMBO J 14:6148–6156

Erickson SL, Desauvage FJ, Kikly K, Carvermoore K, Pittsmeek S, Gillett N, Sheehan K, Schreiber RD, Goeddel DV, Moore MW (1994) Decreased sensitivity to tumour-necrosis factor but normal T-cell development in TNF receptor-2-deficient mice. Nature 372:560–563

Fernandesalnemri T, Armstrong RC, Krebs J, Srinivasula SM, Wang L, Bullrich F, Fritz LC, Trapani JA, Tomaselli KJ, Litwack G, Alnemri ES (1996) In vitro activation of CPP32 and Mch3 by Mch4, a novel human apoptotic cysteine protease containing two FADD-like domains. Proc Natl Acad Sci USA 93:7464–7469

Fisher GH, Rosenberg FJ, Straus SE, Dale JK, Middleton LA, Lin AY, Strober W, Lenardo MJ, Puck JM (1995) Dominant interfering Fas gene mutations impair apoptosis in a human autoimmune lymphoproliferative syndrome. Cell 81:935–946

Gagliardini V, Fernandez PA, Lee RKK, Drexler HCA, Rotello RJ, Fishman MC, Yuan J (1994) Prevention of vertebrate neuronal death by the Crm A gene. Science 263:826–828

Goldberg YP, Nicholson DW, Rasper DM, Kalchman MA, Koide NA, Vaillancourt JP, Hayden MR (1996) Cleavage of huntington by apopain, a proapoptotic cysteine protease, is modulated by the polyglutamine tract. Nat Genet 13:442–449

Grimm LM, Goldberg AL, Poirier GG, Schwartz LM, Osborne BA (1996) Proteasomes play an essential role in thymocyte apoptosis. EMBO J 15:3835–3844

Gschwind M, Huber G (1995) Apoptotic cell death induced by beta-amyloid(1–42) peptide is cell type dependent. J Neurochem 65:292–300

Haldar S, Beatty C, Tsujimoto Y, Croce CM (1989) The bcl-2 gene encodes a novel G protein. Nature 342:195–198

Hawkins CJ, Uren AG, Hacker G, Medcalf RL, Vaux DL (1996) Inhibition of interleukin 1-beta-converting enzyme-mediated apoptosis of mammalian cells by baculovirus IAP. Proc Natl Acad Sci USA 93:13786–13790

Hedgecock EM, Sulston JE, Thomson JN (1983) Mutations affecting programmed cell deaths in the nematode *Caenorhabditis elegans*. Science 220:1277–1279

Henderson S, Huen D, Rowe M, Dawson C, Johnson G, Rickinson A (1993) Epstein-Barr virus coded BHRF1 protein, a viral homolog of bcl-2, protects human B cells from programmed cell death. Proc Natl Acad Sci USA 90:8479–8483

Hengartner MO, Horvitz HR (1994) *C. elegans* cell survival gene ced-9 encodes a functional homolog of the mammalian proto-oncogene bcl-2. Cell 76:665–676

Hengartner MO, Ellis RE, Horvitz HR (1992) *Caenorhabditis elegans* gene ced-9 protects cells from programmed cell death. Nature 356:494–499

Henkart P (1996) ICE family proteases: mediators of all apoptotic cell death? Immunity 4:195–201

Heusel JW, Wesselschmidt RL, Shresta S, Russell JH, Ley TJ (1994) Cytotoxic lymphocytes require granzyme B for the rapid induction of DNA fragmentation and apoptosis in allogeneic target cells. Cell 76:977–987

Hockenbery DM, Oltvai ZN, Yin XM, Milliman CL, Korsmeyer SJ (1993) Bcl-2 functions in an antioxidant pathway to prevent apoptosis. Cell 75:241–251

Hofmann K, Bucher P, Tschopp J (1997) The CARD domain – a new apoptotic signalling motif. Trends Biochem 22:155–156

Irmler M, Hofmann K, Vaux D, Tschopp J (1997) Direct physical interaction between the *Caenorhabditis elegans* death proteins ced-3 and ced-4. FEBS Lett 406:189–190

Kamada S, Shimono A, Shinto Y, Tsujimura T, Takahashi T, Noda T, Kitamura Y, Kondoh H, Tsujimoto Y (1995) Bcl-2 deficiency in mice leads to pleiotropic abnormalities – accelerated lymphoid cell death in thymus and spleen, polycystic kidney, hair hypopigmentation, and distorted small intestine. Cancer Res 55:354-359

Kischkel FC, Hellbardt S, Behrmann I, Germer M, Pawlita M, Krammer PH, Peter ME (1995) Cytotoxicity-dependent APO-1 (Fas/CD95)-associated proteins form a death-inducing signaling complex (DISC) with the receptor. EMBO J 14:5579-5588

Kluck RM, Bossywetzel E, Green DR, Newmeyer DD (1997) The release of cytochrome c from mitochondria - a primary site for bcl-2 regulation of apoptosis. Science 275:1132-1136

Kuida K, Lippke JA, Ku G, Harding MW, Livingston DJ, Su MS, Flavell RA (1995) Altered cytokine export and apoptosis in mice deficient in interleukin-1 beta converting enzyme. Science 267:2000-2003

Kuida K, Zheng TS, Na SQ, Kuan CY, Yang D, Karasuyama H, Rakic P, Flavell RA (1996) Decreased apoptosis in the brain and premature lethality in CPP32-deficient mice. Nature 384:368-372

Lazebnik YA, Kaufmann SH, Desnoyers S, Poirier GG, Earnshaw WC (1994) Cleavage of poly(ADP-ribose) polymerase by a proteinase with properties like ICE. Nature 371:346-347

Lazebnik YA, Takahashi A, Moir RD, Goldman RD, Poirier GG, Kaufmann SH, Earnshaw WC (1995) Studies of the lamin proteinase reveal multiple parallel biochemical pathways during apoptotic execution. Proc Natl Acad Sci USA 92:9042-9046

Li P, Allen H, Banerjee S, Franklin S, Herzog L, Johnston C, Mcdowell J, Paskind M, Rodman L, Salfeld J, Towne E, Tracey D, Wardwell S, Wei FY, Wong W, Kamen R, Seshadri T (1995) Mice deficient in IL-1-beta-converting enzyme are defective in production of mature IL-1-beta and resistant to endotoxic shock. Cell 80:401-411

Liu XS, Zou H, Slaughter C, Wang XD (1997) DFF, a heterodimeric protein that functions downstream of caspase-3 to trigger DNA fragmentation during apoptosis. Cell 89:175-184

Liu ZG, Hsu HL, Goeddel DV, Karin M (1996) Dissection of TNF receptor 1 effector functions - JNK activation is not linked to apoptosis while NF-kappa-B activation prevents cell death. Cell 87:565-576

Lowin B, Hahne M, Mattmann C, Tschopp J (1994) Cytolytic T-cell cytotoxicity is mediated through perforin and Fas lytic pathways. Nature 370:650-652

Malkin D, Li FP, Strong LC, Fraumeni J Jr, Nelson CE, Kim DH, Kassel J, Gryka MA, Bischoff FZ, Tainsky MA, et al. (1990) Germ line p53 mutations in a familial syndrome of breast cancer, sarcomas, and other neoplasms. Science 250:1233-1238

Martin SJ, Amarantemendes GP, Shi LF, Chuang TH, Casiano CA, Obrien GA, Fitzgerald P, Tan EM, Bokoch GM, Greenberg AH, Green DR (1996) The cytotoxic cell protease granzyme B initiates apoptosis in a cell-free system by proteolytic processing and activation of the ICE/ced-3 family protease, CPP32, via a novel two-step mechanism. EMBO J 15:2407-2416

McDonnell TJ, Deane N, Platt FM, Nunez G, Jaeger U, McKearn JP, Korsmeyer SJ (1989) Bcl-2-immunoglobulin transgenic mice demonstrate extended B cell survival and follicular lymphoproliferation. Cell 57:79-88

Medema JP, Scaffidi C, Kischkel FC, Shevchenko A, Mann M, Krammer PH, Peter ME (1997) FLICE is activated by association with the CD95 death-inducing signalling complex (DISC). EMBO J 16:2794-2804

Minn AJ, Velez P, Schendel SL, Liang H, Muchmore SW, Fesik SW, Fill M, Thompson CB (1997) Bcl-x(L) forms an ion channel in synthetic lipid membranes. Nature 385:353-357

Molineaux SM, Casano FJ, Rolando AM, Peterson EP, Limjuco G, Chin J, Griffin PR, Calaycay JR, Ding GJF, Yamin T-T, Palyha OC, Luell S, Fletcher D, Miller DK, Howard AD, Thornberry NA, Kostura MJ (1993) Interleukin 1β (IL-1β) processing in murine macrophages requires a structurally conserved homologue of human IL-1β converting enzyme. Proc Natl Acad Sci USA 90:1809-1813

Motoyama N, Wang FP, Roth KA, Sawa H, Nakayama K, Nakayama K, Negishi I, Senju S, Zhang Q, Fujii S, Loh DY (1995) Massive cell death of immature hematopoietic cells and neurons in bcl-x-deficient mice. Science 267:1506-1510

Muzio M, Chinnaiyan AM, Kischkel FC, Orourke K, Shevchenko A, Ni J, Scaffidi C, Bretz JD, Zhang M, Gentz R, Mann M, Krammer PH, Peter ME, Dixit VM (1996) FLICE, a novel FADD-homologous ICE/ced-3-like protease, is recruited to the CD95 (Fas/Apo-1) death-inducing signaling complex. Cell 85:817–827

Muzio M, Salvesen GS, Dixit VM (1997) FLICE induced apoptosis in a cell-free system – cleavage of caspase zymogens. J Biol Chem 272:2952–2956

Nakayama K, Nakayama K, Nagashi I, Kulda K, Shinkai Y, Louie MC, Fields LE, Lucas PJ, Stewart V, Alt FW, Loh DY (1993) Disappearance of the lymphoid system in bcl-2 homozygous mutant chimeric mice. Science 261:1534–1538

Neilan JG, Lu Z, Afonso CL, Kutish GF, Sussman MD, Rock DL (1993) An African swine fever virus gene with similarity to the proto-oncogene bcl-2 and the Epstein-Barr virus gene BHRF1. J Virol 67:4391–4394

Rieux LF, Le DF, Hivroz C, Roberts IA, Debatin KM, Fischer A, de VJ (1995) Mutations in Fas associated with human lymphoproliferative syndrome and autoimmunity. Science 268:1347–1349

Rothe M, Pan MG, Henzel WJ, Ayres TM, Goeddel DV (1995) The TNFR2-TRAF signaling complex contains two novel proteins related to baculoviral-inhibitor of apoptosis proteins. Cell 83:1243–1252

Roy N, Mahadevan MS, Mclean M, Shutler G, Yaraghi Z, Farahani R, Baird S, Besnerjohnston A, Lefebvre C, Kang XL, Salih M, Aubry H, Tamai K, Guan XP, Ioannou P, Crawford TO, Dejong PJ, Surh L, Ikeda JE, Korneluk RG, Mackenzie A (1995) The gene for neuronal apoptosis inhibitory protein is partially deleted in individuals with spinal muscular atrophy. Cell 80:167–178

Sarid R, Sato T, Bohenzky RA, Russo JJ, Chang Y (1997) Kaposis sarcoma-associated herpesvirus encodes a functional bcl-2 homologue. Nat Med 3:293–298

Shaham S, Horvitz HR (1996) Developing *Caenorhabditis elegans* neurons may contain both cell-death protective and killer activities. Genes Dev 10:578–91

Shi L, Kam CM, Powers JC, Aebersold R, Greenberg AH (1992) Purification of three cytotoxic lymphocyte granule serine proteases that induce apoptosis through distinct substrate and target cell interactions. J Exp Med 176:1521–1529

Spector MS, Desnoyers S, Hoeppner DJ, Hengartner MO (1997) Interaction between the *C. elegans* cell-death regulators ced-9 and ced-4. Nature 385:653–656

Srinivasula SM, Ahmad M, Fernandesalnemri T, Litwack G, Alnemri ES (1996) Molecular ordering of the Fas-apoptotic pathway – the Fas/Apo-1 protease Mch5 is a CrmA-inhibitable protease that activates multiple ced-3/ICE-like cysteine proteases. Proc Natl Acad Sci USA 93:14486–14491

Srivastava S, Zou ZQ, Pirollo K, Blattner W, Chang EH (1990) Germ-line transmission of a mutated p53 gene in a cancer-prone family with Li-Fraumeni syndrome. Nature 348:747–749

Strasser A, Whittingham S, Vaux DL, Bath ML, Adams JM, Cory S, Harris AW (1991) Enforced bcl-2 expression in B-lymphoid cells prolongs antibody responses and elicits autoimmune disease. Proc Natl Acad Sci USA 88:8661–8665

Strasser A, Harris AW, Cory S (1993) E mu-bcl-2 transgene facilitates spontaneous transformation of early pre-B and immunoglobulin-secreting cells but not T cells. Oncogene 8:1–9

Strasser A, Harris AW, Jacks T, Cory S (1994) DNA damage can induce apoptosis in proliferating lymphoid cells via p53-independent mechanisms inhibitable by Bcl-2. Cell 79:329–339

Strasser A, Harris AW, Huang D, Krammer PH, Cory S (1995) Bcl-2 and Fas/Apo-1 regulate distinct pathways to lymphocyte apoptosis. EMBO J 14:6136–6147

Thome M, Schneider P, Hofmann K, Fickenscher H, Meinl E, Neipel F, Mattmann C, Burns K, Bodmer JL, Schroter M, Scaffidi C, Krammer PH, Peter ME, Tschopp J (1997) Viral FLICE-inhibitory proteins (FLIPs) prevent apoptosis induced by death receptors. Nature 386:517–521

Tsujimoto Y, Cossman J, Jaffe E, Croce CM (1985) Involvement of the bcl-2 gene in human follicular lymphoma. Science 228:1440–1443

Uren AG, Pakusch M, Hawkins CJ, Puls KL, Vaux DL (1996) Cloning and expression of apoptosis inhibitory protein homologs that function to inhibit apoptosis and/or bind tumor necrosis factor receptor-associated factors. Proc Natl Acad Sci USA 93:4974–4978

Vaux DL, Hacker G (1995) Hypothesis – apoptosis caused by cytotoxins represents a defensive response that evolved to combat intracellular pathogens. Clin Exp Pharm Physiol 22:861–863

Vaux DL, Cory S, Adams JM (1988) Bcl-2 gene promotes haemopoietic cell survival and cooperates with c-myc to immortalize pre-B cells. Nature 335:440–442

Vaux DL, Weissman IL, Kim SK (1992) Prevention of programmed cell death in *Caenorhabditis elegans* by human bcl-2. Science 258:1955–1957

Veis DJ, Sorenson CM, Shutter JR, Korsmeyer SJ (1993) Bcl-2-deficient mice demonstrate fulminant lymphoid apoptosis, polycystic kidneys, and hypopigmented hair. Cell 75:229–240

Walker NPC, Talanian RV, Brady KD, Dang LC, N.J. B, Ferenz CR, Franklin S, Ghayur T, Hackett MC, Hamill LD, Herzog L, Hugunin M, Houy W, Mankovich JA, McGuiness L, Orlewicz E, Paskind M, Pratt CA, Reis P, Summani A, Terranova M, Welch JP, Xiong L, Möller A, Tracey DE, Kamen R, Wong WW (1994) Crystal structure of the cysteine protease interleukin-1-β-converting enzyme: a (p20/p10)$_2$ homodimer. Cell 78:343–352

Wang ZQ, Auer B, Stingl L, Berghammer H, Haidacher D, Schweiger M, Wagner EF (1995) Mice lacking ADPRT and poly(ADP-ribosyl)ation develop normally but are susceptible to skin disease. Genes Dev 9:509–520

Watanabe FR, Brannan CI, Itoh N, Yonehara S, Copeland NG, Jenkins NA, Nagata S (1992) The cDNA structure, expression, and chromosomal assignment of the mouse Fas antigen. J Immunol 148:1274–1279

Wilson KP, Black J, Thomson JA, Kim EE, Griffith JP, Navia MA, Murcko MA, Chambers SP, Aldape RA, Raybuck SA, Livingston DJ (1994) Structure and mechanism of interleukin-1-beta converting enzyme. Nature 370:270–275

Wu DY, Wallen HD, Nunez G (1997) Interaction and regulation of subcellular localization of ced-4 by ced-9. Science 275:1126–1129

Xue D, Shaham S, Horvitz HR (1996) The *Caenorhabditis elegans* cell-death protein ced-3 is a cysteine protease with substrate specificities similar to those of the human CPP32 protease. Genes Dev 10:1073–1083

Yang J, Liu XS, Bhalla K, Kim CN, Ibrado AM, Cai JY, Peng TI, Jones DP, Wang XD (1997) Prevention of apoptosis by bcl-2 – release of cytochrome c from mitochondria blocked. Science 275:1129–1132

Yonish RE, Resnitzky D, Lotem J, Sachs L, Kimchi A, Oren M (1991) Wild-type p53 induces apoptosis of myeloid leukaemic cells that is inhibited by interleukin-6. Nature 353:345–347

Yuan JY, Shaham S, Ledoux S, Ellis HM, Horvitz HR (1993) The *C. elegans* cell death gene ced 3 encodes a protein similar to mammalian interleukin 1 beta converting enzyme. Cell 75:641–652

Zou H, Henzel WJ, Liu X, Lutschg A, Wang X (1997) Apaf-1, a human protein homologous to *C. elegans* CED-4, participates in cytochrome c-dependent activation of caspase-3. Cell 90:405–413

Zumbansen M, Stoffel W (1997) Tumor necrosis factor alpha activates NF-kappa-B in acid sphingomyelinase-deficient mouse embryonic fibroblasts. J Biol Chem 272:10904–10909

The Death Receptors

Marcus E. Peter[1], Carsten Scaffidi[1], Jan Paul Medema[2],
Frank Kischkel[1] and Peter H. Krammer[1]

1
Introduction

In recent years apoptosis, also called programmed cell death, has been recognized to be the physiological way for a nucleated animal cell to die. Apoptosis takes care of unwanted, injured or virus-infected cells (Farber 1994; Collins 1995). Autoreactive T and B cells that are produced by the immune system by the millions every day are also eliminated by apoptosis. A large number of stimuli can trigger apoptosis. However, only the discovery of the existence of receptors that could trigger apoptosis convinced everyone that a certain substance would not just kill a cell due to its high toxicity but involve special apoptosis-inducing mechanisms. A number of receptors that were first shown to have other functions besides induction of apoptosis could kill cells. These receptors include the T cell receptor/CD3 complex (Smith et al. 1989; Takahashi et al. 1989; Newell et al. 1990), the B cell receptor (Ales-Martinez et al. 1992), CD2 (Merkenschlager and Fisher 1991), CD4 (Wadsworth et al. 1990) and the mouse antigen Thy-1 (Ucker et al. 1989).

In the meantime it has become apparent that most apoptosis-inducing receptors are found in a family of receptors that is structurally related to the tumor necrosis factor (TNF) receptors (Fig. 1). The receptor superfamily is characterized by cysteine-rich extracellular domains present in 2–5 copies. For most members of the family an apoptosis-inducing activity has been reported. However, some of them also have other functions such as induction of proliferation, differentiation and so forth. Receptors with pleiotropic functions include TNF-R1 (Loetscher et al. 1990; Schall et al. 1990; Smith et al. 1990), TNF-R2 (Dembic et al. 1990), CD40 (Stamenkovic et al. 1989), CD30 (Durkop et al. 1992), CD27 (Camerini et al. 1991), OX-40 (Mallett et al. 1990), 4-1BB (Kwon und Weissman 1989), NGF-R (Radeke et al. 1987), DR3 (TRAMP/wsl-1/

[1] Tumor Immunology Program, German Cancer Research Center, Im Neuenheimer Feld 280, 69120 Heidelberg, Germany
[2] Present address: Department of Immunohematology and Bloodbank, University Hospital Leiden, Albinusdreef 2, 2300RC, Leiden, The Netherlands

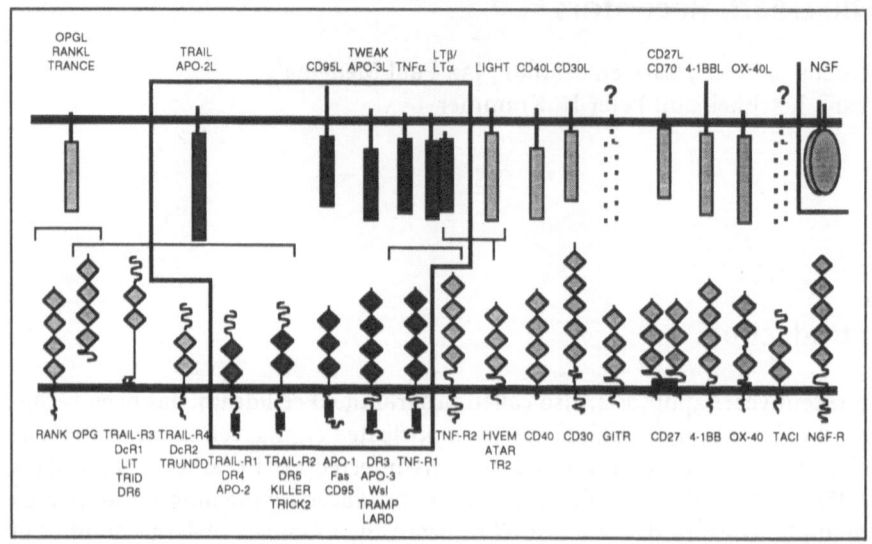

Fig. 1. Members of the tumor necrosis factor receptor superfamily and their ligands. Subfamilies of death receptors and their ligands are *boxed*. For details see text.

APO-3/LARD/AIR) (Chinnaiyan et al. 1996a; Kitson et al. 1996; Marsters et al. 1996b; Bodmer et al. 1997; Degli-Esposti et al. 1997; Screaton et al. 1997), HVEM/ATAR/TR2 (Montgomery et al. 1996; Hsu et al. 1997; Kwon et al. 1997) and GITR (Nocentini et al. 1997).

With the discovery of the first "professional" apoptosis-inducing receptor the importance of apoptosis induction became most obvious. This receptor was called APO-1 or Fas and is now called CD95 (Oehm et al. 1992; Watanabe-Fukunaga et al. 1992b). By comparing CD95 with TNF-R1, a domain was identified in the intracellular tail of both receptors that was essential for transduction of a death signal into cells (Itoh and Nagata 1993; Tartaglia et al. 1993a). It was called the death domain (DD). Surprisingly, it took 4 years to identify other receptors that carry a DD in their intracellular part. DR3 (TRAMP/wsl-1/APO-3/LARD/AIR) is both structurally and functionally similar to TNF-R1 whereas DR4/TRAIL-R1 is functionally similar to CD95 as its main function seems to be to induce apoptosis (Pan et al. 1997). It is the only TNF-R-like molecule possessing only two cysteine-rich domains. Through the recent discovery of DD containing apoptosis-inducing receptors a new subgroup of TNF-R-like receptors has emerged, the death receptors. They are defined by the presence of an intracellular DD.

2
Death Ligands

For most members of the TNF-R family a ligand has been identified. Three of these ligands, CD95L, TNFα and TRAIL, bind to death receptors (Suda et al. 1993; Beutler and van Huffel 1994; Wiley et al. 1995). All ligands that bind to members of the TNF-R superfamily with the exception of NGF form a family of the TNF-related molecules (Fig. 1). They are type II transmembrane proteins. The crystal structure of TNFα and TNFβ alone or TNFβ in complex with the extracellular domain of the TNF-R1 revealed a trimeric structure (Eck and Sprang 1989; Eck et al. 1992; Jones et al. 1992; Banner et al. 1993). Together with a number of biochemical data that also suggested that TNF-like ligands can form a trimer (Karpusas et al. 1995; Pitti et al. 1996) it is believed that all these active ligands have a trimeric structure in solution. It is likely that they bind to their cognate receptors and activate them by trimerization. Data on the CD95 receptor demonstrating that dimerization of CD95 was not sufficient to trigger apoptosis supported this notion (Dhein et al. 1992; Kischkel et al. 1995). Another feature shared by most of these ligands is that they are expressed as transmembrane proteins. However, for most of them soluble forms have been identified. The secreted form of the ligands is generated by the activity of metalloproteinases. This was suggested for the CD95L (Mariani et al. 1995). For TNFα a metalloprotease (TACE) could be cloned. TACE was shown to specifically cleave TNFα (Black et al. 1997; Moss et al. 1997). Either TACE has a broader specificity and processes other TNF-like ligands or each ligand has its own specific protease for its release. We will refer to the ligands binding to the death receptors as the death ligands. This review will focus on the function of death receptors and their ligands with special emphasis on the signal transduction pathways used by these receptors.

3
Biological Functions of the Death Receptors

TNF-R1 and CD95 are the best characterized members of the death receptor family. The study of these receptors and their ligands underlined the biological relevance of apoptotic systems within an organism. In the mature organism apoptosis induced by death receptors plays a role in the homeostasis of tissues. Senescent, injured or mutated cells are being eliminated. Moreover, apoptosis plays an important function in the immune system. Autoreactive cells must be deleted in order to prevent autoimmunity. At the end of an immune response activated lymphocytes are removed to maintain immune homeostasis (peripheral deletion). Furthermore, immune surveillance involves apoptotic processes to remove aberrant cells. Recently, it has become clear that death ligands are responsible for the extraordinary status of certain tissues such as brain, ovary, testis, uterus during pregnancy, placenta, eye and the hamster cheek pouch. These sites are exempt from immune responses and were therefore labeled "immunologically privileged" (Barker and Billingham 1977). Within these sites immune cells are cleared via apoptosis. This has tremendous implication for transplantation medicine, since the problem of graft-versus-host and host-versus-graft diseases is still unsolved. A better understanding of the mechanism to maintain these privileged sites may result in ways to specifically suppress an immune response. The apoptotic machinery utilizing death receptor/ligand systems is very powerful and requires tight regulation. Disturbance of these systems can cause severe disease.

3.1
The CD95/CD95L System

The CD95 receptor is widely expressed on lymphoid and nonlymphoid cells (Watanabe-Fukunaga et al. 1992b; Leithauser et al. 1993; Hiramatsu et al. 1994; Galle et al. 1995), whereas expression of CD95L is more tightly regulated. Originally, expression of the ligand seemed to be restricted to T cells (Suda et al. 1993), but turned out to be expressed in some important nonlymphoid areas as discussed below.

Identification of CD95 and CD95L helped to explain the phenotype of two mutant mice: *lpr* (for *lymphoproliferation*) (Watanabe-Fukunaga et al. 1992a) and *gld* (for *generalized lymphoproliferative disease*) (Takahashi et al. 1994), respectively. These mice display a phenotype of uncontrolled accumulation of CD4$^-$CD8$^-$ T cells, leading to lymphadenopathy, splenomegaly and an autoimmunity that closely resembles systemic lupus erythematosus. The *lpr* mutation results from the insertion of a transposable element into intron 2 of the gene encoding CD95, preventing transcription of full-length mRNA (Adachi et al. 1993). The *gld* defect arises from a point mutation within the gene encoding CD95L, changing an amino acid critical for CD95 binding. In humans a similar disease was reported (Fisher et al. 1995; Rieux-Laucat et al. 1995), with a dysfunction of the CD95/CD95L system. Children with this "autoimmune lymphoproliferative syndrome" (ALPS) have massive, nonmalignant lymphadenopathy, an altered and enlarged T cell population and a massive autoimmune disorder.

Although there are a few reports showing that triggering of CD95 results in secretion of IL-8 (Abreu-Martin et al. 1995) or proliferation (Alderson et al. 1993; Mapara et al. 1993; Aggarwal et al. 1995; Freiberg et al. 1997), the overall *in vitro* and *in vivo* data suggest that CD95 is a receptor which mainly mediates apoptosis. The physiological role of the CD95/CD95L system seems to be strongly connected to the immune system and to the liver (Adachi et al. 1995). It could be shown that after T cell receptor engagement on previously activated or transformed T cells (Alderson et al. 1995; Brunner et al. 1995; Dhein et al.

1995; Ju et al. 1995), CD95L expression is upregulated, and cells then undergo apoptosis due to CD95/CD95L interactions. This activation-induced cell death (AICD) is also involved in peripheral deletion *in vivo*, in which T cells responding to a strong antigenic stimulus decrease in number over time, a process that may be important in immune homeostasis. Peripheral deletion is at least partially defective in animals deficient in CD95 (Singer and Abbas 1994; Mogil et al. 1995), which supports the view that expression of this receptor–ligand pair is important for depleting excess lymphocytes after an immune response. Because the CD95/CD95L system is involved in AICD it seems to be at least partially responsible for the depletion of $CD4^+$ T cells in AIDS. It could be shown that indirect mechanisms lead to a sensitization of noninfected T cells towards AICD after HIV-1 infection (Szawlowski et al. 1993; Zagury et al. 1993; Li et al. 1995a; Westendorp et al. 1995). In addition, T lymphocytes of HIV-1 infected patients expressed increased CD95 on their surface (Debatin et al. 1994; McCloskey et al. 1995). Katsikis et al. (1995) demonstrated that T cells of HIV-1 seropositive individuals were more sensitive towards CD95-mediated apoptosis. A scenario can be drawn in which AICD-sensitive T cells of HIV-patients show an elevated expression of CD95. Stimulation of these cells with antigen or anti-CD95 autoantibodies leads to apoptosis. This mechanism could contribute to the increased depletion of T cells during the course of the AIDS disease.

CD95L seems to be required for some tissues to maintain an immune privileged status. After viral inoculation into the anterior chamber of the eye, lymphocytes and granulocytes that are recruited undergo apoptosis probably due to CD95L expressed on epithelial cells (Griffith et al. 1995, 1996). This apoptosis is not observed in the eyes of animals with defective CD95L (*gld* mice), and the resulting uncontrolled inflammation destroys the tissue. Thus, CD95L is necessary for the maintenance of the privileged status of the eye. It is interesting to note that *gld* and *lpr* mice have no apparent ocular abnormalities and no increased lymphocytic infiltration. Thus, even with aberrant CD95L expression, organs such as the eye maintain their function, probably due to other mechanisms that insure immune privilege, particularly the blood–eye barrier.

The involvement of the CD95/CD95L system in immune privilege gave rise to the idea to use this system in transplantation to assure graft acceptance by ectopic expression of CD95L. It was recently shown that human corneas express functional CD95L (Stuart et al. 1997), raising the possibility that this molecule could act to protect these cornea grafts in humans. Examination of corneal transplants in mice supported this idea; while approximately 45% of allogeneic cornea transplants survived for an extended period, no graft survival was seen with corneas expressing defective CD95L (*gld*) or when the recipient mice had a defect in CD95 expression (*lpr*). Thus, protection of allogeneic grafts was dependent upon the presence of functional CD95L. Other studies on CD95L involvement in graft acceptance are less clear and currently very controversially discussed. A striking protective effect of CD95L expressed in the testis was observed by transplantation of allogenic testis under the kidney capsule of recipient mice by one group (Bellgrau et al. 1995), but this result could not be confirmed by others (Allison et al. 1997). In another example of contradicting results syngeneic myoblasts expressing ectopic CD95L effectively protected allogeneic pancreatic islets coimplanted under the kidney capsule of animals made diabetic by streptozotocin treatment (Lau et al. 1996). These grafts, which were quickly rejected if myoblasts did not express CD95L, maintained function for an extended period of time. Consistent with this was the observation that murine or allogeneic rat islets showed delayed rejection when coimplanted with CD95L-expressing testis tissue in rats (Selawry and Cameron 1993). However, Allison et al. (1997) reported that expression of functional CD95L in the pancreatic islets of transgenic mice failed to protect these islets from allogenic transplant rejection when placed under the kidney capsule of recipient mice. The presence of CD95L induced a granulocytic infiltrate in the transgenic animals themselves, which damaged (but did not destroy) the islets. These findings were consistent with a report showing that CD95L on tumor cells can induce a granulocyte-mediated rejection reaction (Seino et al. 1997).

One other recent example of CD95L-dependent immune privilege was described in a different context. A number of different murine and human tumors, including many nonlymphoid tumors, have been observed to constitutively express functional CD95L (Hahne et al. 1996; O'Connell et al. 1996; Niehans et al. 1997; Shiraki et al. 1997). For example, a CD95L-expressing melanoma was capable of inducing potent antitumor immunity, providing that the host was defective in CD95 expression (Hahne et al. 1996). This suggested that the mechanism responsible for protecting tissues from autoimmune destruction during inflammatory responses, or during graft rejection, also could be used by tumors to escape immune surveillance.

These observations strongly implicate nonlymphoid CD95L in the control of immune responses, via induction of apoptosis in infiltrating lymphocytes and granulocytes. However, CD95L expressed on some tumors resulted in rejection rather than protection of the tumor (EgSeino et al. 1997). In addition, it is known that CD95L can induce tissue damage. In graft-versus-host disease, the ability of the graft effector cells to express functional CD95L contributes to the destructive assault (Baker et al. 1996; Braun et al. 1996). Anti-CD95 antibody induces apoptosis in hepatocytes *in vivo* (Ogasawara et al. 1993), and this has led to the idea that CD95L-induced apoptosis of these cells contributes to some forms of hepatitis, which could be confirmed at least in patients with alcoholic liver damage, in which the hepatocytes express CD95L (Galle et al. 1995).

CD95L-induced apoptosis seems to play a role in some autoimmune diseases. Normal thyrocytes constitutively express functional CD95L, but normally do not express CD95. However, in Hashimoto's thyroiditis patients, the thyrocytes do express CD95, and these cells undergo apoptosis (Giordano et al. 1997). *In vitro*, normal thyrocytes express CD95 after exposure to interleukin-1, and the resulting apoptosis can be blocked by antibodies that disrupt CD95/CD95L interactions. Hence, in Hashimoto's thyroiditis the normally protective function of CD95L on thyrocytes leads to the destruction of the thyroid gland. The cause of this dysfunction is unclear, but it is likely that the CD95R/L system contributes to the disease process.

3.2
The TNF/TNF-R System

The biological function of the TNF/TNF-R system is more complex and less well defined than the CD95/CD95L system, since two different ligands [TNF-α and lymphotoxine (LT)-α] can bind to two receptors TNF-R1 and TNF-R2 (Perez et al. 1990; Beutler and van Huffel 1994; Decoster et al. 1995). In addition, membrane-bound TNF (mTNF) has different binding affinities to the receptors from the soluble form of TNF (sTNF). TNF is a pleiotropic cytokine and is produced by many cell types, including macrophages, monocytes, lymphoid cells, and fibroblasts, in response to inflammation, infection, and other environmental challenges (Tracey and Cerami 1993). TNF elicits a wide spectrum of organismal and cellular responses, including fever, shock, tissue injury, tumor necrosis, anorexia, induction of other cytokines and immunoregulatory molecules, cell proliferation, differentiation, and apoptosis (Tracey and Cerami 1993; Vandenabeele et al. 1995). The corresponding receptors are coexpressed in most cell lines and primary tissues, although expression of TNF-R1 and TNF-R2 is controlled by distinct mechanisms. Typically, TNF-R1 is constitutively expressed at a rather low level, whereas the level of TNF-R2 expression is subject to both transcriptional and posttranscriptional regulation induced by external stimuli (Brockhaus et al. 1990; Thoma et al. 1990).

TNF-R1 is responsible for most of the biological properties of TNF, including programmed cell death, antiviral activity, and activation of the transcription factor NF-κB in a wide variety of cell types (Engelmann et al. 1990; Espevik et al. 1990; Tartaglia et al. 1991, 1993a; Wong et al. 1992). It also plays an essential role in the host defense against microorganisms and bacterial pathogens (Pfeffer et al. 1993; Rothe et al. 1993). The contribution of TNF-R2 to cellular responses induced by sTNF appears to be of a supportive or modulating nature, with two distinct functional properties. First, the proteolytically cleaved extracellular domain may buffer excessive sTNF and, as a consequence, might be effective as a TNF inhibitor (Porteu and Hieblot 1994). Second, TNF-R2-bound ligand may be passed over to TNF-R1 to enhance TNF-R1 signaling, a process termed ligand passing that is favored by the distinct kinetics of ligand association and dissociation of the two receptors (Tartaglia et al. 1993b). However, the prime physiological activator of TNF-R2 seems to be mTNF, since TNF-R2 can be strongly stimulated by mTNF rather than by sTNF. As mTNF also signals via TNF-R1, the resulting cooperativity of both receptors leads to cellular responses much stronger than those achievable with sTNF alone. Moreover, it was shown that upon appropriate activation of TNF-R2, a phenotypic switch of the cellular response pattern to TNF could be observed, such that, as an example, cells fully resistant to the cytotoxic action of sTNF become highly susceptible and are killed upon contact with mTNF (Grell et al. 1995).

Gene targeting and transgene technologies have been used in order to unravel the in vivo role of the TNF/TNF-Rs system and to establish genetically defined models of human diseases. TNF$^{-/-}$ mice show an almost normal phenotype histologically, but have reduced sensitivity to lipopolysoccharide (LPS)-mediated toxicity and increased sensitivity to intracellular pathogens like Listeria, due to severely impaired macrophage function (Pasparakis et al. 1996). Moreover, they lack typical germinal center formation and, although capable of undergoing IgG and IgE class switching, show quantitative differences in the Ig composition of secondary antibody responses. All data suggest that TNF$^{-/-}$ mice have a general defect in the amplification of humoral and probably also cellular immune response. The importance of mTNF *in vivo* has been elegantly demonstrated in several transgenic models. For example, it was reported that the deficiencies of the TNF$^{-/-}$ mice are reconstituted by an ectopically expressed, noncleavable mTNF (Korner and Sedgwick 1996). Further, in a different transgenic model it was shown that mTNF induces multiple sclerosis (MS)-like disease with paralysis and a histopathology resembling experimental autoimmune encephalomyelitis (EAE) when expressed in microglia, but not neuronal cells.

In contrast to TNF-R1$^{-/-}$ mice (Rothe et al. 1993), deletion of TNF-R2 has no apparent influence on lymphoid organ development. TNF-R2 is critically involved in mediating pathogenicity during cerebral malaria (Garcia et al. 1995), is essential for LPS-induced leukostasis and downregulates TNF-R1 dependent neutrophil influx in a lung inflammation model representing farmer's lung. A dominant role of TNF-R1 in mediating TNF's pathogenic activities was evident early on from models of septic shock and arthritis (Hayward and Fiedler-Nagy 1987; Espevik et al. 1988; Shimamoto et al. 1988). The growing knowledge about the pathophysiological role of TNF in acute and especially in chronic diseases calls for strategies to intervene with the deleterious effects of TNF. Clinical therapies employing anti-TNF reagents have been impressively successful in different diseases such as rheumatoid arthritis, septic shock and inflammatory bowel disease (for example, Crohn's disease) (Stokkers et al. 1995; Maini 1996; Glauser 1996).

3.3
The DR3 and DR4/TRAIL System

Since DR3, DR4 and TRAIL-R2 have only recently been cloned, not much is known about their biological function. DR3 seems to be restricted to tissues enriched in lymphocytes (Chinnaiyan et al. 1996a; Bodmer et al. 1997). Therefore, this receptor might play a role in lymphocyte homeostasis. Eventually TRAIL is also involved in the regulation of the immune system. It was reported that TRAIL can induce AICD in activated T cells (Marsters et al. 1996a). Nevertheless, it will be interesting to see how many members of the death receptor family and their ligands are needed to properly regulate apoptosis in all cellular contexts.

4
Death Receptor-Associating Molecules

4.1
Death Domain Proteins

The solution structure of the CD95 death domain (DD) has been determined by NMR spectroscopy. The structure consists of six antiparallel, amphipathic alpha-helices arranged in a novel fold (Huang et al. 1996). The first three nonreceptor DDs identified were human FADD (fas-associated DD protein)/ MORT1 (Boldin et al. 1995; Chinnaiyan et al. 1995), murine RIP (receptor interactive protein) (Stanger et al. 1995) and human TRADD (TNF-R1 related DD protein) (Hsu et al. 1995). The yeast two-hybrid system was employed to clone FADD and RIP with the CD95 cytoplasmic tail (ct) and TRADD with the TNF-R1 ct as bait, respectively. Subsequent cloning of human RIP, however, suggested that RIP would more likely interact with TNF-R1 than with CD95 (Hsu et al. 1996b).

Overexpression of DD proteins causes cell death, indicating that these molecules are involved in apoptosis signaling. Using classical biochemical methods we identified a complex of proteins that associated with stimulated CD95 (Kischkel et al. 1995). Treatment of CD95-positive cells with the agonistic mAb anti-APO-1 and subsequent immunoprecipitation of the receptor with protein A-Sepharose resulted in the identification of four cytotoxicity-dependent APO-1-associated proteins (CAP1–4) on 2D-IEF/SDS gels within seconds after receptor triggering. Together with the receptor, these proteins formed the death-inducing signaling complex (DISC). Using a specific rabbit antiserum CAP1 and 2 could be identified as two different serine phosphorylated species of FADD and demonstrated that FADD bound to CD95 in a stimulation-dependent fashion *in vivo*. Overexpression of the N-terminal half of FADD induced cell death (Chinnaiyan et al. 1995). This part of FADD was therefore termed the death effector domain (DED). Overexpression of the C-terminal DD-containing part (FADD-DN), however, protected cells from CD95-mediated apoptosis and functioned as a dominant negative. This suggested that the N-terminus of FADD coupled to the cytotoxic machinery. In cells stably transfected with FADD-DN the DISC formed and FADD-DN was recruited to CD95 instead of the endogenous FADD. Analysis on 2D gels revealed that CAP3 and 4 were not part of the DISC anymore (Chinnaiyan et al. 1996b). These proteins were therefore prime candidates for the signaling molecules. Using nano-electrospray tandem mass spectrometry, sequence information of CAP3 and 4 was obtained that led to the retrieval of a full-length clone from a cDNA data base that contained all sequenced peptides (Muzio et al. 1996). This protein contained two DEDs at its N-terminus, and at its C-terminus the typical domain structure of an ICE-like protease. It was therefore termed FLICE (FADD-like ICE). FLICE was also cloned by two other groups and named MACH and Mch5 (Boldin et al. 1996; Fernandes-Alnemri et al. 1996). FLICE belongs to the group of cysteine proteases which are now called the caspases (Alnemri et al. 1996; see Sect. 8). FLICE is caspase-8. Identification of caspase-8 as being part of the DISC connected two different levels in apoptosis pathways, the receptor level with the level of the apoptosis executioner, the caspases.

Both TRADD and RIP can induce apoptosis and activate NF-κB when expressed in cells which is a typical feature of TNF-induced signaling (Hsu et al. 1995, 1996a; Park and Baichwal 1996; Ting et al. 1996). For TRADD, *in vivo*-binding to TNF-R1 could be established (Hsu et al. 1996b). RIP, however, does not seem to directly bind to CD95 or TNF-R1. In contrast to the original idea that RIP played a role in CD95 signaling, it was later identified to be crucial for TNF-R1-mediated NF-κB activation. In a mutant cell line that had lost expression of RIP, CD95 signaling was not affected, whereas TNF-R1-mediated NF-κB activation was blocked (Ting et al. 1996). After reconstitution with RIP, NF-κB activation in this cell line was restored. Recently, a new DD protein, MADD, was cloned and shown to bind to TNF-R1 activating ERK2 (Schievella et al. 1997).

4.2
Death Effector Domain Proteins

When it became obvious that caspase-8 contained two regions at its N-terminus that shared a weak sequence homology with the DED of FADD, this information could be used to screen databases for DED-containing proteins that might be involved in death receptor-mediated apoptosis pathways. The first protein to be identified was PEA-15 (Boldin et al. 1996; Muzio et al. 1996), a phosphoprotein of unknown function expressed in astrocytes. So far it has not been possible to link this protein with an apoptosis pathway. Using database searches Fernandes-Alnemri et al. (1996) and later Vincenz and Dixit

(1997) identified a new caspase (caspase-10, Mch4, FLICE2) that contained two DED proteins similar to caspase-8. Its role in apoptosis is unclear as it has not been identified on the protein level, due to the lack of antibody reagents. In the next "wave" of database searches an entire family of DED proteins was identified. They are encoded by herpes viruses of the gamma class such as herpes virus saimiri (HVS), the Kaposi sarcoma virus-associated herpes virus 8 (HHV-8) and two strains of moluscipox viruses (reviewed by Peter et al. 1997b). These proteins were called E8 proteins or v-FLIPs (viral FLICE inhibitory proteins). v-FLIPs have a unique structure, they consist of two DED proteins. Biochemical analysis of v-FLIP transfected cells showed that they bind to the CD95–FADD complex preventing caspase-8 recruitment and complete DISC formation. Interestingly, in v-FLIP transfected cells apoptosis signaling through all death receptors CD95, TNF-R1, DR4 and DR3 was impaired suggesting that they all use similar pathways to kill (Bertin et al. 1997; Hu et al. 1997; Thome et al. 1997). Having now about 20 DED-containing molecules available (Peter et al. 1997b), it should be possible using refined search algorithms to identify more DED proteins that might be involved in the regulation of death receptor-mediated apoptosis signaling.

5
Death Receptor Signaling Complexes

5.1
CD95

Caspase-8 was identified as part of the *in vivo* CD95 DISC (Fig. 2). This suggested that caspase-8 activation occurred at the DISC level. We have recently confirmed that the entire cytoplasmic caspase-8 in cells can be converted into active caspase-8 subunits and this activation occurs at the DISC level (Medema et al. 1997a). After stimulation FADD and caspase-8 are recruited to CD95 within seconds after receptor engagement. Binding of caspase-8 likely causes a structural change resulting in autoproteolytic activation of caspase-8. The active subunits p10 and p18 are released into the cytoplasm and part of the prodomain stays bound to the DISC. Using specific anti-caspase-8 monoclonal antibodies it became clear that from all the eight published caspase-8 isoforms, two were predominantly expressed on the protein level in 13 different cell lines tested (Scaffidi et al. 1997). These isoforms are caspase-8/a and caspase-8/b. Both isoforms are recruited to the DISC and processed with similar kinetics. Recently, two reports demonstrated that recombinant caspase-8 lacking the prodomain could cleave caspase-8 in vitro suggesting an amplification step with caspase-8 at the top of a caspase cascade (Srinivasula et al. 1996a; Muzio et al. 1997). However, using the *in vivo* DISC we could not confirm this observation (Medema et al. 1997a). It is therefore possible that recombinant caspase-8 lacking the prodomain displays a different substrate specificity as

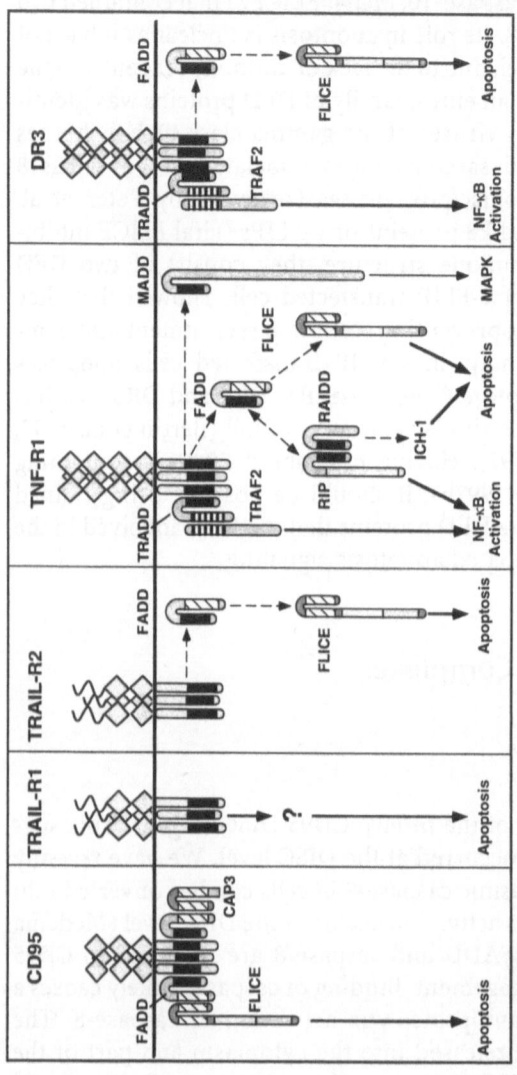

Fig. 2. Putative signaling pathways of death receptors. Depicted are only signaling molecules that have been shown or suggested to directly bind to receptors containing either a death domain, a death effector domain or a TRAF (TNF receptor-associated factor) domain. For associations/activations indicated by a *hatched arrow* no *in vivo* association has been demonstrated to occur in a ligand-dependent fashion

compared to full length caspase-8 *in vivo*. It is assumed right now that the active caspase-8 subunits cleave various death substrates including other caspases such as caspase-3 leading to the execution of apoptosis. Recently, a protein DAXX was identified that binds to the CD95 DD directly activating Jun N-terminal kinases (JNK) (Yang et al. 1997b). JNKs are being controversially discussed to be activated upon CD95 triggering (Cahill et al. 1996; Latinis and Koretzky 1996; Wilson et al. 1996; Goillot et al. 1997; Juo et al. 1997; Lenczowski et al. 1997; Nishina et al. 1997).

5.2
TNF-R1

Both FADD-DN and functional inactive caspase-8 not only blocked CD95 signaling but also inhibited TNF-R1-induced signaling in overexpression systems, suggesting that a system was coupling to TNF-R1 similar to the CD95 signaling complex (Boldin et al. 1996; Chinnaiyan et al. 1996b). However, overexpression of a DED containing molecule such as caspase-8 or FADD could just block any other DED protein binding to the receptor and thereby block the signaling pathway. Consistent with this assumption, we and others have not been able to detect a direct interaction of FADD with TNF-R1 (unpubl. data; H.-B. Shu and D. Goeddel, pers. comm.). Hence, either TNF-R1 uses a different set of signaling molecules (e.g. a FADD-like molecule) or FADD couples indirectly to this receptor (Fig. 2). Therefore, the question whether caspase-8 would be activated by TNF-R1 was believed to provide an answer to this question. Testing cells stimulated with TNFα demonstrated that caspase-8 was processed (unpubl. data) suggesting that TNF-R1 indeed uses a similar signaling pathway for induction of apoptosis. It is unclear at the moment whether caspase-8 is activated by association with TNF-R1 or whether its activation is a result of a secondary activation event. In contrast to CD95, TNF-R1 signaling resulting in the induction of cell death seems to be more complex. A DD containing protein was recently identified that bound to TNF-R1 *in vitro* or when overexpressed in 293 cells. This protein, RAIDD/CRADD (Ahmad et al. 1997; Duan and Dixit 1997), carries the DD at its C-terminus and at its N-terminus it has homologies with the prodomain of caspase-2. It bound to TNF-R1 more efficiently in the presence of RIP. It was therefore suggested that RAIDD would engage TNF-R1 and induce activation of caspase-2. However, no in vivo data are available and the mechanism of caspase-2 activation remains to be determined.

5.3
DR3 and DR4

DR3 has been reported to bind FADD, TRADD, TRAF2, and caspase-8 (Chinnaiyan et al. 1996a; Fig. 2). Together with its TNF-R1-like structure it is therefore expected to have a similar signaling function as TNF-R1. Like TNF-R1 it can induce apoptosis and it activates NF-κB (Chinnaiyan et al. 1996a; Kitson et al. 1996; Marsters et al. 1996b; Bodmer et al. 1997; Degli-Esposti et al. 1997; Screaton et al. 1997). However, no ligand has been cloned and agonistic antibodies have not yet been generated. All information regarding its signaling pathways have been obtained using overexpression systems.

DR4 was the first receptor cloned to bind TRAIL (Pan et al. 1997). Subsequently, it was shown that TRAIL-R signaling may not involve binding of the receptor to FADD since FADD-DN did not inhibit TRAIL-induced apoptosis (Marsters et al. 1996a). However, testing BJAB cells expressing FADD-DN treated with TRAIL we observed a complete block of TRAIL-R signaling (S. M. Mariani, M. E. Peter unpubl. data). In addition, it was shown that TRAIL-R signaling involves activation of caspases (Mariani et al. 1997). Recently, a second TRAIL-R was cloned that was shown to bind to FADD (Walczak et al. 1997; Fig. 2). The existence of two TRAIL-Rs might help to explain the contradicting results with respect to involvement of FADD in the signaling pathway. Testing TRAIL sensitive BJAB cells we have recently detected caspase-8 activation upon addition of TRAIL (unpubl. results). It is likely that caspase-8 among other caspases is involved in the TRAIL-R signaling pathway.

6
Regulation of Apoptotic Signal Transduction Initiated by Death Receptors

Whether triggering of CD95 or TNF-R1 actually results in the onset of apoptosis depends on the cell line tested. Several cell lines display a high level of resistance towards the induction of apoptosis although their receptor expression is sufficient to warrant effective signaling. Often this resistance can be overcome by pretreatment with cycloheximide (Chx) or actinomycin D (Act D), suggesting that the cell maintains this state of apoptosis resistance by actively transcribing/translating one or more resistance factor(s). Several levels have been described at which the induction of apoptosis may be regulated. For CD95 a negative regulatory role for the C-terminus of the receptor has been suggested, since deletion of the last 15 amino acids of CD95 increases the sensitivity towards CD95-induced apoptosis (Itoh et al. 1993). Both in the yeast two-hybrid system and in *in vitro* binding studies this region of CD95 interacts with a protein phosphatase called Fas-associated phosphatase-1 (FAP-1) (Sato et al. 1995). Overexpression of FAP-1 partially inhibits CD95-induced apoptosis, while its expression inversely correlates with the sensitivity towards an anti-CD95 antibody in T helper cell subsets Th1 and Th2 (Zhang et al. 1997). In addition, the region of CD95 that is required for interaction with FAP-1 has recently been narrowed down to the last three amino acids and microinjection of this tripeptide into cells, which blocks binding of FAP-1 to

CD95 *in vitro*, facilitates CD95 signaling (Yanagisawa et al. 1997). These results are consistent with a negative regulatory role for FAP-1 in CD95 signaling. Yet, so far conclusions are solely based on correlations and no in vivo association of FAP-1 with CD95 has been detected nor has a substrate for FAP-1 been identified. Therefore, further investigations are needed to establish the exact function of FAP-1 in CD95 signal transduction. Signaling of TNF-R1 is not affected by FAP-1 nor does TNF-R1 bind FAP-1 (Sato et al. 1995). Recently, a protein, Sentrin, has been identified that binds to the DD of CD95 and TNF-R1, protecting cells against apoptosis (Okura et al. 1996). Sentrin has homologies to ubiquitin but its *in vivo* function is unknown.

In T cells, resistance towards CD95 mediated apoptosis is in part regulated at the level of the DISC. Short-term activated CD95 resistant T cells do not recruit full length caspase-8 into the DISC whereas in the DISC of long-term activated CD95 sensitive T cells pro-caspase-8 can be detected (Peter et al. 1997a).

A zinc finger binding protein called A20 has been identified which inhibits TNF-induced apoptosis and NF-κB activation, but has no effect on CD95 signaling (Opipari et al. 1992; Jäättelä et al. 1996). Expression of A20 is induced by TNF treatment through transcriptional activation of NF-κB (Krikos et al. 1992). Once expressed A20 binds to the TRAF1/TRAF2 complex and thereby likely inhibits NF-κB activation and thus its own expression (Song et al. 1996). However, how A20 inhibits apoptosis, which as mentioned is not mediated by TRAF1/TRAF2 but by TRADD instead, remains to be determined.

NF-κB is another molecule which negatively affects TNF-mediated apoptosis. Activation of NF-κB by treatment with TNF, IL-1β or artificially by overexpression of the transcriptional active NF-κB subunits p65 or c-Rel greatly enhances resistance to TNF, anti-IgM, daunorubicin and ionizing radiation-induced apoptosis (Beg and Baltimore 1996; Z. G. Liu et al. 1996; Van Antwerp et al. 1996; Wang et al. 1996a; Wu et al. 1996). Conversely, expression of the NF-κB inhibitor I-κBα breaks this resistance. Furthermore, embryonic fibroblasts from c-Rel$^{-/-}$ mice readily die upon TNF treatment in the absence of Chx while the c-Rel$^{+/+}$ cells are completely resistant (Beg and Baltimore 1996; Van Antwerp et al. 1996). As described, A20 is a good candidate for such an anti-apoptosis gene since it is driven by NF-κB and inhibits TNF-induced apoptosis (Krikos et al. 1992; Opipari et al. 1992; Song et al. 1996). Yet transfection of A20 in c-Rel$^{-/-}$ embryonic fibroblasts does not rescue these cells from TNF-induced apoptosis (Beg et al. 1996), suggesting that other resistance genes are involved as well. Interestingly, TNF itself is a potent activator of NF-κB and thus blocks its own induction of apoptosis (X. G. Liu et al. 1996). This may explain why in most cases TNF-mediated apoptosis is only observed in the presence of Act D or Chx to block transcription/translation (Kull and Besterman 1990; Leist et al. 1994; Trost and Lemasters 1994). Furthermore, this could also serve as an explanation for the fact that triggering of CD95 results in a more rapid induction of apoptosis than TNF (Nagata and

Golstein 1995). First, in most cells CD95 triggering does not lead to detectable NF-κB activation. This is probably due to the rapid induction of cell death, since in cell lines where apoptosis is blocked NF-κB activation can be detected (Mandal et al. 1996; Ponton et al. 1996). Second, CD95 triggering seems to be insensitive to this NF-κB-driven factor, since even in a cell line (T24) where a block in transcription/translation is required to obtain CD95-induced cell death and where CD95-induced NF-κB activation can be detected, expression of I-κBα does not alleviate the necessity to block RNA and protein synthesis for apoptosis to occur (Van Antwerp et al. 1996). NF-κB thus drives a cellular resistance towards killing by several apoptogenic factors, yet so far does not seem to affect CD95-induced apoptosis.

A separate level by which apoptosis is regulated represents the Bcl-2 family of proteins, which contains pro-apoptotic members like Bax, Bak and Bad and anti-apoptotic members like Bcl-2 and Bcl-x_L. Homo- and heterodimerization between pro- and anti-apoptotic members seem to determine the apoptotic fate of a cell (reviewed by Oltvai and Korsmeyer 1994; also see Chapter by O'Connor and Strasser in this Volume). Indeed, overexpression of Bcl-2/Bcl-x_L abrogates apoptosis induced by ionizing radiation, chemotherapeutic drugs and viral infection (Boise et al. 1993; Cory 1995), while overexpression of Bax promotes apoptosis (Oltvai et al. 1993). However, recent evidence suggest that dimerization of Bax and Bcl-x_L is due to the presence of nonionic detergent in the dimerization assays and that these family members might not dimerize *in vivo* at all (Y. T. Hsu and Youle 1997). In accordance with this is the observation that Bcl-x_L mutants that can no longer heterodimerize with Bax can still inhibit apoptosis (Cheng et al. 1996) and vice versa, mutants of Bax which do not bind Bcl-x_L still promote apoptosis (Simonian et al. 1996). It therefore seems that these Bcl-2 family members, instead of heterodimerizing, compete for a common binding partner. Generally Bcl-2 or Bcl-x_L do not inhibit CD95-induced apoptosis. Either separate pathways are used by the different apoptogenic factors or CD95 utilizes two or more redundant pathways. The latter seems likely since in a few cell lines CD95-induced apoptosis can be blocked by overexpression of Bcl-2/Bcl-x_L (Itoh et al. 1993; Mandal et al. 1996), suggesting that in these cells redundancy is lost. Overexpression of Bcl-2/Bcl-x_L clearly blocks the apoptosis pathway upstream of caspase-3 (Monney et al. 1996). Several models, which are so far mainly focused on the involvement of mitochondria, have evolved to explain the anti-apoptotic effects exerted by Bcl-2/Bcl-x_L. Recent studies showed that in transient transfections, Bcl-x_L can bind to caspase-8 and suggested that this binding would inhibit the activation of caspase-8 (Chinnaiyan et al. 1997). However, in MCF7-Fas cells, which were made resistant towards CD95-mediated apoptosis by overexpression of Bcl-x_L, no association between caspase-8 and Bcl-x_L was detected (Medema et al. 1998). In addition, activation of caspase-8 was also unaffected. X-ray and NMR structure of Bcl-x_L revealed a similarity to the pore-forming domains of bacterial toxins (Muchmore et al. 1996). Both Bcl-2 and Bcl-x_L have now been shown to form channels in artificial lipid layers with a selectivity for K^+ (Minn et al. 1997; Schendel et al. 1997).

Thus, Bcl-2 family members might form channels in mitochondria with different selectivity which could influence the outcome of apoptosis. Sequence comparison reveals a domain (BH4) in the anti-apoptotic members which is not present in the pro-apoptotic members. This domain is responsible for binding of the kinase raf-1 and targeting raf-1 to the mitochondria (Wang et al. 1996b). Raf-1 can phosphorylate the pro-apoptotic member Bad, which then loses its affinity for Bcl-x_L (Zha et al. 1996) leaving Bcl-x_L free to exert its anti-apoptotic activities. Despite the clear effects of Bcl-2/Bcl-x_L, these features do not explain how these Bcl-2 family members can block the activation of caspase-3. Two other models, however, can address this issue. Upon induction of apoptosis two pro-apoptotic proteins, apoptosis inducing factor (AIF) and cytochrome c (Cyt c), have been shown to be released from mitochondria (X. Liu et al. 1996; Susin et al. 1996). Both proteins induce apoptosis in a cell free system. Cyt c in its reduced form induces activation of caspase-3. AIF is a 50 kDa protease itself and is blocked by N-benzyloxycarbonyl-Val-Ala-Asp.fluoromethylketone (zVAD.fmk), an antagonist of caspases that is also an efficient inhibitor of apoptosis in cells. Using mitochondria from Bcl-2/Bcl-x_L overexpressing cells in a cell free system revealed that this expression blocks the release of both these factors and thus could inhibit the induction of apoptosis (Susin et al. 1996; Kluck et al. 1997; Yang et al. 1997a). How Bcl-2/Bcl-x_L can block this release or whether this can be connected to its pore-forming capacities remains to be determined but will give an exciting insight into the action of the Bcl-2 family members.

7
Alternative Death Receptor Signaling Pathways

7.1
The Sphingomyelin Pathway

The second messenger ceramide is produced by hydrolysis of the plasma membrane phospholipid sphingomyelin or via de novo synthesis by the ceramide synthetase (see Chapter by Malisan et al. this Volume). Sphingomyelin degradation is catalyzed by sphingomyelinase (SMase), a sphingomyelin-specific form of phospholipase C. Two forms of SMases have been identified based on their pH optima. Acidic SMase (pH optimum 4.5–5) resides in lysosomes and was also found in plasma membranes. Neutral SMase has a pH optimum of 7.4. Ceramide signaling involves a number of direct targets such as a ceramide-activated protein kinase (CAP kinase), a ceramide-activated protein phosphatase and the protein kinase C isoform ζ. Ceramide production has been reported following a number of apoptotic stimuli including ionizing radiation (Haimovitz-Friedman et al. 1994) or daunorubicin treatment (Bose et al. 1995; Jaffrezou et al. 1996). In addition, triggering of the death receptors CD95 and TNF-R1 has been shown to activate the sphingomyelin pathway. However, although a cell permeable synthetic ceramide analog, C2-ceramide, can induce apoptosis in some cell types (Jarvis et al. 1994; Sweeney et al. 1996), recent reports suggest that ceramide production is likely to be independent of cell death and may not be involved in CD95 signaling (Sillence and Allan 1997; Watts et al. 1997).

TNF-R1 has been shown to activate neutral-SMase through Fan, a protein that interacts with a stretch of nine amino acids upstream of the death domain. A dominant negative mutant of Fan is able to block TNF-R1-mediated neutral-SMase activation completely without affecting cell death (Adam-Klages et al. 1996). However, recent data from two different groups indicate that all functions exerted by TNF-R1 require a functional DD. First, overexpression of a trimerized TNF-R1 DD was sufficient to induce apoptosis and to activate NF-κB (Vandevoorde et al. 1997). Second, TNF-R1 knock-out mice expressing a TNF-R1 transgene lacking the 30 terminal amino acids have the same phenotype as TNF-R1$^{-/-}$ mice. They are resistant to endotoxic shock and susceptible to Listeria infection (T. Plitz and K. Pfeffer, pers. comm.). All these data suggest that the juxtamembrane region of TNF-R1 is not required for the main functions of the receptor. Hence, the relevance of FAN binding and activation of neutral SMase needs to be shown.

Similar to TNF-R1, mutant CD95, which is defective in cell death signaling, is still able to activate neutral SMase (Cifone et al. 1995). Therefore, neutral SMase mediated ceramide production is independent of cell death signaling by CD95 and TNF-R1. In addition, cells from patients with Nieman-Pick disease type A, lacking functional acidic SMase, have only been reported to be resistant to ionizing radiation and not to CD95- or TNF-R1-induced apoptosis (Santana et al. 1996). Therefore, although both neutral and acidic SMase have been implicated in ceramide production and cell death signaling through CD95 and TNF-R1, neither seems to be essential or sufficient for apoptosis induction by these receptors. Moreover, recent publications place ceramide production downstream of caspases, as it can be blocked by caspase inhibitors such as CrmA, zVAD or DEVD (Gamen et al. 1996; Dbaibo et al. 1997; Sillence et al. 1997). Therefore, ceramide production seems to be a more downstream signal which may be independent of apoptosis, since it is also observed after Ca^{2+} ionophore treatment (Sillence and Allan 1997) or in senescent cells (Venable et al. 1995) without being associated with cell death. Alternatively, it could serve as a secondary modulatory pathway.

7.2
Activation of Stress-Activated Protein Kinases (SAPK/JNK)

Stress-activated protein kinases (SAPK), also known as the Jun N-terminal kinases (JNK) are part of an alternative MAP kinase cascade activated by a diverse set of stress inducers like UV irradiation, heat shock and protein synthesis inhibitors (Hibi et al. 1993; Derijard et al. 1994; Meier et al. 1996; Zanke et al. 1996). TNF-R1 has also been shown to activate SAPK/JNK. However, this activation is not linked to cell death, as it is not blocked by a dominant negative FADD mutant, which blocks apoptosis induction by TNF-R1. In addition, SAPK/JNK activation by TNF-R1 was shown to be mediated by TRAF2 and RIP via a noncytotoxic pathway (Z. G. Liu et al. 1996; Natoli et al. 1997). Therefore, the cytotoxic signal of TNF-R1, mediated through TRADD, and the activation of SAPK/JNK, mediated through TRAF2 and RIP, are two separate pathways that diverge at the level of the receptor bound signaling molecules.

The CD95 receptor has also been reported to activate SAPK/JNK, although TRAF2 is not associated with this receptor (Cahill et al. 1996; Latinis and Koretzky 1996). Moreover, SAPK/JNK activation is located downstream of caspases in the CD95 signaling pathway, since it can be blocked by zVAD and CrmA (Cahill et al. 1996; Juo et al. 1997). In addition, SEK1, which is a direct activator of SAPK/JNK in response to environmental stress or mitogenic factors, is able to inhibit SAPK/JNK activation when expressed as a dominant negative mutant without affecting CD95-mediated apoptosis (Lenczowski et al. 1997). This again suggests that the pathway of SAPK/JNK activation is independent from apoptosis induction.

However, there are contradictory results placing SAPK/JNK activation right into the apoptotic pathway: for example, ASK1, a direct activator of SEK1, was found to inhibit TNF mediated cytotoxicity when expressed as a dominant negative mutant, suggesting that ASK1 is a mediator of TNF induced apoptosis (Ichijo et al. 1997). Conversely, SEK1 itself has been reported to inhibit CD95 mediated apoptosis, since SEK1 deficient cells were found to be more susceptible to CD95 triggering (Nishina et al. 1997). In addition, a recent publication reported that SAPK/JNK activation is mediated by DAXX, a protein that is supposed to interact with the death domain of the CD95 receptor (Yang et al. 1997b). Therefore, SAPK/JNK activation would be independent of caspases. Also, it was found that a dominant negative SEK1 mutant is able to block both SAPK/JNK activation and cell death in certain cells (Yang et al. 1997b). Therefore, a secondary apoptotic pathway may exist in certain cells that is dependent on SAPK/JNK activation and that is not mediated by caspases. However, the relevance of this pathway in biological systems needs to be established.

8
The Role of Downstream Caspases in Death Receptor Signaling

Caspase-8 (FLICE/MACH/Mch5) belongs to a growing family of cysteine proteases, recently named caspases (Alnemri et al. 1996).

A direct link between caspase activation and death receptor triggering was established by cloning of caspase-8 as part of the CD95 DISC (Muzio et al. 1996). The proform of caspase-8 is recruited to the multimerized receptor and then likely activated by autoproteolytic cleavage at the DISC (Medema et al. 1997a). Therefore, caspase-8 is the most upstream caspase in the CD95 pathway. Either caspase-8 itself or a caspase-8-like caspase such as caspase-10 may be involved in a similar fashion in the signal transduction of the other death receptors.

Although overexpression of caspases in mammalian or in insect cells induces apoptosis, only caspase-3, caspase-6, caspase-7 and caspase-8 have been shown to be activated *in vivo* upon triggering of the death receptors (Duan et al. 1996; Orth et al. 1996; Schlegel et al. 1996; Faleiro et al. 1997; Medema et al. 1997a; Scaffidi et al. 1997). Caspase-8 has been shown *in vitro* to cleave caspase-3, caspase-4, caspase-7, caspase-9 and caspase-10 directly, whereas caspase-2 and caspase-6 were cleaved indirectly by other caspase-8-activatable caspases present in cellular extracts (Muzio et al. 1997). Therefore, caspase-8 is able to start a cascade of caspases. The order of caspases in this cascade is not clear so far. Orth et al. (1996) place caspase-6 upstream of caspase-3 and caspase-7 whereas it has also been demonstrated that activated caspase-3 can in turn cleave and activate caspase-6, caspase-7 and caspase-9 (Fernandes-Alnemri et al. 1995, 1996; Srinivasula et al. 1996b). However, there is no report so far demonstrating that a single caspase is crucial for apoptosis signaling by death receptors. Caspase-1 has been proposed to be a key molecule in CD95-mediated apoptosis, and thymocytes from caspase-1$^{-/-}$ mice have been reported to be resistant to CD95-induced cell death (Kuida et al. 1995). However, others could not find an impairment of apoptosis in caspase-1$^{-/-}$ mice (Li et al. 1995b; Smith et al. 1997) or failed to demonstrate activation of caspase-1 upon CD95 triggering (Muzio et al. 1997). Therefore, either caspase-1 does not play a role in apoptosis signaling through the death receptors, or another caspase-1-like caspase can substitute for its function in different cellular contexts. Similarly, mice deficient for caspase-3 showed only impairment of brain development, whereas thymocytes were not affected at all (Kuida et al. 1996). However, due to the redundancy of different caspases with caspase-3-like activity, the knockout technology may not be suitable to investigate the role of a single caspase in apoptosis signaling.

Recently, a direct link between caspase-3 and DNA fragmentation was found by cloning a heterodimering factor, DFF, that was activated to induce DNA fragmentation on isolated nuclei after cleavage of its 45-kDa component by caspase-3 (Liu et al. 1997). Therefore, the theoretical possibility exists of a death receptor signaling pathway involving caspases only. CD95 could activate caspase-8 that cleaves caspase-3 which in turn could activate DFF. This scenario would not require mitochondria.

Further studies are necessary to unravel the caspase cascade induced by the different death receptors and to identify crucial targets for caspases that establish the link between caspase activation and more downstream events in apoptosis.

9
Viral Inhibitors of Apoptosis

Multicellular organisms are constantly under attack by more or less harmful agents such as bacteria, viruses or other pathogens. To survive these attacks the immune system utilizes apoptosis as a defense strategy. For example, virus-specific T-lymphocytes express the CD95 ligand to kill CD95 expressing infected cells or initiate the apoptotic machinery through the perforin/granzyme system which leads to the direct activation of caspase-8 by granzyme B (Medema et al. 1997b). Alternatively, infected cells may sense the presence of viruses intracellularly and respond by committing apoptotic suicide.

Viruses have developed different strategies to avoid the host's apoptotic response. For example, human cytomegalovirus (CMV) expresses several genes that solely function to avoid presentation of viral peptides by MHC class I and thereby prevent activation of cytotoxic T-lymphocytes. Interestingly, due to the absence of endogenous MHC class I molecules on the surface of the infected cell recognition by natural killer cells is also inhibited by expression of viral MHC class I-like molecules of CMV (Farrell et al. 1997; Reyburn et al. 1997).

Some viruses have adapted to the host's apoptotic defense by expressing genes that are able to directly interfere with the apoptotic signaling machinery, and thereby have the advantage of increased viral replication.

9.1
Viral Inhibitors of Caspases

Cowpox viruses encode different genes that are able to affect the host cytokine response, for example by synthesizing virus encoded cytokines that antagonize the effect of host cytokines mediating antiviral processes or by interfering with the interaction between a cytokine and its receptor, for example by secreting soluble receptors for tumor necrosis factor. Another strategy is to inhibit the synthesis and release of cytokines from infected cells (Pickup 1994). One of these inhibitors is crmA which was found to inhibit the release of mature IL-1β by inhibiting caspase-1 (Ray et al. 1992). As caspase-1 and other caspases were recognized as being important for apoptosis it was found that crmA inhibited not only IL-1β maturation, but also apoptosis induced by CD95 or TNF-R1 (Enari et al. 1995; Los et al. 1995; Tewari et al. 1995a). CrmA belongs to a family of serine protease inhibitors (serpins) and it can form a stable complex with the cysteine protease caspase-1 (Komiyama et al. 1994) and also inhibit caspase-3 (Tewari et al. 1995b). CrmA protects host cells by inhibiting direct

CTL mediated apoptosis through the CD95 dependent pathway (Tewari et al. 1995c).

Other viruses also code for crmA related proteins such as the vaccinia virus B13R (SPI-2) (Dobbelstein and Shenk 1996; Kettle et al. 1997) or the rabbitpox-virus (SPI-2) (Brooks et al. 1995). These proteins have been demonstrated to protect from apoptosis by a similar mechanism as crmA.

Another viral inhibitor of caspases is the baculovirus p35 protein which was discovered to inhibit apoptosis in infected insect cells (Clem et al. 1991). p35 is an early gene and is targeted to the cytosol of infected cells (Hershberger et al. 1994). It was found to be a competitive inhibitor of CED-3 and of other mammalian caspases (Xue and Horvitz 1995) and forms a stable complex with the caspase after being cleaved by the very same caspase (Bump et al. 1995). Therefore, the cleavage site within p35 is essential for its function (Bertin et al. 1996). Similar to crmA, p35 was also found to inhibit mammalian cell death induced through CD95 or TNF-R1 (Beidler et al. 1995). Interestingly, p35 is necessary for the efficient replication of baculovirus, as p35 mutant strains show an approximately 1000 times reduced replication efficiency and induce massive apoptosis in infected cells (Clem et al. 1991; Clem and Miller 1993; Lerch and Friesen 1993).

9.2
Viral Bcl-2 Homologues

Bcl-2 and its family members are proteins that can inhibit apoptosis induction by a variety of stimuli through a mechanism that is not completely understood so far (see Chapter by O'Connor and Strasser, this Volume). However, several viruses code for genes that show homology to Bcl-2 not only at the protein level but also in functional terms. One of the best-studied examples is the adenovirus E1B 19kDa protein. Adenovirus infection leads to the induction of host cell apoptosis that is dependent on the adenovirus E1A protein and inhibited by E1B 19kDa. E1A induced cell death results in accumulation of wildtype p53 and subsequent upregulation of Bax and other apoptosis inducing factors (Debbas and White 1993; McCurrach et al. 1997) as well as in activation of caspase-3 and subsequent cleavage of lamin and PARP (Boulakia et al. 1996; Rao et al. 1996). It has been shown that E1A induced apoptosis as well as the above described effects can be blocked by E1B 19kDa or by Bcl-2 demonstrating the functional homology of these two proteins (Boulakia et al. 1996; Rao et al. 1996). Therefore, adenovirus needs E1B or Bcl-2 expression in order to lead to high frequency transformation of cells (Rao et al. 1992). As the E1B 19kDa protein has been shown to be a functional equivalent of the mammalian inhibitor Bcl-2 or Bcl-x_L, the ability of E1B 19kDa to inhibit CD95- or TNF-induced cell death is questionable and reports span from complete inhibition (Hashimoto et al. 1991; White et al. 1995) through partial inhibition (Memon et al. 1995) to no inhibition (Moreno et al. 1996; Huang et al. 1997).

Recently, the CELO adenovirus protein GAM-1 has also been shown to inhibit apoptosis in a similar fashion to E1B 19kDa or Bcl-2 although no homology of GAM-1 and Bcl-2 exists (Chiocca et al. 1997). Therefore, GAM-1 may represent another unknown type of viral apoptosis inhibitor.

Other viruses have also been found to encode a Bcl-2 homologous protein that inhibits different kinds of apoptosis. For example, the Epstein–Barr virus immediate early protein BHRF1 inhibits apoptosis induced by DNA damaging agents, CD95 stimulation or TNFα and can substitute for E1B 19kDa in adenovirus replication (Tarodi et al. 1994; Kawanishi 1997). The African swine fever virus also contains two Bcl-2 homologous genes designated LMW5-HL and A179L that both have anti-apoptotic function (Neilan et al. 1993; Afonso et al. 1996; Brun et al. 1996). Another example is the group of γ-herpes viruses that code for Bcl-2-like proteins such as KSbcl-2 of human herpes virus-8 or the ORF16 of herpes virus saimiri (Cheng et al. 1997).

9.3
The Family of IAPs

In a search for p35-like proteins in other baculoviruses a new class of apoptosis inhibitors, named IAPs (inhibitors of apoptosis) was discovered. Two baculovirus IAPs named Cp-IAP and Op-IAP were found to be functionally analogous to p35, but act at a different level of apoptosis inhibition (Crook et al. 1993; Birnbaum et al. 1994). IAPs contain a zinc finger-like domain at their C-terminus, named RING-finger, and a Cys/His motif at their N-terminus, named BIR motif (for baculovirus IAP repeat), both necessary for apoptosis protection (Clem and Miller 1994). Recently, Drosophila IAP-like proteins, designated DIAP1 and DIAP2 (DILP/DIAP) were cloned that inhibit cell death in Drosophila (Hay et al. 1995; Duckett et al. 1996; Liston et al. 1996). Interestingly, also several human IAP homologues were cloned and found to be preferentially expressed in T cells (ITA) (Digby et al. 1996), to bind to the TNF-R2 signaling complex (cIAP1 (HIAP2) and cIAP2 (HIAP1)) through interaction with TRAF1 and TRAF2 (Rothe et al. 1995; Liston et al. 1996), to inhibit neuronal apoptosis (NAIP) (Liston et al. 1996) or caspase-1 induced cell death (HILP/xIAP) (Ducket et al. 1996; Liston et al. 1996). However, the mechanism by which the mammalian IAP-like proteins function as apoptosis inhibitors remains elusive.

9.4
Other Viral Anti-Apoptotic Proteins

Several other viruses have developed strategies to inhibit host cell apoptosis by known or unknown proteins. One mechanism is the expression of soluble receptors that compete for binding of cytotoxic molecules such as TNFα, which is one of the strategies of cowpox viruses used to affect the host cytokine response, as described in Section 9.1. Another example is the myxoma virus that also codes for a TNF-R homologue which is expressed as a secreted glycoprotein (Schreiber et al. 1997). Similarly, human herpes virus 2 inhibits the cell surface expression of CD95 ligand and avoids suicide of infected CD95 positive cells (Sieg et al. 1996). Interestingly, vaccinia virus WR was found to inhibit CD95-induced cell death of infected cells; however, the mechanism or the responsible gene product is unknown so far (Heinkelein et al. 1996).

9.5
DED Containing Viral Proteins

DED containing proteins such as FADD or caspase-8 are known key players in the first steps of apoptosis induction (see Sect. 4.2). Recently, the search for other DED containing proteins led to the discovery of a new family of anti-apoptotic viral proteins. Members of this family are the E8 protein of equine herpes virus 2, the ORF71 of human herpes virus 8 and herpes virus saimiri, a protein encoded by the bovine herpes virus 4 and two genes (ORF159L and ORF160L) of human moluscipox virus. All these proteins were found to contain two DED motives (Bertin et al. 1997; Hu et al. 1997; Thome et al. 1997). Overexpression of these proteins results in protection from CD95-, TNF-R1-, DR3- and DR4-induced apoptosis. The mechanism by which these viral proteins interfere with the apoptotic machinery of the host is by association with FADD or a homologous protein involved in the signal transduction of the other death receptors distinct from CD95 and thereby inhibiting the association as well as the activation of caspase-8 or a related caspase by the receptor signaling complex (Thome et al. 1997). Therefore, this family of viral proteins was named v-FLIPs (for *viral FLICE-inhibitory* proteins). v-FLIPs are the first known anti-apoptotic viral proteins that interfere with the most upstream events in apoptosis signaling of all death receptors known so far. For the human herpes virus saimiri (ORF71)-FLIP it was shown that the protein is expressed late in the lytic cycle and thereby renders the cells resistant to CD95-induced apoptosis. Therefore, HVS (ORF71)-FLIP may protect infected cells from premature apoptosis induced by viral overload (Thome et al. 1997).

Interestingly, several viruses code for more than one gene that inhibits apoptosis at various levels. For example, in addition to the FLIP protein the family of γ-herpes viruses also contains a Bcl-2-like protein as mentioned above (see Sect. 9.2). Similarly, the African swine fewer virus encodes two Bcl-2 homologues and one IAP-like molecule (Neilan et al. 1993; Chacon et al. 1995; Brun et al. 1996; Afonso et al. 1996). Therefore some viruses may take advantage of several different anti-apoptotic functions that are able to protect the infected cell from apoptosis induced by a variety of stimuli.

In contrast to the viral Bcl-2-like molecules, the mammalian homologues of IAPs were discovered after the viral apoptosis inhibitory proteins were found. This is also the case for the family of v-FLIPs, where mammalian counterparts have recently been discovered. Therefore, the study of viral strategies to inhibit apoptosis may lead to the discovery of unknown mammalian apoptosis modulators. As viruses have developed strategies to interfere with all major steps in apoptosis signaling known so far it can be postulated that by resolving unknown mechanisms of viral anti-apoptotic proteins, new signaling events in mammalian apoptosis can be found.

10
Conclusions

This overview describes the variety of death receptors and the signaling pathways they use to induce apoptosis, as far as they are known at the present time. In addition, we describe checkpoints in these signaling pathways at which cellular proteins or proteins from infectious agents such as viruses interfere with and block apoptosis. Our present view of the different functions of the various death systems is still fairly restricted and it can be anticipated that more death receptors and elements of their signaling pathways will be discovered in the future. Therefore, it will be the main object of research to study the biology of such systems and to put order into the respective signaling cascades. Thus, some of the major questions such as why does the organism need so many different death systems, when and how do they operate and what assures their specificity will be solved.

References

Abreu-Martin MT, Vidrich A, Lynch DH, Targan SR (1995) Divergent induction of apoptosis and IL-8 secretion in HT-29 cells in response to TNF-alpha and ligation of Fas antigen. J Immunol 155:4147–4154

Adachi M, Watanabe-Fukunaga R, Nagata S (1993) Aberrant transcription caused by the insertion of an early transposable element in an intron of the Fas antigen gene of lpr mice. Proc Natl Acad Sci USA 90:1756–1760

Adachi M, Suematsu S, Kondo T, Ogasawara J, Tanaka T, Yoshida N, Nagata S (1995) Targeted mutation in the Fas gene causes hyperplasia in peripheral lymphoid organs and liver. Nat Genet 11:294–300

Adam-Klages S, Adam D, Wiegmann K, Struve S, Kolanus W, Schneider-Mergener J, Kronke M (1996) FAN, a novel WD-repeat protein, couples the p55 TNF-receptor to neutral sphingomyelinase. Cell 86:937–947

Afonso CL, Neilan JG, Kutish GF, Rock DL (1996) An African swine fever virus Bcl-2 homolog, 5-HL, suppresses apoptotic cell death. J Virol 70:4858–4863

Aggarwal BB, Singh S, LaPushin R, Totpal K (1995) Fas antigen signals proliferation of normal human diploid fibroblast and its mechanism is different from tumor necrosis factor receptor. FEBS Lett 364:5–8

Ahmad M, Srinivasula SM, Wang L, Talanian RV, Litwack G, Fernandes-Alnemri T, Alnemri ES (1997) CRADD, a novel human apoptotic adaptor molecule for caspase-2, and FasL/tumor necrosis factor receptor-interacting protein RIP. Cancer Res 57:615–619

Alderson MR, Armitage RJ, Maraskovsky E, Tough TW, Roux E, Schooley K, Ramsdell F, Lynch DH (1993) Fas transduces activation signals in normal human T lymphocytes. J Exp Med 178:2231–2235

Alderson MR, Tough TW, Davis Smith T, Braddy S, Falk B, Schooley KA, Goodwin RG, Smith CA, Ramsdell F, Lynch DH (1995) Fas ligand mediates activation-induced cell death in human T lymphocytes. J Exp Med 181:71–77

Ales Martinez JE, Scott DW, Phipps RP, Casnellie JE, Kroemer G, Martinez C, Pezzi L (1992) Cross-linking of surface IgM or IgD causes differential biological effects in spite of overlap in tyrosine (de)phosphorylation profile. Eur J Immunol 22:845–850

Allison J, Georgiou HM, Strasser A, Vaux DL (1997) Transgenic expression of CD95 ligand on islet beta cells induces a granulocytic infiltration but does not confer immune privilege upon islet allografts. Proc Natl Acad Sci USA 94:3943–3947

Alnemri ES, Livingston DJ, Nicholson DW, Salvesen G, Thornberry NA, Wong WW, Yuan J (1996) Human ICE/CED-3 protease nomenclature. Cell 87:171

Baker MB, Altman NH, Podack ER, Levy RB (1996) The role of cell-mediated cytotoxicity in acute GVHD after MHC-matched allogeneic bone marrow transplantation in mice. J Exp Med 183:2645–2656

Banner DW, D'Arcy A, Janes W, Gentz R, Schoenfeld HJ, Broger C, Loetscher H, Lesslauer W (1993) Crystal structure of the soluble human 55 kD TNF receptor–human TNF beta complex: implications for TNF receptor activation. Cell 73:431–445

Barker CF, Billingham RE (1977) Immunologically privileged sites. Adv Immunol 25:1–54

Beg AA, Baltimore D (1996) An essential role for NF-kappaB in preventing TNF-alpha-induced cell death. Science 274:782–784

Beidler DR, Tewari M, Friesen PD, Poirier G, Dixit VM (1995) The baculovirus p35 protein inhibits Fas- and tumor necrosis factor-induced apoptosis. J Biol Chem 270:16526–16528

Bellgrau D, Gold D, Selawry H, Moore J, Franzusoff A, Duke RC (1995) A role for CD95 ligand in preventing graft rejection. Nature 377:630–632

Bertin J, Mendrysa SM, LaCount DJ, Gaur S, Krebs JF, Armstrong RC, Tomaselli KJ, Friesen PD (1996) Apoptotic suppression by baculovirus p35 involves cleavage by and inhibition of a virus-induced CED-3/ICE-like protease. J Virol 70:6251–6259

Bertin J, Armstrong RC, Ottilie S, Martin DA, Wang Y, Banks S, Wang GH, Senkevich TG, Alnemri ES, Moss B, Lenardo MJ, Tomaselli KJ, Cohen JI (1997) Death effector domain-containing herpesvirus and poxvirus proteins inhibit both Fas- and TNFR1-induced apoptosis. Proc Natl Acad Sci USA 94:1172–1176

Beutler B, van Huffel C (1994) Unraveling function in the TNF ligand and receptor families. Science 264:667–8

Birnbaum MJ, Clem RJ, Miller LK (1994) An apoptosis-inhibiting gene from a nuclear polyhedrosis virus encoding a polypeptide with Cys/His sequence motifs. J Virol 68:2521–2528

Black RA, Rauch CT, Kozlosky CJ, Peschon JJ, Slack JL, Wolfson MF, Castner BJ, Stocking KL, Reddy P, Srinivasan S, Nelson N, Boiani N, Schooley KA, Gerhart M, Davis R, Fitzner JN, Johnson RS, Paxton RJ, March CJ, Cerretti DP (1997) A metalloproteinase disintegrin that releases tumour-necrosis factor-alpha from cells. Nature 385:729–733

Bodmer JL, Burns K, Schneider P, Hofmann K, Steiner V, Thome M, Bornand T, Hahne M, Schroter M, Becker K, Wilson A, French LE, Browning JL, MacDonald HR, Tschopp J (1997) TRAMP, a novel apoptosis-mediating receptor with sequence homology to tumor necrosis factor receptor 1 and Fas (Apo-1/CD95). Immunity 6:79–88

Boise LH, Gonzalez Garcia M, Postema CE, Ding L, Lindsten T, Turka LA, Mao X, Nunez G, Thompson CB (1993) Bcl-x, a bcl-2-related gene that functions as a dominant regulator of apoptotic cell death. Cell 74:597–608

Boldin MP, Varfolomeev EE, Pancer Z, Mett IL, Camonis JH, Wallach D (1995) A novel protein that interacts with the death domain of Fas/APO1 contains a sequence motif related to the death domain. J Biol Chem 270:7795–7798

Boldin MP, Goncharov TM, Goltsev YV, Wallach D (1996) Involvement of MACH, a novel MORT1/FADD-interacting protease, in Fas/APO-1- and TNF receptor-induced cell death. Cell 85:803–815

Bose R, Verheij M, Haimovitz-Friedman A, Scotto K, Fuks Z, Kolesnick R (1995) Ceramide synthase mediates daunorubicin-induced apoptosis: an alternative mechanism for generating death signals. Cell 82:405–414

Boulakia CA, Chen G, Ng FW, Teodoro JG, Branton PE, Nicholson DW, Poirier GG, Shore GC (1996) Bcl-2 and adenovirus E1B 19 kDa protein prevent E1A-induced processing of CPP32 and cleavage of poly(ADP-ribose) polymerase. Oncogene 12:529–535

Braun MY, Lowin B, French L, Acha Orbea H, Tschopp J (1996) Cytotoxic T cells deficient in both functional fas ligand and perforin show residual cytolytic activity yet lose their capacity to induce lethal acute graft-versus-host disease. J Exp Med 183:657–661

Brockhaus M, Schoenfeld HJ, Schlaeger EJ, Hunziker W, Lesslauer W, Loetscher H (1990) Identification of two types of tumor necrosis factor receptors on human cell lines by monoclonal antibodies. Proc Natl Acad Sci USA 87:3127–3131

Brooks MA, Ali AN, Turner PC, Moyer RW (1995) A rabbitpox virus serpin gene controls host range by inhibiting apoptosis in restrictive cells. J Virol 69:7688–7698

Brun A, Rivas C, Esteban M, Escribano JM, Alonso C (1996) African swine fever virus gene A179L, a viral homologue of bcl-2, protects cells from programmed cell death. Virology 225:227–230

Brunner T, Mogil RJ, LaFace D, Yoo NJ, Mahboubi A, Echeverri F, Martin SJ, Force WR, Lynch DH, Ware CF, et al. (1995) Cell-autonomous Fas (CD95)/Fas-ligand interaction mediates activation-induced apoptosis in T-cell hybridomas. Nature 373:441–444

Bump NJ, Hackett M, Hugunin M, Seshagiri S, Brady K, Chen P, Ferenz C, Franklin S, Ghayur T, Li P, et al. (1995) Inhibition of ICE family proteases by baculovirus antiapoptotic protein p35. Science 269:1885–1888

Cahill MA, Peter ME, Kischkel FC, Chinnaiyan AM, Dixit VM, Krammer PH, Nordheim A (1996) CD95 (APO-1/Fas) induces activation of SAP kinases downstream of ICE-like proteases. Oncogene 13:2087–2096

Camerini D, Walz G, Loenen WA, Borst J, Seed B (1991) The T cell activation antigen CD27 is a member of the nerve growth factor/tumor necrosis factor receptor gene family. J Immunol 147:3165–3169

Chacon MR, Almazan F, Nogal ML, Vinuela E, Rodriguez JF (1995) The African swine fever virus IAP homolog is a late structural polypeptide. Virology 214:670–674

Cheng EH, Levine B, Boise LH, Thompson CB, Hardwick JM (1996) Bax-independent inhibition of apoptosis by Bcl-XL. Nature 379:554–556

Cheng EH, Nicholas J, Bellows DS, Hayward GS, Guo HG, Reitz MS, Hardwick JM (1997) A Bcl-2 homolog encoded by Kaposi sarcoma-associated virus, human herpesvirus 8, inhibits apoptosis but does not heterodimerize with Bax or Bak. Proc Natl Acad Sci USA 94:690–694

Chinnaiyan AM, O'Rourke K, Tewari M, Dixit VM (1995) FADD, a novel death domain-containing protein, interacts with the death domain of Fas and initiates apoptosis. Cell 81:505–512

Chinnaiyan AM, O'Rourke K, Yu GL, Lyons RH, Garg M, Duan DR, Xing L, Gentz R, Ni J, Dixit VM (1996a) Signal transduction by DR3, a death domain-containing receptor related to TNFR-1 and CD95. Science 274:990–992

Chinnaiyan AM, Tepper CG, Seldin MF, O'Rourke K, Kischkel FC, Hellbardt S, Krammer PH, Peter ME, Dixit VM (1996b) FADD/MORT1 is a common mediator of CD95 (Fas/APO-1) and tumor necrosis factor receptor-induced apoptosis. J Biol Chem 271:4961–4965

Chinnaiyan AM, O'Rourke K, Lane BR, Dixit VM (1997) Interaction of CED-4 with CED-3 and CED-9: a molecular framework for cell death. Science 275:1122–1126

Chiocca S, Baker A, Cotten M (1997) Identification of a novel antiapoptotic protein, GAM-1, encoded by the CELO adenovirus. J Virol 71:3168–77

Cifone MG, Roncaioli P, De Maria R, Camarda G, Santoni A, Ruberti G, Testi R (1995) Multiple pathways originate at the Fas/APO-1 (CD95) receptor: sequential involvement of phosphatidylcholine-specific phospholipase C and acidic sphingomyelinase in the propagation of the apoptotic signal. EMBO J 14:5859–5868

Clem RJ, Miller LK (1993) Apoptosis reduces both the in vitro replication and the in vivo infectivity of a baculovirus. J Virol 67:3730–3738

Clem RJ, Miller LK (1994) Control of programmed cell death by the baculovirus genes p35 and iap. Mol Cell Biol 14:5212–5222

Clem RJ, Fechheimer M, Miller LK (1991) Prevention of apoptosis by a baculovirus gene during infection of insect cells. Science 254:1388–1390

Collins M (1995) Potential roles of apoptosis in viral pathogenesis. Am J Respir Crit Care Med 152:S20–S24

Cory S (1995) Regulation of lymphocyte survival by the BCL-2 family. Annu Rev Immunol 13:513–543

Crook NE, Clem RJ, Miller LK (1993) An apoptosis-inhibiting baculovirus gene with a zinc finger-like motif. J Virol 67:2168–2174

Dbaibo GS, Perry DK, Gamard CJ, Platt R, Poirier GG, Obeid LM, Hannun YA (1997) Cytokine response modifier A (CrmA) inhibits ceramide formation in response to tumor necrosis factor (TNF)-alpha: CrmA and Bcl-2 target distinct components in the apoptotic pathway. J Exp Med 185:481–490

Debatin KM, Fahrig Faissner A, Enenkel Stoodt S, Kreuz W, Benner A, Krammer PH (1994) High expression of APO-1 (CD95) on T lymphocytes from human immunodeficiency virus-1-infected children. Blood 83:3101–3103

Debbas M, White E (1993) Wild-type p53 mediates apoptosis by E1A, which is inhibited by E1B. Genes Dev 7:546–554

Decoster E, Vanhaesebroeck B, Vandenabeele P, Grooten J, Fiers W (1995) Generation and biological characterization of membrane-bound, uncleavable murine tumor necrosis factor. J Biol Chem 270:18473–18478

Degli-Esposti MA, Din WS, Cosman D, Smith CA, Goodwin RG (1997) AIR, a novel member of the TNF receptor family, is a strong inducer of apoptosis (submitted)

Dembic Z, Loetscher H, Gubler U, Pan YC, Lahm HW, Gentz R, Brockhaus M, Lesslauer W (1990) Two human TNF receptors have similar extracellular, but distinct intracellular, domain sequences. Cytokine 2:231–237

Derijard B, Hibi M, Wu IH, Barrett T, Su B, Deng T, Karin M, Davis RJ (1994) JNK1: a protein kinase stimulated by UV light and Ha-Ras that binds and phosphorylates the c-Jun activation domain. Cell 76:1025–1037

Dhein J, Daniel PT, Trauth BC, Oehm A, Moller P, Krammer PH (1992) Induction of apoptosis by monoclonal antibody anti-APO-1 class switch variants is dependent on cross-linking of APO-1 cell surface antigens. J Immunol 149:3166–3173

Dhein J, Walczak H, Baumler C, Debatin KM, Krammer PH (1995) Autocrine T-cell suicide mediated by APO-1/(Fas/CD95). Nature 373:438–441

Digby MR, Kimpton WG, York JJ, Connick TE, Lowenthal JW (1996) ITA, a vertebrate homologue of IAP that is expressed in T lymphocytes. DNA Cell Biol 15:981–988

Dobbelstein M, Shenk T (1996) Protection against apoptosis by the vaccinia virus SPI-2 (B13R) gene product. J Virol 70:6479–6485

Duan H, Orth K, Chinnaiyan AM, Poirier GG, Froelich CJ, He WW, Dixit VM (1996) ICE-LAP6, a novel member of the ICE/Ced-3 gene family, is activated by the cytotoxic T cell protease granzyme B. J Biol Chem 271:16720–16724

Duan H, Dixit VM (1997) RAIDD is a new "death" adaptor molecule. Nature 385:86–89

Duckett CS, Nava VE, Gedrich RW, Clem RJ, Van Dongen JL, Gilfillan MC, Shiels H, Hardwick JM, Thompson CB (1996) A conserved family of cellular genes related to the baculovirus iap gene and encoding apoptosis inhibitors. EMBO J 15:2685–2694

Durkop H, Latza U, Hummel M, Eitelbach F, Seed B, Stein H (1992) Molecular cloning and expression of a new member of the nerve growth factor receptor family that is characteristic for Hodgkin's disease. Cell 68:421–427

Eck MJ, Sprang SR (1989) The structure of tumor necrosis factor-alpha at 2.6 Å resolution. Implications for receptor binding. J Biol Chem 264:17595–17605

Eck MJ, Ultsch M, Rinderknecht E, de Vos AM, Sprang SR (1992) The structure of human lymphotoxin (tumor necrosis factor-beta) at 1.9-Å resolution. J Biol Chem 267:2119–2122

EgSeino K, Kayagaki N, Okumura K, Yagita H (1997) Anti tumor effect of locally produced CD95 ligand. Nat Med 3:165–170

Enari M, Hug H, Nagata S (1995) Involvement of an ICE-like protease in Fas-mediated apoptosis. Nature 375:78–81

Engelmann H, Holtmann H, Brakebusch C, Avni YS, Sarov I, Nophar Y, Hadas E, Leitner O, Wallach D (1990) Antibodies to a soluble form of a tumor necrosis factor (TNF) receptor have TNF-like activity. J Biol Chem 265:14497–14504

Espevik T, Waage A (1988) The involvement of tumor necrosis factor-alpha (TNF-alpha) in immunomodulation and in septic shock. Dev Biol Stand 69:139–142

Espevik T, Brockhaus M, Loetscher H, Nonstad U, Shalaby R (1990) Characterization of binding and biological effects of monoclonal antibodies against a human tumor necrosis factor receptor. J Exp Med 171:415–426

Faleiro L, Kobayashi R, Fearnhead H, Lazebnik Y (1997) Multiple species of CPP32 and Mch2 are the major active caspases present in apoptotic cells. EMBO J 16:2271–2281

Farber E (1994) Programmed cell death: necrosis versus apoptosis. Mod Pathol 7:605–609

Farrell HE, Vally H, Lynch DM, Fleming P, Shellam GR, Scalzo AA, Davis-Poynter NJ (1997) Inhibition of natural killer cells by a cytomegalovirus MHC class I homologue in vivo. Nature 386:510–514

Fernandes-Alnemri T, Takahashi A, Armstrong R, Krebs J, Fritz L, Tomaselli KJ, Wang L, Yu Z, Croce CM, Salveson G, et al. (1995) Mch3, a novel human apoptotic cysteine protease highly related to CPP32. Cancer Res 55:6045–6052

Fernandes-Alnemri T, Armstrong RC, Krebs J, Srinivasula SM, Wang L, Bullrich F, Fritz LC, Trapani JA, Tomaselli KJ, Litwack G, Alnemri ES (1996) In vitro activation of CPP32 and Mch3 by Mch4, a novel human apoptotic cysteine protease containing two FADD-like domains. Proc Natl Acad Sci USA 93:7464–7469

Fisher GH, Rosenberg FJ, Straus SE, Dale JK, Middleton LA, Lin AY, Strober W, Lenardo MJ, Puck JM (1995) Dominant interfering Fas gene mutations impair apoptosis in a human autoimmune lymphoproliferative syndrome. Cell 81:935–946

Freiberg RA, Spencer DM, Choate KA, Duh HJ, Schreiber SL, Crabtree GR, Khavari PA (1997) Fas signal transduction triggers either proliferation or apoptosis in human fibroblasts. J Invest Dermatol 108:215–219

Galle PR, Hofmann WJ, Walczak H, Schaller H, Otto G, Stremmel W, Krammer PH, Runkel L (1995) Involvement of the CD95 (APO-1/Fas) receptor and ligand in liver damage. J Exp Med 182:1223–1230

Gamen S, Marzo I, Anel A, Pineiro A, Naval J (1996) CPP32 inhibition prevents Fas-induced ceramide generation and apoptosis in human cells. FEBS Lett 390:232–237

Garcia I, Miyazaki Y, Araki K, Araki M, Lucas R, Grau GE, Milon G, Belkaid Y, Montixi C, Lesslauer W, et al. (1995) Transgenic mice expressing high levels of soluble TNF-R1 fusion protein are protected from lethal septic shock and cerebral malaria, and are highly sensitive to Listeria monocytogenes and Leishmania major infections. Eur J Immunol 25:2401–2407

Giordano C, Stassi G, De Maria R, Todaro M, Richiusa P, Papoff G, Ruberti G, Bagnasco M, Testi R, Galluzzo A (1997) Potential involvement of Fas and its ligand in the pathogenesis of Hashimoto's thyroiditis. Science 275:960–963

Glauser MP (1996) The inflammatory cytokines. New developments in the pathophysiology and treatment of septic shock. Drugs 52 Suppl 2:9–17

Goillot E, Raingeaud J, Ranger A, Tepper RI, Davis RJ, Harlow E, Sanchez I (1997) Mitogen-activated protein kinase-mediated Fas apoptotic signaling pathway. Proc Natl Acad Sci USA 94:3302–3307

Grell M, Douni E, Wajant H, Lohden M, Clauss M, Maxeiner B, Georgopoulos S, Lesslauer W, Kollias G, Pfizenmaier K, et al. (1995) The transmembrane form of tumor necrosis factor is the prime activating ligand of the 80 kDa tumor necrosis factor receptor. Cell 83:793–802

Griffith TS, Brunner T, Fletcher SM, Green DR, Ferguson TA (1995) Fas ligand-induced apoptosis as a mechanism of immune privilege. Science 270:1189–1192

Griffith TS, Yu X, Herndon JM, Green DR, Ferguson TA (1996) CD95-induced apoptosis of lymphocytes in an immune privileged site induces immunological tolerance. Immunity 5:7–16

Gu Y, Kuida K, Tsutsui H, Ku G, Hsiao K, Fleming MA, Hayashi N, Higashino K, Okamura H, Nakanishi K, Kurimoto M, Tanimoto T, Flavell RA, Sato V, Harding MW, Livingston DJ, Su MS (1997) Activation of interferon-gamma inducing factor mediated by interleukin-1beta converting enzyme. Science 275:206–209

Hahne M, Rimoldi D, Schroter M, Romero P, Schreier M, French LE, Schneider P, Bornand T, Fontana A, Lienard D, Cerottini J, Tschopp J (1996) Melanoma cell expression of Fas (Apo-1/CD95) ligand: implications for tumor immune escape. Science 274:1363–1366

Haimovitz Friedman A, Kan CC, Ehleiter D, Persaud RS, McLoughlin M, Fuks Z, Kolesnick RN (1994) Ionizing radiation acts on cellular membranes to generate ceramide and initiate apoptosis. J Exp Med 180:525–535

Hashimoto S, Ishii A, Yonehara S (1991) The E1B oncogene of adenovirus confers cellular resistance to cytotoxicity of tumor necrosis factor and monoclonal anti-Fas antibody. Int Immunol 3:343–351

Hay BA, Wassarman DA, Rubin GM (1995) Drosophila homologs of baculovirus inhibitor of apoptosis proteins function to block cell death. Cell 83:1253–1262

Hayward M, Fiedler-Nagy C (1987) Mechanisms of bone loss: rheumatoid arthritis, periodontal disease and osteoporosis. Agents Actions 22:251–254

Heinkelein M, Pilz S, Jassoy C (1996) Inhibition of CD95 (Fas/Apo1)-mediated apoptosis by vaccinia virus WR. Clin Exp Immunol 103:8–14

Hershberger PA, LaCount DJ, Friesen PD (1994) The apoptotic suppressor p35 is required early during baculovirus replication and is targeted to the cytosol of infected cells. J Virol 68:3467–3477

Hibi M, Lin A, Smeal T, Minden A, Karin M (1993) Identification of an oncoprotein- and UV-responsive protein kinase that binds and potentiates the c-Jun activation domain. Genes Dev 7:2135–2148

Hiramatsu N, Hayashi N, Katayama K, Mochizuki K, Kawanishi Y, Kasahara A, Fusamoto H, Kamada T (1994) Immunohistochemical detection of Fas antigen in liver tissue of patients with chronic hepatitis C. Hepatology 19:1354–1359

Hsu H, Xiong J, Goeddel DV (1995) The TNF receptor 1-associated protein TRADD signals cell death and NF-kappa B activation. Cell 81:495–504

Hsu H, Huang J, Shu HB, Baichwal V, Goeddel DV (1996a) TNF-dependent recruitment of the protein kinase RIP to the TNF receptor-1 signaling complex. Immunity 4:387–396

Hsu H, Shu HB, Pan MG, Goeddel DV (1996b) TRADD-TRAF2 and TRADD-FADD interactions define two distinct TNF receptor 1 signal transduction pathways. Cell 84:299–308

Hsu H, Solovyev I, Colombero A, Elliott R, Kelley M, Boyle WJ (1997) ATAR, a novel tumor necrosis factor receptor family member, signals through TRAF2 and TRAF5. J Biol Chem 272:13471–13474

Hsu YT, Youle RJ (1997) Nonionic detergents induce dimerization among members of the Bcl-2 family. J Biol Chem 272:13829–13834

Hu S, Vincenz C, Buller M, Dixit VM (1997) A novel family of viral death effector domain-containing molecules that inhibit both CD-95- and tumor necrosis factor receptor-1-induced apoptosis. J Biol Chem 272:9621–9624

Huang B, Eberstadt M, Olejniczak ET, Meadows RP, Fesik SW (1996) NMR structure and mutagenesis of the Fas (APO-1/CD95) death domain. Nature 384:638-641

Huang DC, Cory S, Strasser A (1997) Bcl-2, Bcl-XL and adenovirus protein E1B19kD are functionally equivalent in their ability to inhibit cell death. Oncogene 14:405-414

Ichijo H, Nishida E, Irie K, ten-Dijke P, Saitoh M, Moriguchi T, Takagi M, Matsumoto K, Miyazono K, Gotoh Y (1997) Induction of apoptosis by ASK1, a mammalian MAPKKK that activates SAPK/JNK and p38 signaling pathways. Science 275:90-94

Itoh N, Nagata S (1993) A novel protein domain required for apoptosis. Mutational analysis of human Fas antigen. J Biol Chem 268:10932-10937

Itoh N, Tsujimoto Y, Nagata S (1993) Effect of bcl-2 on Fas antigen-mediated cell death. J Immunol 151:621-627

Jäättelä M, Mouritzen H, Elling F, Bastholm L (1996) A20 zinc finger protein inhibits TNF and IL-1 signaling. J Immunol 156:1166-1173

Jaffrezou JP, Levade T, Bettaieb A, Andrieu N, Bezombes C, Maestre N, Vermeersch S, Rousse A, Laurent G (1996) Daunorubicin-induced apoptosis: triggering of ceramide generation through sphingomyelin hydrolysis. EMBO J 15:2417-2424

Jarvis WD, Fornari FA Jr, Browning JL, Gewirtz DA, Kolesnick RN, Grant S (1994) Attenuation of ceramide-induced apoptosis by diglyceride in human myeloid leukemia cells. J Biol Chem 269:31685-31692

Jones EY, Stuart DI, Walker NP (1992) Crystal structure of TNF. Immunol Ser 56:93-127

Ju ST, Panka DJ, Cui H, Ettinger R, el Khatib M, Sherr DH, Stanger BZ, Marshak Rothstein A (1995) Fas(CD95)/FasL interactions required for programmed cell death after T-cell activation. Nature 373:444-448

Juo P, Kuo CJ, Reynolds SE, Konz RF, Raingeaud J, Davis RJ, Biemann HP, Blenis J (1997) Fas activation of the p38 mitogen-activated protein kinase signalling pathway requires ICE/CED-3 family proteases. Mol Cell Biol 17:24-35

Karpusas M, Hsu YM, Wang JH, Thompson J, Lederman S, Chess L, Thomas D (1995) 2 A crystal structure of an extracellular fragment of human CD40 ligand. Structure 3:1031-1039

Katsikis PD, Wunderlich ES, Smith CA, Herzenberg LA (1995) Fas antigen stimulation induces marked apoptosis of T lymphocytes in human immunodeficiency virus-infected individuals. J Exp Med 181:2029-2036

Kawanishi M (1997) Epstein-Barr virus BHRF1 protein protects intestine 407 epithelial cells from apoptosis induced by tumor necrosis factor alpha and anti-Fas antibody. J Virol 71:3319-3322

Kettle S, Alcami A, Khanna A, Ehret R, Jassoy C, Smith GL (1997) Vaccinia virus serpin B13R (SPI-2) inhibits interleukin-1beta-converting enzyme and protects virus-infected cells from TNF- and Fas-mediated apoptosis, but does not prevent IL-1beta-induced fever. J Gen Virol 78:677-685

Kischkel FC, Hellbardt S, Behrmann I, Germer M, Pawlita M, Krammer PH, Peter ME (1995) Cytotoxicity-dependent APO-1 (Fas/CD95)-associated proteins form a death-inducing signaling complex (DISC) with the receptor. EMBO J 14:5579-5588

Kitson J, Raven T, Jiang YP, Goeddel DV, Giles KM, Pun KT, Grinham CJ, Brown R, Farrow SN (1996) A death-domain-containing receptor that mediates apoptosis. Nature 384:372-375

Kluck RM, Bossy Wetzel E, Green DR, Newmeyer DD (1997) The release of cytochrome c from mitochondria: a primary site for Bcl-2 regulation of apoptosis. Science 275:1132-1136

Komiyama T, Ray CA, Pickup DJ, Howard AD, Thornberry NA, Peterson EP, Salvesen G (1994) Inhibition of interleukin-1 beta converting enzyme by the cowpox virus serpin CrmA. An example of cross-class inhibition. J Biol Chem 269:19331-19337

Korner H, Sedgwick JD (1996) Tumour necrosis factor and lymphotoxin: molecular aspects and role in tissue-specific autoimmunity. Immunol Cell Biol 74:465-472

Krikos A, Laherty CD, Dixit VM (1992) Transcriptional activation of the tumor necrosis factor alpha-inducible zinc finger protein, A20, is mediated by kappa B elements. J Biol Chem 267:17971-17976

Kuida K, Lippke JA, Ku G, Harding MW, Livingston DJ, Su MS, Flavell RA (1995) Altered cytokine export and apoptosis in mice deficient in interleukin-1 beta converting enzyme. Science 267:2000–2003

Kuida K, Zheng TS, Na S, Kuan C, Yang D, Karasuyama H, Rakic P, Flavell RA (1996) Decreased apoptosis in the brain and premature lethality in CPP32-deficient mice. Nature 384:368–372

Kull FC Jr, Besterman JM (1990) Drug-induced alterations of tumor necrosis factor-mediated cytotoxicity: discrimination of early versus late stage action. J Cell Biochem 42:1–12

Kwon BS, Weissman SM (1989) cDNA sequences of two inducible T cell genes. Proc Natl Acad Sci USA 86:1963–1968

Kwon BS, Tan KB, Ni J, Lee KO, Kim KK, Kim YJ, Wang S, Gentz R, Yu GL, Harrop J, Lyn SD, Silverman C, Porter TG, Truneh A, Young PR (1997) A newly identified member of the tumor necrosis factor receptor superfamily with a wide tissue distribution and involvement in lymphocyte activation. J Biol Chem 272:14272–14276

Lahti JM, Xiang J, Heath LS, Campana D, Kidd VJ (1995) PITSLRE protein kinase activity is associated with apoptosis. Mol Cell Biol 15:1–11

Latinis KM, Koretzky GA (1996) Fas ligation induces apoptosis and Jun kinase activation independently of CD45 and Lck in human T cells. Blood 87:871–875

Lau HT, Yu M, Fontana A, Stoeckert CJJ (1996) Prevention of islet allograft rejection with engineered myoblasts expressing FasL in mice. Science 273:109–112

Leist M, Gantner F, Bohlinger I, Germann PG, Tiegs G, Wendel A (1994) Murine hepatocyte apoptosis induced in vitro and in vivo by TNF-alpha requires transcriptional arrest. J Immunol 153:1778–1788

Leithauser F, Dhein J, Mechtersheimer G, Koretz K, Bruderlein S, Henne C, Schmidt A, Debatin KM, Krammer PH, Moller P (1993) Constitutive and induced expression of APO-1, a new member of the nerve growth factor/tumor necrosis factor receptor superfamily, in normal and neoplastic cells. Lab Invest 69:415–429

Lenczowski JM, Dominguez L, Eder AM, King LB, Zacharchuk CM, Ashwell JD (1997) Lack of a role for Jun kinase and AP-1 in Fas-induced apoptosis. Mol Cell Biol 17:170–181

Lerch RA, Friesen PD (1993) The 35-kilodalton protein gene (p35) of Autographa californica nuclear polyhedrosis virus and the neomycin resistance gene provide dominant selection of recombinant baculoviruses. Nucleic Acids Res 21:1753–1760

Li CJ, Friedman DJ, Wang C, Metelev V, Pardee AB (1995a) Induction of apoptosis in uninfected lymphocytes by HIV-1 Tat protein. Science 268:429–431

Li P, Allen H, Banerjee S, Franklin S, Herzog L, Johnston C, McDowell J, Paskind M, Rodman L, Salfeld J, et al. (1995b) Mice deficient in IL-1 beta-converting enzyme are defective in production of mature IL-1 beta and resistant to endotoxic shock. Cell 80:401–411

Lippke JA, Gu Y, Sarnecki C, Caron PR, Su MS (1996) Identification and characterization of CPP32/Mch2 homolog 1, a novel cysteine protease similar to CPP32. J Biol Chem 271:1825–1828

Liston P, Roy N, Tamai K, Lefebvre C, Baird S, Cherton Horvat G, Farahani R, McLean M, Ikeda JE, MacKenzie A, Korneluk RG (1996) Suppression of apoptosis in mammalian cells by NAIP and a related family of IAP genes. Nature 379:349–353

Liu X, Kim CN, Yang J, Jemmerson R, Wang X (1996) Induction of apoptotic program in cell-free extracts: requirement for dATP and cytochrome c. Cell 86:147–157

Liu X, Zou H, Slaughter C, Wang X (1997) DFF, a heterodimeric protein that functions downstream of caspase 3 to trigger DNA fragmentation during apoptosis. Cell 89:175–184

Liu ZG, Hsu H, Goeddel DV, Karin M (1996) Dissection of TNF receptor 1 effector functions: JNK activation is not linked to apoptosis while NF-kappaB activation prevents cell death. Cell 87:565–576

Loetscher H, Pan YC, Lahm HW, Gentz R, Brockhaus M, Tabuchi H, Lesslauer W (1990) Molecular cloning and expression of the human 55 kD tumor necrosis factor receptor. Cell 61:351–359

Los M, Van de Craen M, Penning LC, Schenk H, Westendorp M, Baeuerle PA, Droge W, Krammer PH, Fiers W, Schulze Osthoff K (1995) Requirement of an ICE/CED-3 protease for Fas/APO-1-mediated apoptosis. Nature 375:81-83

Maini RN (1996) The role of cytokines in rheumatoid arthritis. The Croonian Lecture 1995. J R Coll Phys Lond 30:344-351

Mallett S, Fossum S, Barclay AN (1990) Characterization of the MRC OX40 antigen of activated CD4 positive T lymphocytes - a molecule related to nerve growth factor receptor. EMBO J 9:1063-1068

Mandal M, Maggirwar SB, Sharma N, Kaufmann SH, Sun SC, Kumar R (1996) Bcl-2 prevents CD95 (Fas/APO-1)-induced degradation of lamin B and poly(ADP-ribose) polymerase and restores the NF-kappaB signaling pathway. J Biol Chem 271:30354-30359

Mapara MY, Bargou R, Zugck C, Dohner H, Ustaoglu F, Jonker RR, Krammer PH, Dorken B (1993) APO-1 mediated apoptosis or proliferation in human chronic B lymphocytic leukemia: correlation with bcl-2 oncogene expression. Eur J Immunol 23:702-708

Mariani SM, Matiba B, Baumler C, Krammer PH (1995) Regulation of cell surface APO-1/Fas (CD95) ligand expression by metalloproteases. Eur J Immunol 25:2303-2307

Mariani SM, Matiba B, Armandola EA, Krammer PH (1997) Interleukin 1 beta-converting enzyme related proteases/caspases are involved in TRAIL-induced apoptosis of myeloma and leukemia cells. J Cell Biol 137:221-229

Marsters SA, Pitti RM, Donahue CJ, Ruppert S, Bauer KD, Ashkenazi A (1996a) Activation of apoptosis by Apo-2 ligand is independent of FADD but blocked by CrmA. Curr Biol 6:750-752

Marsters SA, Sheridan JP, Donahue CJ, Pitti RM, Gray CL, Goddard AD, Bauer KD, Ashkenazi A (1996b) Apo-3, a new member of the tumor necrosis factor receptor family, contains a death domain and activates apoptosis and NF-kappa B. Curr Biol 6:1669-1676

McCloskey TW, Oyaizu N, Kaplan M, Pahwa S (1995) Expression of the Fas antigen in patients infected with human immunodeficiency virus. Cytometry 22:111-114

McCurrach ME, Connor TM, Knudson CM, Korsmeyer SJ, Lowe SW (1997) Bax-deficiency promotes drug resistance and oncogenic transformation by attenuating p53-dependent apoptosis. Proc Natl Acad Sci USA 94:2345-2349

Medema JP, Scaffidi C, Kischkel FC, Shevchenko A, Mann M, Krammer PH, Peter ME (1997a) FLICE is activated by association with the CD95 death-inducing signaling complex (DISC). EMBO J 16:2794-2804

Medema JP, Toes REM, Scaffidi C, Zheng TS, Flavell RA, Melief CJM, Peter ME, Offringa R, Krammer PH (1997b) FLICE activation by granzyme B during CTL mediated apoptosis. Eur J Immunol 27:3492-3498

Medema JP, Scaffidi C, Krammer PH, Peter ME (1998) Bcl-x$_L$ acts downstream of caspase-8 activation by the death-inducing signaling complex. J Biol Chem 273:3388-3393

Meier R, Rouse J, Cuenda A, Nebreda AR, Cohen P (1996) Cellular stresses and cytokines activate multiple mitogen-activated-protein kinase homologues in PC12 and KB cells. Eur J Biochem 236:796-805

Memon SA, Moreno MB, Petrak D, Zacharchuk CM (1995) Bcl-2 blocks glucocorticoid- but not Fas- or activation-induced apoptosis in a T cell hybridoma. J Immunol 155:4644-4652

Merkenschlager M, Fisher AG (1991) CD45 isoform switching precedes the activation-driven death of human thymocytes by apoptosis. Int Immunol 3:1-7

Minn AJ, Velez P, Schendel SL, Liang H, Muchmore SW, Fesik SW, Fill M, Thompson CB (1997) Bcl-x(L) forms an ion channel in synthetic lipid membranes. Nature 385:353-357

Mogil RJ, Radvanyi L, Gonzalez Quintial R, Miller R, Mills G, Theofilopoulos AN, Green DR (1995) Fas (CD95) participates in peripheral T cell deletion and associated apoptosis in vivo. Int Immunol 7:1451-1458

Monney L, Otter I, Olivier R, Ravn U, Mirzasaleh H, Fellay I, Poirier GG, Borner C (1996) Bcl-2 overexpression blocks activation of the death protease CPP32/Yama/apopain. Biochem Biophys Res Commun 221:340-345

Montgomery RI, Warner MS, Lum BJ, Spear PG (1996) Herpes simplex virus-1 entry into cells mediated by a novel member of the TNF/NGF receptor family. Cell 87:427–436

Moreno MB, Memon SA, Zacharchuk CM (1996) Apoptosis signaling pathways in normal T cells: differential activity of Bcl-2 and IL-1beta-converting enzyme family protease inhibitors on glucocorticoid- and Fas-mediated cytotoxicity. J Immunol 157:3845–3849

Moss ML, Jin SL, Milla ME, Burkhart W, Carter HL, Chen WJ, Clay WC, Didsbury JR, Hassler D, Hoffman CR, Kost TA, Lambert MH, Leesnitzer MA, McCauley P, McGeehan G, Mitchell J, Moyer M, Pahel G, Rocque W, Overton LK, Schoenen F, Seaton T, Su JL, Warner J, Becherer JD, et al. (1997) Cloning of a disintegrin metalloproteinase that processes precursor tumour-necrosis factor-alpha. Nature 385:733–736

Muchmore SW, Sattler M, Liang H, Meadows RP, Harlan JE, Yoon HS, Nettesheim D, Chang BS, Thompson CB, Wong SL, Ng SL, Fesik SW (1996) X-ray and NMR structure of human Bcl-xL, an inhibitor of programmed cell death. Nature 381:335–341

Muzio M, Chinnaiyan AM, Kischkel FC, O'Rourke K, Shevchenko A, Ni J, Scaffidi C, Bretz JD, Zhang M, Gentz R, Mann M, Krammer PH, Peter ME, Dixit VM (1996) FLICE, a novel FADD-homologous ICE/CED-3-like protease, is recruited to the CD95 (Fas/APO-1) death-inducing signaling complex. Cell 85:817–827

Muzio M, Salvesen GS, Dixit VM (1997) FLICE induced apoptosis in a cell-free system. Cleavage of caspase zymogens. J Biol Chem 272:2952–2956

Nagata S, Golstein P (1995) The Fas death factor. Science 267:1449–1456

Natoli G, Costanzo A, Ianni A, Templeton DJ, Woodgett JR, Balsano C, Levrero M (1997) Activation of SAPK/JNK by TNF receptor 1 through a noncytotoxic TRAF2-dependent pathway. Science 275:200–203

Neilan JG, Lu Z, Afonso CL, Kutish GF, Sussman MD, Rock DL (1993) An African swine fever virus gene with similarity to the proto-oncogene bcl-2 and the Epstein–Barr virus gene BHRF1. J Virol 67:4391–4394

Newell MK, Haughn LJ, Maroun CR, Julius MH (1990) Death of mature T cells by separate ligation of CD4 and the T-cell receptor for antigen. Nature 347:286–289

Nicholson DW, Ali A, Thornberry NA, Vaillancourt JP, Ding CK, Gallant M, Gareau Y, Griffin PR, Labelle M, Lazebnik YA, et al. (1995) Identification and inhibition of the ICE/CED-3 protease necessary for mammalian apoptosis. Nature 376:37–43

Niehans GA, Brunner T, Frizelle SP, Liston JC, Salerno CT, Knapp DJ, Green DR, Kratzke RA (1997) Human lung carcinomas express Fas ligand. Cancer Res 57:1007–1012

Nishina H, Fischer KD, Radvanyi L, Shahinian A, Hakem R, Rubie EA, Bernstein A, Mak TW, Woodgett JR, Penninger JM (1997) Stress-signalling kinase Sek1 protects thymocytes from apoptosis mediated by CD95 and CD3. Nature 385:350–353

Nocentini G, Giunchi L, Ronchetti S, Krausz LT, Bartoli A, Moraca R, Migliorati G, Riccardi C (1997) A new member of the tumor necrosis factor/nerve growth factor receptor family inhibits T cell receptor-induced apoptosis. Proc Natl Acad Sci USA 94:6216–6221

O'Connell J, O'Sullivan GC, Collins JK, Shanahan F (1996) The Fas counterattack: Fas-mediated T cell killing by colon cancer cells expressing Fas ligand. J Exp Med 184:1075–1082

Oehm A, Behrmann I, Falk W, Pawlita M, Maier G, Klas C, Li Weber M, Richards S, Dhein J, Trauth BC, et al. (1992) Purification and molecular cloning of the APO-1 cell surface antigen, a member of the tumor necrosis factor/nerve growth factor receptor superfamily. Sequence identity with the Fas antigen. J Biol Chem 267:10709–10715

Ogasawara J, Watanabe-Fukunaga R, Adachi M, Matsuzawa A, Kasugai T, Kitamura Y, Itoh N, Suda T, Nagata S (1993) Lethal effect of the anti-Fas antibody in mice. Nature 364:806–809

Okura T, Gong L, Kamitani T, Wada T, Okura I, Wei CF, Chang HM, Yeh ET (1996) Protection against Fas/APO-1- and tumor necrosis factor-mediated cell death by a novel protein, sentrin. J Immunol 157:4277–4281

Oltvai ZN, Korsmeyer SJ (1994) Checkpoints of dueling dimers foil death wishes. Cell 79:189–192

Oltvai ZN, Milliman CL, Korsmeyer SJ (1993) Bcl-2 heterodimerizes in vivo with a conserved homolog, Bax, that accelerates programmed cell death. Cell 74:609–619

Opipari AW Jr, Hu HM, Yabkowitz R, Dixit VM (1992) The A20 zinc finger protein protects cells from tumor necrosis factor cytotoxicity. J Biol Chem 267:12424–12427

Orth K, O'Rourke K, Salvesen GS, Dixit VM (1996) Molecular ordering of apoptotic mammalian CED-3/ICE-like proteases. J Biol Chem 271:20977–20980

Pan G, O'Rourke K, Chinnaiyan AM, Gentz R, Ebner R, Ni J, Dixit VM (1997) The receptor for the cytotoxic ligand TRAIL. Science 276:111–113

Park A, Baichwal VR (1996) Systematic mutational analysis of the death domain of the tumor necrosis factor receptor 1-associated protein TRADD. J Biol Chem 271:9858–9862

Pasparakis M, Alexopoulou L, Episkopou V, Kollias G (1996) Immune and inflammatory responses in TNF alpha-deficient mice: a critical requirement for TNF alpha in the formation of primary B cell follicles, follicular dendritic cell networks and germinal centers, and in the maturation of the humoral immune response. J Exp Med 184:1397–1411

Perez C, Albert I, DeFay K, Zachariades N, Gooding L, Kriegler M (1990) A nonsecretable cell surface mutant of tumor necrosis factor (TNF) kills by cell-to-cell contact. Cell 63:251–258

Peter ME, Kischkel FC, Scheuerpflug CG, Medema JP, Debatin KM, Krammer PH (1997a) Resistance of cultured peripheral T cells towards activation-induced cell death involves a lack of recruitment of FLICE (MACH/caspase 8) to the CD95 death-inducing signaling complex. Eur J Immunol 27:1207–1212

Peter ME, Medema JP, Krammer PH (1997b) Does the *Caenorhabditis elegans* protein CED-4 contain a region of homology to the mammalian death effector domain? Cell Death Differ 4:523–525

Pfeffer K, Matsuyama T, Kundig TM, Wakeham A, Kishihara K, Shahinian A, Wiegmann K, Ohashi PS, Kronke M, Mak TW (1993) Mice deficient for the 55 kD tumor necrosis factor receptor are resistant to endotoxic shock, yet succumb to *L. monocytogenes* infection. Cell 73:457–467

Pickup DJ (1994) Poxviral modifiers of cytokine responses to infection. Infect Agents Dis 3:116–127

Pitti RM, Marsters SA, Ruppert S, Donahue CJ, Moore A, Ashkenazi A (1996) Induction of apoptosis by Apo-2 ligand, a new member of the tumor necrosis factor cytokine family. J Biol Chem 271:12687–12690

Ponton A, Clement MV, Stamenkovic I (1996) The CD95 (APO-1/Fas) receptor activates NF-kappaB independently of its cytotoxic function. J Biol Chem 271:8991–8995

Porteu F, Hieblot C (1994) Tumor necrosis factor induces a selective shedding of its p75 receptor from human neutrophils. J Biol Chem 269:2834–2840

Radeke MJ, Misko TP, Hsu C, Herzenberg LA, Shooter EM (1987) Gene transfer and molecular cloning of the rat nerve growth factor receptor. Nature 325:593–597

Rao L, Debbas M, Sabbatini P, Hockenbery D, Korsmeyer S, White E (1992) The adenovirus E1A proteins induce apoptosis, which is inhibited by the E1B 19-kDa and Bcl-2 proteins. Proc Natl Acad Sci USA 89:7742–7746

Rao L, Perez D, White E (1996) Lamin proteolysis facilitates nuclear events during apoptosis. J Cell Biol 135:1441–1455

Ray CA, Black RA, Kronheim SR, Greenstreet TA, Sleath PR, Salvesen GS, Pickup DJ (1992) Viral inhibition of inflammation: cowpox virus encodes an inhibitor of the interleukin-1 beta converting enzyme. Cell 69:597–604

Reyburn HT, Mandelboim O, Vales Gomez M, Davis DM, Pazmany L, Strominger JL (1997) The class I MHC homologue of human cytomegalovirus inhibits attack by natural killer cells. Nature 386:514–517

Rieux-Laucat F, Le Deist F, Hivroz C, Roberts IA, Debatin KM, Fischer A, de Villartay JP (1995) Mutations in Fas associated with human lymphoproliferative syndrome and autoimmunity. Science 268:1347–1349

Rothe J, Lesslauer W, Lotscher H, Lang Y, Koebel P, Kontgen F, Althage A, Zinkernagel R, Steinmetz M, Bluethmann H (1993) Mice lacking the tumour necrosis factor receptor 1 are resistant to TNF-mediated toxicity but highly susceptible to infection by *Listeria monocytogenes*. Nature 364:798–802

Rothe M, Pan MG, Henzel WJ, Ayres TM, Goeddel DV (1995) The TNFR2-TRAF signaling complex contains two novel proteins related to baculoviral inhibitor of apoptosis proteins. Cell 83:1243–1252

Santana P, Pena LA, Haimovitz Friedman A, Martin S, Green D, McLoughlin M, Cordon Cardo C, Schuchman EH, Fuks Z, Kolesnick R (1996) Acid sphingomyelinase-deficient human lymphoblasts and mice are defective in radiation-induced apoptosis. Cell 86:189–99

Sato T, Irie S, Kitada S, Reed JC (1995) FAP-1: a protein tyrosine phosphatase that associates with Fas. Science 268:411–415

Scaffidi C, Medema JP, Krammer PH, Peter ME (1997) FLICE is predominantly expressed as two functionally active isoforms, caspase-8/a and caspase-8/b. J Biol Chem 272:26953–26958

Schall TJ, Lewis M, Koller KJ, Lee A, Rice GC, Wong GH, Gatanaga T, Granger GA, Lentz R, Raab H, et al. (1990) Molecular cloning and expression of a receptor for human tumor necrosis factor. Cell 61:361–370

Schendel SL, Xie Z, Montal MO, Matsuyama S, Montal M, Reed JC (1997) Channel formation by antiapoptotic protein Bcl-2. Proc Natl Acad Sci USA 94:5113–5118

Schievella AR, Chen JH, Graham JR, Lin LL (1997) MADD, a novel death domain protein that interacts with the type 1 tumor necrosis factor receptor and activates mitogen-activated protein kinase. J Biol Chem 272:12069–12075

Schlegel J, Peters I, Orrenius S, Miller DK, Thornberry NA, Yamin TT, Nicholson DW (1996) CPP32/apopain is a key interleukin 1 beta converting enzyme-like protease involved in Fas-mediated apoptosis. J Biol Chem 271:1841–1844

Schreiber M, Sedger L, McFadden G (1997) Distinct domains of M-T2, the myxoma virus tumor necrosis factor (TNF) receptor homolog, mediate extracellular TNF binding and intracellular apoptosis inhibition. J Virol 71:2171–2181

Screaton GR, Xu XN, Olsen AL, Cowper AE, Tan R, McMichael AJ, Bell JI (1997) LARD: a new lymphoid-specific death domain containing receptor regulated by alternative pre-mRNA splicing. Proc Natl Acad Sci USA 94:4615–4619

Seino K, Kayagaki N, Okumura K, Yagita H (1997) Antitumor effect of locally produced CD95 ligand. Nat Med 3:165–170

Selawry HP, Cameron DF (1993) Sertoli cell-enriched fractions in successful islet cell transplantation. Cell Transplant 2:123–129

Shimamoto Y, Chen RL, Bollon A, Chang A, Khan A (1988) Monoclonal antibodies against human recombinant tumor necrosis factor: prevention of endotoxic shock. Immunol Lett 17:311–317

Shiraki K, Tsuji N, Shioda T, Isselbacher KJ, Takahashi H (1997) Expression of fas ligand in liver metastases of human colonic adenocarcinomas. Proc Natl Acad Sci USA 94:6420–6425

Sieg S, Yildirim Z, Smith D, Kayagaki N, Yagita H, Huang Y, Kaplan D (1996) Herpes simplex virus type 2 inhibition of Fas ligand expression. J Virol 70:8747–8751

Sillence DJ, Allan D (1997) Evidence against an early signalling role for ceramide in Fas-mediated apoptosis. Biochem J 324:29–32

Simonian PL, Grillot DA, Andrews DW, Leber B, Nunez G (1996) Bax homodimerization is not required for Bax to accelerate chemotherapy-induced cell death. J Biol Chem 271:32073–32077

Singer GG, Abbas AK (1994) The fas antigen is involved in peripheral but not thymic deletion of T lymphocytes in T cell receptor transgenic mice. Immunity 1:365–371

Smith CA, Williams GT, Kingston R, Jenkinson EJ, Owen JJ (1989) Antibodies to CD3/T-cell receptor complex induce death by apoptosis in immature T cells in thymic cultures. Nature 337:181–184

Smith CA, Davis T, Anderson D, Solam L, Beckmann MP, Jerzy R, Dower SK, Cosman D, Goodwin RG (1990) A receptor for tumor necrosis factor defines an unusual family of cellular and viral proteins. Science 248:1019–1023

Smith DJ, McGuire MJ, Tocci MJ, Thiele DL (1997) IL-1 beta convertase (ICE) does not play a requisite role in apoptosis induced in T lymphoblasts by Fas-dependent or Fas-independent CTL effector mechanisms. J Immunol 158:163–170

Song HY, Rothe M, Goeddel DV (1996) The tumor necrosis factor-inducible zinc finger protein A20 interacts with TRAF1/TRAF2 and inhibits NF-kappaB activation. Proc Natl Acad Sci USA 93:6721–6725

Srinivasula SM, Ahmad M, Fernandes-Alnemri T, Litwack G, Alnemri ES (1996a) Molecular ordering of the Fas-apoptotic pathway: the Fas/APO-1 protease Mch5 is a CrmA-inhibitable protease that activates multiple Ced-3/ICE-like cysteine proteases. Proc Natl Acad Sci USA 93:14486–14491

Srinivasula SM, Fernandes-Alnemri T, Zangrilli J, Robertson N, Armstrong RC, Wang L, Trapani JA, Tomaselli KJ, Litwack G, Alnemri ES (1996b) The Ced-3/interleukin 1beta converting enzyme-like homolog Mch6 and the lamin-cleaving enzyme Mch2alpha are substrates for the apoptotic mediator CPP32. J Biol Chem 271:27099–27106

Stamenkovic I, Clark EA, Seed B (1989) A B-lymphocyte activation molecule related to the nerve growth factor receptor and induced by cytokines in carcinomas. EMBO J 8:1403–1410

Stanger BZ, Leder P, Lee TH, Kim E, Seed B (1995) RIP: a novel protein containing a death domain that interacts with Fas/APO-1 (CD95) in yeast and causes cell death. Cell 81:513–523

Stokkers PC, Camoglio L, van Deventer SJ (1995) Tumor necrosis factor (TNF) in inflammatory bowel disease: gene polymorphisms, animal models, and potential for anti-TNF therapy. J Inflamm 47:97–103

Stuart PM, Griffith TS, Usui N, Pepose J, Yu X, Ferguson TA (1997) CD95 ligand (FasL)-induced apoptosis is necessary for corneal allograft survival. J Clin Invest 99:396–402

Suda T, Takahashi T, Golstein P, Nagata S (1993) Molecular cloning and expression of the Fas ligand, a novel member of the tumor necrosis factor family. Cell 75:1169–1178

Susin SA, Zamzami N, Castedo M, Hirsch T, Marchetti P, Macho A, Daugas E, Geuskens M, Kroemer G (1996) Bcl-2 inhibits the mitochondrial release of an apoptogenic protease. J Exp Med 184:1331–1341

Sweeney EA, Sakakura C, Shirahama T, Masamune A, Ohta H, Hakomori S, Igarashi Y (1996) Sphingosine and its methylated derivative N,N-dimethylsphingosine (DMS) induce apoptosis in a variety of human cancer cell lines. Int J Cancer 66:358–366

Szawlowski PW, Hanke T, Randall RE (1993) Sequence homology between HIV-1 gp120 and the apoptosis mediating protein Fas. AIDS 7:1018

Takahashi S, Maecker HT, Levy R (1989) DNA fragmentation and cell death mediated by T cell antigen receptor/CD3 complex on a leukemia T cell line. Eur J Immunol 19:1911–1919

Takahashi T, Tanaka M, Brannan CI, Jenkins NA, Copeland NG, Suda T, Nagata S (1994) Generalized lymphoproliferative disease in mice, caused by a point mutation in the Fas ligand. Cell 76:969–976

Tarodi B, Subramanian T, Chinnadurai G (1994) Epstein-Barr virus BHRF1 protein protects against cell death induced by DNA-damaging agents and heterologous viral infection. Virology 201:404–407

Tartaglia LA, Weber RF, Figari IS, Reynolds C, Palladino MA Jr, Goeddel DV (1991) The two different receptors for tumor necrosis factor mediate distinct cellular responses. Proc Natl Acad Sci USA 88:9292–9296

Tartaglia LA, Ayres TM, Wong GH, Goeddel DV (1993a) A novel domain within the 55kD TNF receptor signals cell death. Cell 74:845–853

Tartaglia LA, Pennica D, Goeddel DV (1993b) Ligand passing: the 75-kDa tumor necrosis factor (TNF) receptor recruits TNF for signaling by the 55-kDa TNF receptor. J Biol Chem 268:18542–18548

Tewari M, Dixit VM (1995a) Fas- and tumor necrosis factor-induced apoptosis is inhibited by the poxvirus crmA gene product. J Biol Chem 270:3255–3260

Tewari M, Quan LT, O'Rourke K, Desnoyers S, Zeng Z, Beidler DR, Poirier GG, Salvesen GS, Dixit VM (1995b) Yama/CPP32 beta, a mammalian homolog of CED-3, is a CrmA-inhibitable protease that cleaves the death substrate poly(ADP-ribose) polymerase. Cell 81:801–809

Tewari M, Telford WG, Miller RA, Dixit VM (1995c) CrmA, a poxvirus-encoded serpin, inhibits cytotoxic T-lymphocyte-mediated apoptosis. J Biol Chem 270:22705–22708

Thoma B, Grell M, Pfizenmaier K, Scheurich P (1990) Identification of a 60-kD tumor necrosis factor (TNF) receptor as the major signal transducing component in TNF responses. J Exp Med 172:1019–1023

Thome M, Schneider P, Hofmann K, Fickenscher H, Meinl E, Neipel F, Mattmann C, Burns K, Bodmer JL, Schroter M, Scaffidi C, Krammer PH, Peter ME, Tschopp J (1997) Viral FLICE-inhibitory proteins (FLIPs) prevent apoptosis induced by death receptors. Nature 386:517–521

Ting AT, Pimentel Muinos FX, Seed B (1996) RIP mediates tumor necrosis factor receptor 1 activation of NF-kappaB but not Fas/APO-1-initiated apoptosis. EMBO J 15:6189–6196

Tracey KJ, Cerami A (1993) Tumor necrosis factor, other cytokines and disease. Annu Rev Cell Biol 9:317–343

Trost LC, Lemasters JJ (1994) A cytotoxicity assay for tumor necrosis factor employing a multiwell fluorescence scanner. Anal Biochem 220:149–153

Ucker DS, Ashwell JD, Nickas G (1989) Activation-driven T cell death. I. Requirements for de novo transcription and translation and association with genome fragmentation. J Immunol 143:3461–3469

Van Antwerp DJ, Martin SJ, Kafri T, Green DR, Verma IM (1996) Suppression of TNF-alpha-induced apoptosis by NF-kappaB. Science 274:787–789

Vandenabeele P, Declercq W, Vanhaesebroeck B, Grooten J, Fiers W (1995) Both TNF receptors are required for TNF-mediated induction of apoptosis in PC60 cells. J Immunol 154:2904–2913

Vandevoorde V, Haegeman G, Fiers W (1997) Induced expression of trimerized intracellular domains of the human tumor necrosis factor (TNF) p55 receptor elicits TNF effects. J Cell Biol 137:1627–1638

Venable ME, Lee JY, Smyth MJ, Bielawska A, Obeid LM (1995) Role of ceramide in cellular senescence. J Biol Chem 270:30701–30708

Vincenz C, Dixit VM (1997) Fas-associated death domain protein interleukin-1beta-converting enzyme 2 (FLICE2), an ICE/Ced-3 homologue, is proximally involved in CD95- and p55-mediated death signaling. J Biol Chem 272:6578–6583

Wadsworth S, Yui K, Siegel RM, Tenenholz DE, Hirsch JA, Greene MI (1990) Origin and selection of peripheral CD4-CD8-T cells bearing alpha/beta T cell antigen receptors in autoimmune gld mice. Eur J Immunol 20:723–730

Walczak H, Degli-Esposti MA, Johnson RS, Smolak PJ, Waugh JY, Boiani N, Timour MS, Gerhart MJ, Schooley KA, Smith CA, Goodwin RG, Rauch CT (1997) TRAIL-R2: a novel apoptosis-mediating receptor for TRAIL. EMBO J 16:5386–5397

Wang CY, Mayo MW, Baldwin ASJ (1996a) TNF- and cancer therapy-induced apoptosis: potentiation by inhibition of NF-kappaB. Science 274:784–787

Wang HG, Rapp UR, Reed JC (1996b) Bcl-2 targets the protein kinase Raf-1 to mitochondria. Cell 87:629–638

Watanabe-Fukunaga R, Brannan CI, Copeland NG, Jenkins NA, Nagata S (1992a) Lymphoproliferation disorder in mice explained by defects in Fas antigen that mediates apoptosis. Nature 356:314–317

Watanabe-Fukunaga R, Brannan CI, Itoh N, Yonehara S, Copeland NG, Jenkins NA, Nagata S (1992b) The cDNA structure, expression, and chromosomal assignment of the mouse Fas antigen. J Immunol 148:1274–1279

Watts JD, Gu M, Polverino AJ, Patterson SD, Aebersold R (1997) Fas-induced apoptosis of T cells occurs independently of ceramide generation. Proc Natl Acad Sci USA 94:7292–7296

Westendorp MO, Frank R, Ochsenbauer C, Stricker K, Dhein J, Walczak H, Debatin KM, Krammer PH (1995) Sensitization of T cells to CD95-mediated apoptosis by HIV-1 Tat and gp120. Nature 375:497–500

White E, Sabbatini P, Debbas M, Wold WS, Kusher DI, Gooding LR (1995) The 19-kilodalton adenovirus E1B transforming protein inhibits programmed cell death and prevents cytolysis by tumor necrosis factor alpha. Mol Cell Biol 12:2570–2580

Wiley SR, Schooley K, Smolak PJ, Din WS, Huang CP, Nicholl JK, Sutherland GR, Smith TD, Rauch C, Smith CA (1995) Identification and characterization of a new member of the TNF family that induces apoptosis. Immunity 3:673–682

Wilson DJ, Fortner KA, Lynch DH, Mattingly RR, Macara IG, Posada JA, Budd RC (1996) JNK, but not MAPK, activation is associated with Fas-mediated apoptosis in human T cells. Eur J Immunol 26:989–994

Wong GH, Tartaglia LA, Lee MS, Goeddel DV (1992) Antiviral activity of tumor necrosis factor is signaled through the 55-kDa type I TNF receptor. J Immunol 149:3350–3353

Wu M, Lee H, Bellas RE, Schauer SL, Arsura M, Katz D, FitzGerald MJ, Rothstein TL, Sherr DH, Sonenshein GE (1996) Inhibition of NF-kappaB/Rel induces apoptosis of murine B cells. EMBO J 15:4682–4690

Xue D, Horvitz HR (1995) Inhibition of the *Caenorhabditis elegans* cell-death protease CED-3 by a CED-3 cleavage site in baculovirus p35 protein. Nature 377:248–251

Yanagisawa J, Takahashi M, Kanki H, Yano-Yanagisawa H, Tazunoki T, Sawa E, Nishitoba T, Kamishohara M, Kobayashi E, Kataoka S, Sato T (1997) The molecular interaction of Fas and FAP-1. A tripeptide blocker of human Fas interaction with FAP-1 promotes Fas-induced apoptosis. J Biol Chem 272:8539–8545

Yang J, Liu X, Bhalla K, Kim CN, Ibrado AM, Cai J, Peng TI, Jones DP, Wang X (1997a) Prevention of apoptosis by Bcl-2: release of cytochrome c from mitochondria blocked. Science 275:1129–1132

Yang X, Khoravi Far R, Chang HY, Baltimore D (1997b) Daxx, a novel Fas-binding protein that activates JNK and apoptosis. Cell 89:1067–1076

Zagury JF, Cantalloube H, Achour A, Cho YY, Fall L, Lachgar A, Chams V, Astgen A, Biou D, Picard O, et al. (1993) Striking similarities between HIV-1 Env protein and the apoptosis mediating cell surface antigen Fas. Role in the pathogenesis of AIDS. Biomed Pharmacother 47:331–335

Zanke BW, Boudreau K, Rubie E, Winnett E, Tibbles LA, Zon L, Kyriakis J, Liu FF, Woodgett JR (1996) The stress-activated protein kinase pathway mediates cell death following injury induced by *cis*-platinum, UV irradiation or heat. Curr Biol 6:606–613

Zha J, Harada H, Yang E, Jockel J, Korsmeyer SJ (1996) Serine phosphorylation of death agonist BAD in response to survival factor results in binding to 14-3-3 not BCL-X(L). Cell 87:619–628

Zhang X, Brunner T, Carter L, Dutton RW, Rogers P, Bradley L, Sato T, Reed JC, Green D, Swain SL (1997) Unequal death in T helper cell (Th)1 and Th2 effectors: Th1, but not Th2, effectors undergo rapid Fas/FasL-mediated apoptosis. J Exp Med 185:1837–1849

Lipid and Glycolipid Mediators in CD95-Induced Apoptotic Signaling

Florence Malisan, Maria Rita Rippo, Ruggero De Maria and Roberto Testi

1
Introduction

Ceramides play an important role mediating different cell responses such as proliferation, differentiation, growth arrest and apoptosis. They are released upon sphingomyelin hydrolysis which occurs after triggering of a number of cell-surface receptors including CD95. Ceramides generation also regulates glycosphingolipids and gangliosides metabolism. In particular, ganglioside GD3 biosynthesis may represent an important event for the progression of apoptotic signals generated by CD95 in hematopoietic cells.

2
Ceramide and the Sphingomyelinase Pathway

Sphingomyelin (SM) breakdown initiates an evolutionary conserved and ubiquitous intracellular signaling pathway (Hannun and Obeid 1995; Kolesnick and Fuks 1995). Sphingomyelin is a membrane phospholipid consisting of an aliphatic core, composed of sphingosine and an amide-linked fatty acid, and a phosphocholine polar head group. Sphingosine is an amino alcohol that contains a long, unsaturated hydrocarbon chain. The primary hydroxyl group of sphingosine is esterified to phosphorylcholine. Both sphingosine and the amide-linked fatty acid constitute the lipid component of the sphingomyelin moiety called ceramide (Kolesnick and Fuks 1995; Pushkarava et al. 1995; Testi 1996). As a major phospholipid, sphingomyelin is largely confined to the outer leaflet of cellular membranes and is hydrolyzed by sphingomyelinases, which are sphingomyelin-specific type C phospholipases, in ceramide and phosphocholine (Spence 1993). Different pools of sphingomyelin are potentially available to sphingomyelinases during membrane biogenesis and degradation. After being generated in the early Golgi compartments, sphingomyelin is transported along the exocytic pathway to the extracellular membrane where it is exposed (Rosenwald and Pagano 1993). As little flip-flop occurs in biologi-

Department of Experimental Medicine and Biochemical Sciences, University of Rome "Tor Vergata", Via di Tor Vergata, 135, 00133 Rome, Italy

cal membranes, it can be suggested that sphingomyelin is present mostly at the outer leaflet of both intracellular and extracellular membranes; however, the existence of a discrete inner membrane sphingomyelin pool has been described (Linardic and Hannun 1994). Based on their cellular localization and activation requirements, at least three major mammalian sphingomyelinase activities have been identified. An acidic sphingomyelinase (ASM), having a pH optimum of about 4.5–5.5, is the product of a single gene, and requires diacylglycerol (DAG) to be activated (Kolesnick 1987; Schuchman et al. 1991). Genetic mutations of ASM, impairing the capacity of SM degradation, are responsible for some forms of Niemann-Pick disease, a group of lethal diseases characterized by sphingomyelin accumulation in different tissues. ASM activity has been found in every tissue examined and is localized primarily in caveolae, endosomes and lysosomes (Liu and Anderson 1995). A neutral Mg^{2+}-dependent sphingomyelinase (NSM), produced by a different and unidentified gene, has been described in several mammalian tissues, but the highest activity has been found in the gray matter of the adult brain. It shows a neutral (7–7.5) pH optimum and is located on the external leaflet of cellular membranes (Spence 1993). Another Mg^{2+}-independent NSM has been characterized. This phospholipase is present in the cytosol (Okazaki et al. 1994), is activated by arachidonic acid (Jayadev et al. 1994) and is active on the inner leaflet pool of sphingomyelin.

3
Sphingomyelin Breakdown in Signal Transduction

Sphingomyelin breakdown represents the most relevant source of ceramide in receptor-mediated signal transduction. Different stimuli have been shown to induce sphingomyelin degradation and ceramide generation. Among these are vitamin D_3 (Kolesnick 1987), dexamethasone (Nelson et al. 1982), ionizing radiation (Haimovitz-Friedman et al. 1994), as well as ligation of different membrane receptors, such as the TNF-R1 (Dressler et al. 1992; Schütze et al. 1992), CD95 (Cifone et al. 1994), CD40 (Sallusto et al. 1996), the nerve growth factor receptor (Dobrowsky et al. 1994), CD28 (Boucher et al. 1995) and the interleukin-1 receptor (Wiegmann et al. 1994). Moreover, sphingomyelin hydrolysis is associated with the cellular response to stresses such as nutrient withdrawal and radiations (Haimovitz-Friedman et al. 1994; Jayadev et al. 1995).

Ceramide has been shown to activate multiple pathways. Different major routes have been so far characterized. A sequential activation of different kinases initiated by a ceramide-activated protein kinase (CAPK), involving phosporylation of Raf-1, MEK and ERK-2, with consequent PLA2 activation, may generate a number of arachidonic acid metabolites having disparate functions (Yao et al. 1995). Importantly, ceramide has been proven to act as a catalyst for the stress response kinase cascade through the consecutive involvement of MEKK1, SEK1, SAPK and c-Jun (Verheij et al. 1996). A ceramide-induced PKCζ phosphorylation may contribute to NFκB translocation, resulting in enhanced expression of multiple genes (Lozano et al. 1994). Finally, along with protein kinase activation, ceramide has been shown to activate a heterotrimeric protein phosphatase of the PP2A family (Dobrowsky et al. 1993), which might be responsible for ceramide-induced downregulation of c-myc and cell cycle arrest (Nickels and Broach 1996). Most of the knowledge concerning the biochemical signaling downstream of ceramide comes from studies in which cell-permeable ceramides have been used to mimic the action of endogenous ceramides. These synthetic ceramides diffuse inside the cell, reaching virtually all the subcellular compartments, mimicking the events which follow the activation of different sphingomyelinases. The pleiotropic biological effects of sphingomyelin breakdown result from the extreme complexity in the biochemical signaling induced by ceramide itself.

4
Early Apoptotic Signaling Through the CD95 Receptor

CD95 is a widely distributed surface receptor whose crosslinking by its specific ligand triggers apoptotic cell death (Krammer et al. 1994; Nagata and Suda 1995; also see Chapter by Peter et al. this Volume). Crosslinking of CD95 results in the activation of two main pathways involved in the early propagation of apoptotic signals, both requiring integrity of the receptor death domain: a proteolytic cascade mediated by cysteine proteases of the ICE/CED-3 family (caspases) and the activation of the acidic sphingomyelinase.

4.1
Activation of the Caspase Cascade

As described in an earlier chapter in this volume by Peter et al., upon ligand-induced receptor oligomerization, the 23-kDa adaptor molecule FADD/MORT1 binds directly to CD95 via its own C-terminal death domain (Boldin et al. 1995; Chinnaiyan et al. 1995). Concomitantly, the N-terminal "effector domain" of FADD/MORT1 interacts with a homologous domain of caspase-8 (FLICE/MACH), a CED3-like cysteine protease (Kumar 1995; Boldin et al. 1996; Muzio et al. 1996). FADD/MORT1-mediated recruitment of caspase-8 to the receptor might trigger its proteolytic activity. Subsequent cleavage and downstream activation of other cysteine proteases in a proteolytic cascade may eventually lead to the hydrolysis of different cytosolic and nuclear substrates (Enari et al. 1995; Los et al. 1995).

4.2
Activation of the Acidic Sphingomyelinase

The activation of SM turnover has been suggested to play an important role in initiating the biochemical pathway leading to active cell death, since synthetic ceramide analog C_2-ceramide has been shown to be directly responsible for apoptosis induction in different cell lines. In human cutaneous T cell lymphoma HuT78 cells, susceptible to CD95-mediated apoptosis, CD95 crosslinking induces SM breakdown and ceramide accumulation (Cifone et al. 1994). This sphingomyelinase activity is transient and peaks at 5–15 min. Both neutral and acidic sphingomyelinases contribute to CD95-induced sphingomyelin breakdown. NSM is not dependent on receptor death domain, activates ERK-2 and phospholipase A2 (PLA2) and is not sufficient to induce apoptosis (Cifone et al. 1995). ASM, on the other hand, is dependent on receptor death domain and on the upstream activation of PC-PLC. Cell pretreatment with the phosphatidylcholine-specific phospholipase C (PC-PLC) inhibitor D609, in fact, prevents CD95-induced SM hydrolysis by ASM but not NSM. PC-PLC activity produces diacylglycerol (DAG) and DAG is required for ASM activation. This suggests that ASM activation requires an upstream PC-PLC activation in vivo and that PC-PLC and ASM activation are two sequentially related steps of the same pathway. Therefore, at least two independent lipid pathways are induced upon CD95 stimulation: (1) a series of events, starting from regions of the receptor close to the membrane, which appear insufficient to induce apoptosis, leading to the activation of membrane NSM and PLA2; (2) another series of events, starting from the death domain, required to allow the progression of the apoptotic signal, i.e. the sequential activation of PC-PLC and ASM (Cifone et al. 1995). These data support the involvement of ASM in generating ceramide as mediator of apoptotic cell death. Accordingly, cells from individuals affected by Niemann-Pick disease, genetically deficient in ASM activity, or from mice in which the gene coding for ASM has been targeted, show remarkable defects in executing the apoptotic program in response to radiations (Santana et al. 1996).

A second wave of ceramide accumulation, dependent on cytosolic NSM activation, occurs several hours after CD95 crosslinking. This increase of ceramide intracellular concentration is prolonged and persistent and is also important for the later progression of apoptotic signals (Hannun 1996).

Mitochondrial permeability transition with breakdown of mitochondrial inner membrane potential ($\Delta\Psi$m) is a possible target of sphingomyelin-derived ceramide (Pastorino et al. 1996). Collapse of $\Delta\Psi$m may contribute to the apoptotic pathway generated by CD95 and TNF-RI crosslinking through a mitochondrial-dependent activation of caspases (Kroemer et al. 1997). Again, the activation of NSM and ASM seems to differently contribute to this process. CD95 stimulation in death domain-defective HuT78 cell clones results in the early activation of the neutral Mg^{2+}-dependent NSM with transient generation of ceramide. This mutated CD95 receptor is unable to activate the ASM pathway, and the membrane NSM-derived ceramide does not induce $\Delta\Psi$m alterations (unpubl. observation). It is likely, therefore, that ceramide produced by ASM plays an important role in $\Delta\Psi$m changes which follow CD95 crosslinking.

Compartmentalization of the relevant downstream effectors of ceramide is likely to play a crucial role, as we were unable to find structural differences, in terms of length or saturation of the acyl chain or of the sphingoid base, between ceramide released after CD95 crosslinking by the activation of NSM and ASM. However, the relevant targets of ceramides, and their metabolites, involved in downstream propagation of apoptotic signals remain poorly characterized.

5
Ceramide as a Source of Gangliosides

It is known that newly synthesized or released ceramides are targeted to the Golgi apparatus and regulate sphingolipid and glycosphingolipid metabolism, including the rate of ganglioside biosynthesis within the Golgi (Allan and Kallen 1993; van Echten and Sandhoff 1993). We therefore recently investigated whether changes in ganglioside metabolism could be detected upon CD95 signaling.

5.1
Ganglioside Biosynthetic Pathways

Gangliosides are amphipatic constituents of the outer leaflet of cellular membranes, composed of a common hydrophobic ceramide moiety, which acts as a membrane anchor, and a hydrophilic oligosaccharide chain. Depending on the type, number and linkage of sugars which compose the oligosaccharide chain, a wide variety of different gangliosides arises, whose expression may vary in relation to the cell type, as well as to the cell cycle and the cell differentiation stage. Ganglioside neosynthesis initiates within the biogenic membranes of the early Golgi by stepwise glycosylation of ceramide (Sandhoff and van Echten 1994). Neoformed ceramides coming from the endoplasmic reticulum, or free ceramides coming from degradative subcellular compartments, enter the early Golgi via vesicular membrane flow to initiate ganglioside synthesis (van Meer 1993; Sandhoff and Kolter 1996). The addition of glucose by the action of a cytosolic β-1,1-glucosyltransferase turns ceramide into glucosylceramide (Fig. 1). Subsequently, a cytosolic β-1,4-galactosyltransferase mediates the attachment of a galactose residue to the glucose, forming lactosylceramide. Lactosylceramide then flips into the luminal side of early Golgi cisternae, where the attachment of a sialic acid residue over the galactose, mediated by α-2,3-sialyltransferase, generates

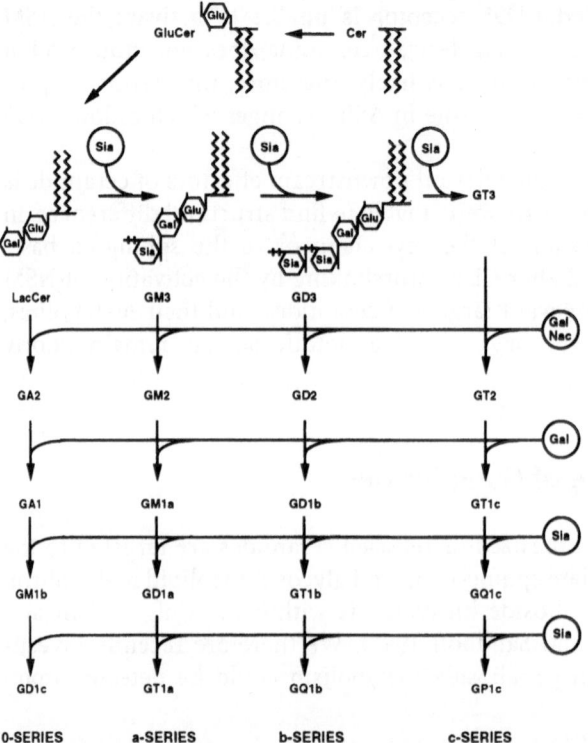

Fig. 1. General scheme for ganglioside biosynthesis. *Glu* Glucose; *GluCer* glucosylceramide; *Cer* ceramide; *Sia* sialic acid; *Gal*, galactose; *LacCer* lactosylceramide; *GalNac* N-acetylgalactosamine; *GM* monosialogangliosides; *GD* disialogangliosides; *GT* trisialogangliosides, etc

GM3, the first sialylated ganglioside. GM3 may give rise to either the monosialylated ganglioside series (GM2, GM1, etc.) or GD3, GT3, GQ3, etc., by the sequential addition of more sialic acid residues.

5.2
GD3 is a Mediator of Apoptosis

We found that GD3 ganglioside is involved in CD95-induced apoptosis (De Maria et al. 1997). GD3 results from the addition of a second sialic acid to the one present on GM3, mediated by the action of GD3 synthase (α-2,8-sialyltransferase), a transmembrane type II protein of about 40 kDa resident in the early Golgi (Haraguchi et al. 1994; Nara et al. 1994; Sasaki et al. 1994). GD3 is in turn metabolized to either GT3, by the same α-2,8-sialyltransferase, or GD2, by the addition of acetylgalactosamine to the galactose residue.

We observed that GD3 transiently accumulates after apoptotic stimulation via CD95 in HuT78 and U937 tumor cells, but not in cell variants which are resistant to CD95 crosslinking. Moreover, C2-ceramide, but not C2-dihydroceramide which is ineffective in inducing apoptosis, is sufficient to induce GD3 accumulation. GD3 is synthesized de novo from GM3 by the action of GD3 synthase, as fluorescent GM3 is converted to GD3 following both CD95 crosslinking and C8-ceramide exposure in intact cells. Moreover, CD95-induced GD3 accumulation is prevented by specific GD3 synthase antisense oligonucleotides. Therefore, triggering the apoptotic program in HuT78 and U937 tumor cells accelerates GD3 synthase function and GD3 neosynthesis. Similarly to ceramides, GD3 can efficiently induce cell death in both HuT78 and U937 cells, while other gangliosides such as GD1a, GT1b and GM1 do not, when added exogenously. Interestingly, CD95-resistant HuT78 and U937 cell lines, which failed to accumulate GD3 after CD95 crosslinking, are as sensitive as their respective wild type cell lines to GD3, suggesting that exogenous GD3 restores the ability to trigger the apoptotic program in cells with upstream signaling defects which result in lack of GD3 accumulation. Furthermore, the transient overexpression of the GD3 synthase induces apoptosis in HuT78 cells. CD95-induced cell death is substantially prevented using antisense oligonucleotides which block GD3 synthase expression. Thus, the ganglioside GD3 mediates cell death induced by CD95 crosslinking and ceramide (De Maria et al. 1997).

Inhibitors of caspases, involved in CD95-induced apoptosis, prevent GD3 accumulation upon CD95 crosslinking, indicating that one or more cysteine proteases may control GD3 synthesis. Glycosphingolipids may have deleterious effects on the conformation of the mitochondrial membrane and the oxidative phosphorylation (Strasberg 1986). Attempts to identify potential targets in the GD3 apoptotic pathway revealed that $\Delta\Psi m$ disruption is a rapid event that precedes all the major apoptotic features induced by GD3. Thus, mitochondria may represent the elective target of the apoptotic signals induced by ceramide and GD3.

6
Conclusions and Perspectives

The acidic sphingomyelinase-mediated pathway therefore contributes to apoptotic signals originated from CD95 death domains. The analysis of genetic models of ASM deficiency in humans and targeted disruption of the ASM gene in mice has established the central role of ASM in the progression of apoptotic signals (Santana et al. 1996). The involvement of GD3 in the apoptotic program helps to explain why the early activation of ASM, and not NSM, is required for the progression of CD95-generated death signal. In fact, CD95-resistant HuT78 cell variants which express death domain-defective CD95 receptors and are unable to activate the ASM after CD95 crosslinking (Cifone et al. 1995; Cascino et al. 1996) fail to accumulate GD3 following CD95 stimulation. This is consistent with the notion that only ceramides produced by ASM in the endocytic degradative compartments come into close functional and topological proximity to the membranogenic organelles, and engage the biosynthetic pathways, while those released at the outer cell surface by NSM do not. Thus, one of the roles of ASM-released ceramide in apoptosis would be serving as a precursor of ganglioside GD3, as transient GD3 accumulation appears intimately associated with ASM activation and is required for the optimal completion of the apoptotic program in hematopoietic cells. In addition, since cysteine protease inhibitors could completely prevent GD3 accumulation following CD95 crosslinking, the pathway which eventually leads to GD3 accumulation might diverge from the cysteine protease cascade at an upstream common proteolytic event triggered within the very first minutes after CD95

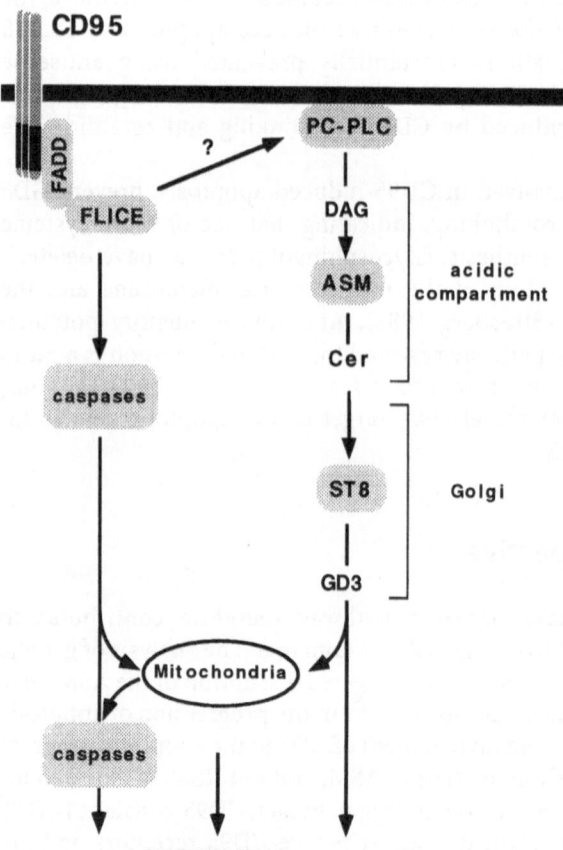

Fig. 2. CD95-induced signaling. *PC-PLC* Phosphatidylcholine-specific phospholipase C; *DAG* diacylglycerol; *ASM* acidic sphingomyelinase; *Cer* ceramide; *ST8* α-2,8-sialyltransferase

crosslinking. Cysteine proteases which express FADD-like domains, such as caspase-8 (Boldin et al. 1996; Muzio et al. 1996), which after ligand engagement are brought in close proximity to the oligomerized receptor, are likely, but not exclusive, candidates for mediating early cleavage of substrates relevant for the activation of the pathway which eventually leads to GD3 accumulation (Fig. 2).

Interestingly, GD3 is abundant in developing tissues (Yu et al. 1988) and in adult tissues displaying relatively high apoptotic rates, such as the thymus and the gastrointestinal tract (Dippold et al. 1985). While a minor ganglioside in most normal tissues, GD3 is also highly expressed in a variety of tumors, particularly of neuroectodermal origin (Hersey et al. 1988; Hakomori 1994), where it may serve as a critical adhesion molecule at the cell membrane (Cheresh et al. 1986; Sariola et al. 1988; Tettamanti and Riboni 1994). It will be of interest to investigate whether some tumors have developed antiapoptotic strategies interfering with a GD3-mediated apoptotic pathway, thus allowing GD3 accumulation and expression at the cell surface.

Mitochondrial damage induced by transient GD3 ganglioside accumulation, moreover, helps to explain how ceramides, specifically those produced in degradative compartments by ASM, might contribute to the apoptotic program induced by a variety of apoptotic stimuli, including cytokines and ionizing radiations (Santana et al. 1996). They also point towards the early Golgi, where GD3 is generated by GD3 synthase, as a critical subcellular compartment for the progression of apoptotic signals. Additional studies are needed to clarify the fate of neosynthesized GD3, its subcellular redistribution, metabolism and molecular targets, in cells undergoing apoptosis.

References

Allan D, Kallen KJ (1993) Transport of lipids to the plasma membrane in animal cells. Prog Lipid Res 32:195–219

Boldin MP, Varfolomeev EE, Pancer Z, Mett I, Camonis JH, Wallach D (1995) A novel protein which interacts with the death domain of Fas/APO-1 contains a sequence motif related to the death domain. J Biol Chem 270:7795–7798

Boldin MP, Goncharov TM, Goltsev YV, Wallach D (1996) Involvement of MACH, a novel MORT1/FADD-interacting protease, in Fas/APO-1- and TNF receptor induced cell death. Cell 85:803–815

Boucher LM, Wiegman K, Fütterer A, Pfeffer K, Machleidt T, Schültze S, Mak TW, Krönke M (1995) CD28 signals through acidic sphingomyelinase. J Exp Med 181:2059–2068

Cascino I, papoff G, De Maria R, Testi R, Ruberti G (1996) Fas/APO-1/CD95 receptor lacking the intracytoplasmic signaling domain protects tumor cells from Fas-mediated apoptosis. J Immunol 156:13–17

Cheresh DA, Pierschbacher MD, Herzig MA, Mujoo K (1986) Disialogangliosides GD2 and GD3 are involved in the attachment of human melanoma and neuroblastoma cells to extracellular matrix proteins. J Cell Biol 102:688–696

Chinnaiyan AM, O'Rourke K, Tewari M, Dixit VM (1995) FADD, a novel death domain-containing protein, interacts with the death domain of Fas and initiates apoptosis. Cell 81:505–512

Cifone MG, De Maria R, Roncaioli P, Rippo MR, Azuma M, Lanier LL, Santoni A, Testi R (1994) Apoptotic signaling through CD95 (Fas/APO-1) activates an acidic sphingomyelinase. J Exp Med 180:1547-1552

Cifone MG, Roncaioli P, De Maria R, Camarda G, Santoni A, Ruberti G, Testi R (1995) Multiple signaling originates at the Fas/Apo-1 (CD95) receptor: sequential involvement of phosphatidylcholine-specific phospholipase C and acidic sphingomyelinase in the propagation of the apoptotic signal. EMBO J 14:5859-5868

De Maria R, Lenti L, Malisan F, d'Agostino F, Tomassini B, Zeuner A, Rippo MR, Testi R (1997) Requirement for GD3 ganglioside in CD95- and ceramide-induced apoptosis. Science 277:1652-1655

Dippold WG, Dienes HP, Knuth A, Meyer zum Buschenfelde KH (1985) Immunohistochemical localization of ganglioside GD3 in human malignant melanoma, epithelial tumors, and normal tissues. Cancer Res 45:3699-3705

Dobrowsky RT, Kamibayashi C, Mumby MC, Hannun YA (1993) Ceramide activates heterotrimeric protein phasphatase 2A. J Biol Chem 268:15523-15530

Dobrowsky RT, Werner MH, Castellino AM, Chao MV, Hannun YA (1994) Activation of the sphingomyelin cycle through the low-affinity neurotrophin receptor. Science 265:1596-1599

Dressler KA, Mathias S, Kolesnick RN (1992) Tumor necrosis factor-α activates the sphingomyelin signal transduction pathway in a cell free system. Science 255:1715-1718

Enari M, Hug H, Nagata S (1995) Involvement of an ICE-like protease in Fas-mediated apoptosis. Nature 375:78-81

Haimovitz-Friedman A, Kan CC, Ehleiter D, Persaud RS, McLoughlin M, Fuks Z, Kolesnick RN (1994) Ionizing radiation acts on cellular membranes to generate ceramide and initiate apoptosis. J Exp Med 180:525-535

Hakomori SI (1994) Role of gangliosides in tumor progression. Prog Brain Res 101:241-250

Hannun YA (1996) Functions of ceramide in coordinating cellular responses to stress. Science 274:1855-1859

Hannun YA, Obeid LM (1995) Ceramide: an intracellular signal for apoptosis. Trends Biochem Sci 20:73-77

Haraguchi M, Yamashiro S, Yamamoto A, Furukawa K, Takamiya K, Lloyd KO, Shiku H, Furukawa K (1994) Isolation of GD3 synthase gene by expression cloning of GM3 α-2,8-sialyltransferase cDNA using anti-GD2 monoclonal antibody. Proc Natl Acad Sci USA 91:10455-10459

Hersey P, Jamal O, Henderson C, Zardawi I, D'Alessandro G (1988) Expression of the gangliosides GM3, GD3 and GD2 in tissue sections of normal skin, naevi, primary and metastatic melanoma. Int J Cancer 41:336-343

Jayadev S, Linardic CM, Hannun YA (1994) Identification of arachidonic acid as a mediator of sphingomyelin hydrolysis in response to tumor necrosis factor. J Biol Chem 269:5757-5763

Jayadev S, Liu B, Bielawska AE, Lee JY, Nazaire F, Pushkareva MY, Obeid LM, Hannun YA (1995) Role for ceramide in cell cycle arrest. J Biol Chem 270:2047-2052

Kolesnick R, Fuks Z (1995) Ceramide: a signal for apoptosis or mitogenesis? J Exp Med 181:1949-1952

Kolesnick RN (1987) 1,2-Diacylglycerols but not phorbol ester stimulate sphingomyelin hydrolysis in GH3 pituitary cells. J Biol Chem 262:16759-16762

Krammer PH, Behrmann I, Daniel P, Dhein J, Debatin KM (1994) Regulation of apoptosis in the immune system. Curr Opin Immunol 6:279-289

Kroemer G, Zamzami N, Susin SA (1997) Mitochondrial control of apoptosis. Immunol Today 18:44-51

Kumar S (1995) ICE-like proteases in apoptosis. Trends Biochem Sci 20:198-202

Linardic CM, Hannun YA (1994) Identification of a distinct pool of sphingomyelin in the sphingomyelin cycle. J Biol Chem 269:23530-23537

Liu P, Anderson RGW (1995) Compartmentalized production of ceramide at the cell surface. J Biol Chem 270:27179-27185

Los M, Van de Craen Penning LC, Schenk H, Westendorp M, Bauerle PA, Dröge W, Krammer PH, Fiers W, Schulze-Osthoff K (1995) Requirement of an ICE/CED-3 protease for Fas/APO-1-mediated apoptosis. Nature 375:81–83

Lozano J, Berra E, Municio MM, Diaz-Meco M, Dominguez I, Sanz L, Moscat J (1994) Protein kinase C ζ isoform is critical for κB-dependent promoter activation by sphingomyelinase. J Biol Chem 269:19200–19202

Muzio M, Chinnaiyan AM, Kischkel FC, O'Rourke K, Shevchenko A, Carsten Scaffidi JN, Bretz JD, Zhang M, Gentz R, Mann M, Krammer PH, Peter ME, Dixit VM (1996) FLICE, a novel FADD homologous ICE/CED-3 – like protease, is recruited to the CD95 (Fas/Apo-1) death-inducing signaling complex. Cell 85:817–827

Nagata S, Suda T (1995) Fas and Fas ligand: Ipr and gld mutations. Immunol Today 16:39–43

Nara K, Watanabe Y, Maruyama K, Kasahara K, Nagai Y, Sanai Y (1994) Expression cloning of a CMP-NeuAc:NeuAc α2-3Gal β1-4Glc β1-1'Cer alpha 2,8-sialyltransferase (GD3 synthase) from human melanoma cells. Proc Natl Acad Sci USA 91:7952–7256

Nelson DH, Murray DK, Brady RO (1982) Dexamethasone-induced change in the sphingomyelin content of human polymorphonuclear leukocytes in vitro. J Clin Endocrinol Metab 54:292–295

Nickels JT, Broach JR (1996) A ceramide-activated protein phosphatase mediates ceramide-induced G1 arrest of *Saccaromyces cerevisiae*. Genes Dev 10:382–394

Okazaki T, Bielawska A, Domae N, Bell R, Hannun YA (1994) Characterization and partial purification of a novel cytosolic, magnesium-independent, neutral sphingomyelinase activated in the early signal transduction of 1α,25-dihydroxyvitamin D$_3$-induced HL-60 differentiation. J Biol Chem 269:4070–4077

Pastorino JG, Simbula G, Yamamoto K, Glascott PA, Rothman R, Farber JL (1996) The cytotoxicity of tumor necrosis factor depends on induction of the mitochondrial permeability transition. J Biol Chem 271:29792–29798

Pushkarava M, Obeid LM, Hannun YA (1995) Ceramide: an endogenous regulator of apoptosis and growth suppression. Immunol Today 16:294–297

Rosenwald AG, Pagano RE (1993) Intracellular transport of ceramide and its metabolites at the Golgi complex: insights from short-chain analogs. Adv Lipid Res 26:101–118

Sallusto F, Nicolo' C, De Maria R, Corinti S, Testi R (1996) Ceramide inhibits antigen uptake and presentation by dendritic cells. J Exp Med 184:2411–2416

Sandhoff K, Kolter T (1996) Topology of glycosphingolipid degradation. Trends Cell Biol 6:98–103

Sandhoff K, van Echten G (1994) Ganglioside metabolism: enzymology, topology and regulation. Prog Brain Res 101:17–29

Santana P, Peña L, Haimovitz-Friedman A, Martin S, Green D, McLaughlin M, Cordon-Cardo C, Schuchman EH, Fuks Z, Kolesnick R (1996) Acid sphingomyelinase-deficient human lymphoblasts and mice are defective in radiation-induced apoptosis. Cell 86:189–199

Sariola H, Aufderheide E, Bernhard H, Henke-Fahle S, Dippold W, Ekblom P (1988) Antibodies to cell surface ganglioside GD3 perturb inductive epithelial–mesenchymal interactions. Cell 54:235–245

Sasaki K, Kurata K, Kojima N, Kurosawa N, Ohta S, Hanai N, Tsuji S, Nishi T (1994) Expression cloning of a GM3-sepcific α-2,8-sialyltransferase (GD3 synthase). J Biol Chem 269:15950–15956

Schuchman EH, Suchi M, Takahashi T, Sandhoff K, Desnick R (1991) Human acidic sphingomyelinase. Isolation, nucleotide sequence and expression of the full length and alternatively spliced cDNA. J Biol Chem 266:8531–8539

Schütze S, Potthof K, Machleidt T, Berkovic D, Wiegmann K, Krönke M (1992) TNF activates NF-κB by phosphatidylcholine-specific phospholipase C-induced "acidic" sphingomyelin breakdown. Cell 71:765–776

Spence MW (1993) Sphingomyelinases. Adv Lipid Res 26:3–23

Strasberg P (1986) Cerebrosides and psychosine disrupt mitochondrial functions. Biochem Cell Biol 64:485–489

Testi R (1996) Sphingomyelin Breakdown and cell fate. Trends Biochem Sci 21:468–471

Tettamanti G, Riboni L (1994) Ganglioside turnover and neural cell function: a new perspective. Prog Brain Res 101:77–100

van Echten G, Sandhoff K (1993) Ganglioside metabolism. Enzymology, topology and regulation. J Biol Chem 268:5341–5344

van Meer G (1993) Transport and sorting of membrane lipids. Curr Opin Cell Biol 5:661–673

Verheij M, Bose R, Lin XH, Yao B, Jarvis WD, Grant S, Birrer MJ, Szabo E, Zon LI, Kyriakis JM, Haimovitz-Friedman A, Fuks Z, Kolesnick RN (1996) Requirement for ceramide-initiated SAPK/JNK signalling in stress-induced apoptosis. Nature 380:75–79

Wiegmann K, Schütze S, Machleidt T, Witte D, Krönke M (1994) Functional dichotomy of neutral and acidic sphingomyelinases in tumor necrosis factor signaling. Cell 78:1005–1015

Yao B, Zhang Y, Delikat S, Mathias S, Basu S, Kolesnick R (1995) Phosphorylation of Raf by ceramide-activated protein kinase. Nature 378:307–310

Yu RK, Macala LJ, Taki T, Weinfield HM, Yu FS (1988) Developmental changes in ganglioside composition and synthesis in embryonic rat brain. J Neurochem 50:1825–1829

Lymphocyte-Mediated Cytolysis: Dual Apoptotic Mechanisms with Overlapping Cytoplasmic and Nuclear Signalling Pathways

Joseph A. Trapani[1] and David A. Jans[2]

1
Introduction

Cytotoxic T lymphocytes (CTL) and natural killer (NK) cells together comprise cytotoxic lymphocytes (CL), which are the principal basis for the immune system's detection and destruction of virus-infected or transformed cells. Although CTL and NK cells recognize foreign antigens and become activated in very different ways, both cell types employ the same two contact-dependent cytolytic mechanisms. In the first, the synergy of two granule-bound factors, a calcium-dependent pore-forming protein, perforin, and a collection of proteases ("granzymes"), results in the entry of effector proteases into the target cell cytoplasm and nucleus. The second mechanism involves trimerization of Fas (CD95/APO-1) molecules on susceptible target cells by binding to Fas ligand (FasL/CD95L) expressed on effector lymphocytes, but has no requirement for calcium. Both cytolytic mechanisms can activate a ubiquitous signalling pathway that involves sequential proteolytic events (the caspase cascade) that lead to apoptosis, with the cleavage of structural and enzymatic targets in both the cytoplasm and nucleus.

While Fas-based cytolysis depends entirely on the generation of activated caspases, the granule-based mechanism may have evolved to circumvent potential blocks in this generic death pathway that are characteristic of many viral infections. Viruses utilize host cell mechanisms for replication, and exploit the host cell environment to evade antibodies and the phagocytic cells of the host immune system. In so doing, they gain time to multiply and invade uninfected cells after replication and packaging are completed. Clearly, it is desirable for the host that infected or otherwise damaged cells do not replicate, and that the passage of such cells through the cell cycle is prevented. Instead, cells may trigger their own programmed demise and die by apoptosis, a rapid process of cellular collapse and involution. Such "altruistic" death is advantageous to the host as it limits the spread of virus. In response to this strategy,

[1] Cellular Cytotoxicity Laboratory, The Austin Research Institute, Studley Road, Heidelberg, 3084, Australia
[2] Nuclear Signalling Laboratory, The John Curtin School of Medical Research, Australian National University, P.O. Box 334, Canberra City, 2601, Australia

many viruses have learnt to delay apoptosis, and in such an event, the immune system faces a potentially serious problem of recognition and elimination of a hidden pathogen. With this in mind, this chapter will compare and contrast the granule exocytosis and FasL pathways, with special emphasis on the differential use of the two pathways by the different types of CL, and the ways in which the death signal is generated and translated into the apoptotic phenotype.

2
Dual Mechanisms of CL-Mediated Cell Death

2.1
Granule Exocytosis

This model proposed that exocytosis of specific mediators from CL could induce death of the target cell, while the CTL could survive the confrontation and recycle to other targets (Zagury et al. 1975). Firstly, cytoplasmic granules of the CL become reoriented close to the site of cell–cell contact (Bykovskaja et al. 1978). Discrete lesions first observed by Henkart and his colleagues then specifically appear on the target cell membrane as a result of encountering the CL (Dourmashkin et al. 1980), and are highly reminiscent of complement lesions, both ultrastructurally and functionally (Podack and Dennert 1983). A protein which Podack and Dennert purified from cytolytic cells and called "perforin" was present in high concentration within the presynaptic granules and could by itself induce changes in permeability of the target cell membrane. Thus arose the granule exocytosis model, embodying the hypothesis that lytic mediators stored within the granule could be released by the CL in a vectorial fashion following the formation of a stable conjugate with a target cell (Dennert and Podack 1983; see also Chapter by Darmon et al., this Vol.).

As elegant as this mechanism appeared, however, the underlying validity of this hypothesis soon came under question (see for example Berke 1991; Krahenbuhl and Tschopp 1991). Some CTL [particularly peritoneal exudate lymphocytes (PEL)] could kill independently of calcium ions, which are necessary both for granule exocytosis and pore-formation by perforin. Also, some potent CTL produced little perforin, were non-granulated, and could kill without exocytosis (Dennert et al. 1987; Trenn et al. 1987; Allbritton et al. 1988; Berke and Rosen, 1988; Ostergaard et al. 1987; Berke 1989; Ostergaard and Clark 1989). Furthermore, purified perforin could kill cells by virtue of its potent lytic potential and induced a type of necrosis; however, most investigators were unable to bring about classic apoptotic changes such as DNA fragmentation and nuclear collapse with perforin alone (Russell 1983; Schmid et al. 1986; Duke et al. 1989). The resolution of this paradox involved two broad areas of study. In the first instance, it was demonstrated that synergy of perforin with other granule proteins can bring about apoptosis, rather than membrane damage alone. The second body of work centred on the finding that T cells (particularly CD4+ Th1 cells) can utilize an alternative but equally potent, granule- and calcium-independent mechanism to kill cells (Podack 1995). The observation that CTLs also kill via the Fas ligand (CD95L)/Fas (CD95) mechanism both solved the conundrum of cytolysis in the absence of perforin, and brought CTL-mediated cell death mechanisms into the broader field of apoptosis research.

2.2
The FasL/Fas Pathway

Fas (APO-1/CD95) is an integral membrane protein with a single trans-membrane domain first identified independently by two groups as the target of mouse antibodies of the IgM (anti-Fas) and IgG3 (anti-APO-1) subclasses that were cytolytic for cells in the absence of complement (Trauth et al. 1989; Yonehara et al. 1989). Fas is a member of the tumour necrosis factor (TNF) family of molecules, other members of which control cell proliferation and apoptosis [e.g. CD40, OX40, CD27, 4-1BB and CD30; (Nagata and Golstein 1995; also see Chap. by Peter et al., this Vol.)]. Although Fas was initially suspected of ligating an unidentified cytokine, experiments from the laboratory of Golstein demonstrated that it interacts with a ligand molecule induced on the surface of certain CTL lines (Rouvier et al. 1993). Although the tenth serial subpassage of a subline of the mouse–rat hybridoma cell line PC60 (PC60-d10s) was deficient in granule-mediated killing, it efficiently killed cells transfected with Fas, but not parental cells or control transfectants. A soluble fusion protein incorporating Fas (Fas-Fc), but not one incorporating TNF receptor I, could block this cell death, implying that cell surface FasL played a major role in inducing Fas-mediated apoptosis (Ju et al. 1994; Stalder et al. 1994; Suda and Nagata 1994). Furthermore, repeated sorting of the highest expressing cells resulted in a subline that could lyse Fas-expressing target cells far more efficiently than the parental cells. Also, pretreatment of d10S with phorbol 12-myristate 13-acetate (PMA) and ionomycin induced high level expression of FasL on the effector cells, and facilitated cytolysis.

CL-mediated cytotoxicity, conventionally assessed using 4-h assays, is accounted for in toto by the perforin and FasL mechanisms. The FasL mechanism functions normally in perforin-deficient CL, and accounts for the remaining cytotoxicity, indicating that the two cytolytic mechanisms can operate independently of one another. Conversely, mice which have mutations of FasL/Fas have apparently normal cytolytic granules and express perforin and granzymes in normal quantities. Interestingly, different subsets of CL have specific preference for the individual pathways (Table 1). The perforin/granzyme mechanism is predominantly used by CD8+ CTL and NK cells, and this is consistent with the central role of these cells in eliminating virus-infected cells. These "professional killer cells" also use the Fas pathway under

Table 1. Mechanisms of cytotoxicity and their relation to CL phenotype. (Adapted from Henkart 1994)

CL phenotype	Mechanism of cytotoxicity	
	FasL/Fas	Granzyme/perforin
CD8+ CTL	+	++
NK	−	++
CD4+ Th1	++	−
CD4+ Th2	−	++

CL, cytotoxic lymphocyte; *CTL*, cytotoxic *T* lymphocyte; *NK*, natural killer.

other circumstances. By contrast, CD4+ Th1 cells readily upregulate FasL expression following conjugate formation, and can lyse in a Fas-dependent fashion more readily than Th2 cells. Deficiencies of the FasL/Fas mechanism arise spontaneously, and mice with deficient receptor (*lpr*) or ligand (*gld*) function have essentially indistinguishable phenotypes. In contrast to abnormalities of granule exocytosis, these mice are not overtly immunodeficient, but die of lymphoproliferation and/or autoimmune diseases. This suggests that the central role of FasL/Fas is to regulate the amplitude and duration of the T cell response to pathogenic (or certain "normal") stimuli. Similar defects have only recently been described in humans who present with an autoimmune lymphoproliferative syndrome (ALPS; Illum et al. 1991; Sneller et al. 1992; Drappa et al. 1996).

2.3
Perforin and Granzymes Synergize to Induce Apoptosis

A causal role for perforin in target cell death was unequivocally established in perforin-deficient gene knock-out mice, reported in rapid succession by four different groups (Kagi et al. 1994; Kojima et al. 1994; Lowin et al. 1994; Walsh et al. 1994). The animals have normal numbers of T cells in their lymphoid organs, normal CD4/CD8 ratios and cytokine production (Kagi et al. 1994). Despite normal activation following infection with lymphocytic choriomeningitis virus (LCMV), in vitro cytotoxic activity was markedly deficient against LCMV-infected targets. The cytolytic deficiency also occurred in primary in vitro-stimulated T cells and against alloreactive tumour cell lines (Kagi et al. 1994). NK activity was also virtually absent in these mice. These studies strongly asserted that the perforin pathway is vital for protection against virus infection, for eliminating alloreactive cells and in NK function (Table 2). The observation that purified perforin could not cause apoptosis, however, confirmed that granule molecules other than perforin were also essential.

Table 2. Properties of the two mechanisms of CTL/NK-mediated apoptosis

	Granule-dependent	FasL/Fas-dependent
Principal effector cells	CD8+ T, CD4+ Th2, NK	CD8+ T, CD4+ Th1
Principal function	Clearing certain viruses	T cell homeostatic mechanisms
	Tumour/allograft rejection	Bystander target lysis
	NK function	
Calcium dependence	Yes	No
Natural mutation	Not described	ALPS (human);
		gld (mouse FasL);
		lpr (mouse Fas)
Effects of mutation	Greater viral susceptibility	Lymphoproliferation
/gene disruption	Defective NK	Autoimmunity
Kinetics of cytolysis	Rapid (<4h)	Rapid (<4h)
Caspase dependence	Nuclear: yes;	Completely dependent
	non-nuclear: no	
Mechanism of cytolysis	Apoptosis	Apoptosis

NK, natural killer; *ALPS*, autoimmune lymphoproliferative syndrome.

Apart from perforin, the second prominent constituent of CTL/NK granules is a family of serine proteases termed "granzymes" (Masson and Tschopp 1987; see also Chapter by Darmon et al., this Vol.), which are biochemically and genetically related to the serine proteases of mast cells and polymorphs (Smyth and Trapani 1995; Smyth et al. 1996). Granzymes have diverse specificities (most of them are trypsin- or chymotrypsin-like) and probably participate in many lymphocyte functions in addition to cytolysis (Table 3). The importance of proteolysis during cell death had previously been pointed out, based on the inhibition of cell death observed in the presence of protease inhibitors (Chang and Eisen 1980; see also Hudig et al. 1991). Like perforin, granzymes are not cytolytic when used by themselves, although granzyme A can induce DNA fragmentation provided intracellular access is provided by partial disruption of the cell membrane (Hayes et al. 1989). Unfractionated cytolytic granules can also mediate DNA release from ^{125}I-UdR-labelled cells, but purified perforin only induces ^{51}Cr release, indicative of membrane damage (Hayes et al. 1989; Sutton et al. 1997). The description of a DNA-fragmenting activity (fragmentin-2) in rat NK cells and its identification as granzyme B, which cleaves substrates at Asp residues ("aspase" activity), greatly strengthened the case for granzyme induced cell death (Shi et al. 1992a). Fascinatingly, fragmentin-2 was unable to induce apoptosis unless sublytic quantities of perforin were also added to the assay. Apoptosis then proceeded rapidly at 37°C, with >90% maximal DNA fragmentation in under

Table 3. Functions of granzymes other than in apoptosis

Granzyme	Putative function	Reference
Granzyme A	B cell mitogen	Simon et al. (1986)
	Cleavage of extracellular matrix proteins	Simon et al. (1987)
	including proteoglycans, collagen,	Simon et al. (1991)
	laminin, fibronectin and thrombin	Simon and Kramer (1994)
	receptor	Suidan et al. (1994)
	Thymic development	Garcia-Sanz et al. (1990)
	IL-6 and IL-8 secretion	Sower et al. (1996b)
	Monocyte activation	Sower et al. (1996a)
	Neurite retraction	Suidan et al. (1996)
	Activation of IL-1	Irmler et al. (1995)
	Direct control of viral replication	Mullbacher et al. (1996)
		Simon et al. (1987)
Granzyme B	Thymic development	Garcia-Sanz et al. (1990)
	Inhibition of anchorage-dependent	Sayers et al. (1992)
	cell growth	

IL, interleukin.

2h. Two other fragmentins (designated fragmentin-1 and -3) also induced DNA fragmentation in combination with perforin, but at a much slower rate than granzyme B. These proteases corresponded to rat granzyme A and tryptase-2, both of which have a trypsin-like activity (cleavage at Lys or Arg) (Shi et al. 1992b). Furthermore, aprotinin, which blocks tryptase activity, could block apoptosis when loaded into the cytoplasm of target cells (Nakajima and Henkart 1994), providing evidence that granzymes have to act on *intracellular* substrates. Finally, inactivation of granzyme B by a synthetic tripeptide (Shi et al. 1992a) or a macromolecular granzyme B inhibitor produced by CL (Sun et al. 1996) blocks apoptosis in vitro, indicating that proteolytic activity is essential for cytolysis.

In further studies, transfection of rat basophilic leukaemia (RBL) cells with the gene for rat perforin conferred efficient cytolytic function towards erythrocytes but not nucleated cells (Shiver and Henkart 1991). These cells only became lytic for nucleated targets when they additionally expressed granzymes A and/or B (Shiver et al. 1992; Nakajima et al. 1995). Expression of antisense granzyme A mRNA by a mouse CTL cell line also partially inhibited cytolysis (Talento et al. 1992). Finally, alloreactive lymphocytes from mice lacking a functional granzyme B gene still induced target cell DNA fragmentation, but far more slowly than granzyme B expressing cells (Heusel et al. 1994). Interestingly, these deficient mice were not overtly immunocompromised and the degree of apoptosis could be rescued with longer incubation, suggesting that other proteases (such as granzyme A) can substitute for granzyme B, albeit less efficiently. The function of perforin, on the other hand, is clearly unique and its absence cannot be compensated, as indicated by the pronounced T cell immundeficiency in perforin-deficient mice.

2.4
Dual Pathways to an Endogenous Apoptotic Cascade in Target Cells

Granzyme B is the only known mammalian serine protease that cleaves at residues with acidic sidechains, with a special predilection for aspartic acid (Poe et al. 1991). This observation is highly significant, since cysteine proteases very similar to the product of the cell death gene *ced-3* of *Caenorhabditis elegans* are potent inducers of apoptosis in mammalian cells through their similar ability to cleave substrates at specific Asp residues (Yuan et al. 1993; for review see Kumar and Lavin 1996, and accompanying chapters, this Vol.). Three genes, *ced-3*, *ced-4* and *ced-9*, regulate apoptosis in *C. elegans* (Yuan and Horvitz 1992), with *ced-3* and *ced-4* permissive and *ced-9* inhibitory for apoptosis. Ced-9 plays a similar role to the mammalian Bcl-2 protein (Vaux et al. 1992). The Ced-3 protein is highly similar to mammalian interleukin-1β (IL-1β) converting enzyme (ICE), a cysteine protease first identified as being responsible for cleavage and activation of pro-IL-1β at Asp116 (Kostura et al. 1989; Thornberry et al. 1992). Multiple mammalian ICE-like proteases have now been identified, cloned, their role in apoptosis characterized (Kumar and Lavin 1996), and each has recently been designated a "caspase" with a numerical suffix, 1 to 10 (e.g. CPP32 is caspase-3; Alnemri et al. 1996). The caspases have been categorized into three groups based on sequence homologies, and each falls into ICE-like, CPP32-like or Nedd-2-like categories. All are synthesized as single chains that require further processing to become active, always by cleavage after specific Asp residues. The typical active caspase is a heterodimer with a heavy chain of ~20 kDa non-covalently bound to a light chain of ~10 kDa. The crystal structure of ICE (Wilson et al. 1994) indicates it to be a heterotetramer of two heavy and two light chains (Thornberry et al. 1992). The residues flanking the active site cysteine (residue 285 of ICE) (Gln-Ala-Cys-Arg/Gln-Gly) are tightly conserved in all caspases, while the imidazole ring of His237 is also essential for catalytic activity (Wilson et al. 1994).

The unique ability of caspases and granzyme B to cleave at Asp residues strongly suggested that an important function of granzyme B is to activate and amplify the caspase cascade (Vaux et al. 1994). In the following sections, we shall see that the FasL/Fas mechanism also activates the caspase cascade. Thus, despite significant functional differences, both the granzyme B/perforin and FasL-mediated apoptotic signalling pathways intersect at the level of caspase activation.

3
Signal Transduction Through the Fas Pathway

3.1
Signalling Through the Fas Receptor

The binding of Fas by FasL exemplifies the growth-modulatory effects of
the burgeoning TNF family of receptor/ligand pairs (Chapter by Peter et al.,
this Vol.). The signal transduction pathways leading on from receptor ligation
have recently been substantially clarified. Ligation of TNFR1 results in
receptor trimerization (Banner et al. 1993), and for both TNFR1 and Fas, there
follows recruitment of a molecular complex (the death inducing signalling
complex, DISC) to the inner leaflet of the cell membrane formed from latent
cytoplasmic proteins (Cleveland and Ihle 1995; Kischkel et al. 1997). The two
receptors have homologous cytoplasmic "death domains" (Itoh and Nagata
1993), which are also related to the drosophila protein reaper, a 65 residue
polypeptide expressed prior to apoptosis (White et al. 1994). The protein
FADD (Fas-associated protein with a death domain; Boldin et al. 1995), also
known as MORT-1 (Chinnaiyan et al. 1995a,b), can bind the death domain of
Fas, while the analogous protein TRADD (TNFR1 related death domain protein) binds to TNFR1 (Hsu et al. 1995), and receptor-interacting protein (RIP;
Stanger et al. 1995) is capable of binding to both receptors (Hsu et al. 1996;
Varfolomeev et al. 1996). All three proteins are equipped with corresponding
death domain homology (DDH) regions that bind to the receptors. In
addition to its carboxy-terminal DDH regions, FADD also has an amino-
terminal death effector domain (DED) which allows FLICE (FADD-like ICE;
Muzio et al. 1996; also known as MACH; Boldin et al. 1996) to attach to the
DISC. Like FADD, pro-FLICE has two distinct functions: it contains two
amino-terminal DEDs, one of which is for interaction with the DED of
FADD, and a carboxy-terminal ICE-like cysteine protease domain (Boldin
et al. 1996; Muzio et al. 1996). Recruitment of pro-FLICE to the DISC
results in autocatalytic detachment of the amino-terminal prodomain that
includes the two DEDs, and liberation of active FLICE protease from the
cell membrane complex. FLICE and the similar molecule Mch4 are the
critical ICE-like proteases that sit at the "apex" of the caspase cascade which is
responsible for transducing the death signal, initiated at the cell membrane,
into a cytosolic signal ultimately leading to apoptosis. It is also likely that
Fas can activate the caspase cascade by a second, FADD-independent
mechanism that operates through the Jun N-terminal kinase (JNK) pathway (Goillot et al. 1997). Recently, Yang et al. (1997) isolated and characterized a novel protein, Daxx, that binds to the Fas death domain, but lacks a
death domain of its own. Overexpression of Daxx leads to JNK activation
and potentiated Fas-mediated apoptosis, while overexpression of the Fas-
binding domain alone acts as a dominant negative inhibitor of Fas-induced

apoptosis. The connection between Jun and caspase activation has not yet been elucidated.

3.2
The ICE-Like Cytoplasmic Protease Cascade and Apoptosis

It is postulated that FLICE activates downstream caspases to bring about the cleavage of structural and enzymatic proteins required to induce apoptosis. Initial indications that activated caspases are apoptotic effector molecules have arisen by analogy to ced-3 and through the demonstration that specific caspase inhibitors could block cell death. The cytokine response modifier A (CrmA; Ray et al. 1992) protein of cowpox virus was first identified as blocking IL-1β production by binding to and inhibiting ICE, but it is now known to block apoptosis induced through ICE (Gagliardini et al. 1994), Fas (Strasser et al. 1995; Tewari and Dixit 1995) and TNF (Miura et al. 1995; Tewari and Dixit 1995). Peptidyl inhibitors of the caspases such as Tyr-Val-Ala-Asp-CMK, Asp-Glu-Val-Asp-CMK and z-Val-Ala-Asp-FMK are potent inhibitors of many types of apoptosis, suggesting that caspase activation is a common feature of diverse forms of apoptosis (Zhivotovsky et al. 1995; Sarin et al. 1997a,b). Furthermore, overexpression of ICE-like proteases can induce apoptosis (Miura et al. 1993; Munday et al. 1995), and expression of antisense mRNA for certain caspases can inhibit it in some cases (Kumar 1995).

3.3
Cleavage by Activated Caspases Contributes to Apoptotic Morphology

Specific structural and catalytic proteins are known to be cleaved early in apoptosis, and this process is fully outlined in Chap. 12, this Vol. The first to be described was the DNA repair enzyme poly(ADP-ribose) polymerase (PARP; Kaufmann 1989; Kaufmann et al. 1993), which can be cleaved by several of the caspases, including CPP32 (Lazebnik et al. 1994; Nicholson et al. 1995; Tewari et al. 1995), Mch3 (Fernandes-Alnemri et al. 1995a), Mch2 and much less efficiently by ICE (Gu et al. 1995). Although PARP is a DNA repair enzyme and can suppress nuclease activity (Lindahl et al. 1995), its degradation is not absolutely required for apoptosis, as mice with a targeted disruption of the PARP gene have normal apoptosis (Wang et al. 1995). Among other proteins, ICE-like proteases also cleave the cell cycle regulatory and anti-apoptotic protein pRb (Janicke et al. 1996), the catalytic subunit of DNA-dependent protein kinase (Casciola-Rosen et al. 1996, Song et al. 1996), the sterol regulatory element-binding proteins 1 and 2 (Wang et al. 1996), nuclear lamins (Lazebnik et al. 1995), actin [thereby reducing its DNAse I binding activity (Kayalar et al. 1996)], and U1 associated 70-kDa protein (Casciola-Rosen et al. 1996). Importantly, some caspases also activate others in a hierarchical manner to amplify the apoptotic cascade. Caspases such as ICE are autocatalytic, while others contribute to their own processing (Fernandes-Alnemri et al. 1996). For example, pro-CPP32 can initially be cleaved at Asp175 by multiple proteases (including granzyme B), and then autocatalyses further cleavage at Asp9 and Asp28. Pro-NEDD-2 can be activated by ICE (Harvey et al. 1996) and Mch4, which like FLICE (Mch5) has two DED elements in its prodomain and can activate pro-CPP32 and pro-Mch3 (Fernandes-Alnemri et al. 1996). At least three activated caspases capable of cleaving nuclear PARP and lamins, namely CPP32, Mch3 and Mch2, are thus generated in the cytoplasm. How/whether these caspases come into contact with their nuclear substrates and thereby contribute to nucleolysis is unclear (see Sect. 5).

4
The Synergy Between Perforin and Granzyme B

4.1
Entry of Granzyme B into the Cell and Generation of the Perforin Signal

Cells exposed to granzyme B in the absence of perforin do not undergo apoptosis (Shi et al. 1992a,b; Jans et al. 1996; Sutton et al. 1997; Trapani et al. 1998). It is clear from a number of observations, however, that perforin does not simply act as a pore through which apoptotic proteases such as granzyme B can diffuse into the cytoplasm. The minimal amounts of perforin required for apoptosis in synergy with granzyme B do not cause appreciable leakage of markers such as ^{51}Cr from the cells when used alone (Shi et al. 1992a). We have also tried to replace the function of perforin with other membrane disruptive/pore-forming agents such as mild detergent treatment (Shi et al. 1997) and complement (J. A. Trapani, unpubl. observ.), but neither can functionally replace perforin. In some of these experiments, even severe membrane disruption (>50% specific ^{51}Cr release, sufficient to permit macromolecules much larger than granzyme B access to the cytoplasm) was insufficient for apoptosis induction by granzyme B. It is now indeed apparent that perforin is not necessary for internalization of granzyme B, and several studies have shown it can cross the plasma membrane in the absence of perforin (Froelich et al. 1996b; Jans et al. 1996; Shi et al. 1997; Trapani et al. 1998). In one study, the amount of fluoresceinated granzyme B entering Yac-1 cells was not altered by the presence or absence of perforin (Shi et al. 1997), whilst studies using HTC (Jans et al. 1996) and FDC-P1 cells (Trapani et al. 1998) demonstrated significantly increased granzyme B cellular uptake when perforin was present. That this increased cellular uptake was specific for granzyme B was indicated by the fact that uptake of small, freely diffusible molecules such as a 20-kDa dextran was negligible in the presence of perforin (Jans et al. 1996, 1997; Trapani et al. 1998).

4.2
Nuclear Targeting of Granzymes in Intact Cells and in Vitro

An important finding from these latter studies was that the co-application of perforin and granzyme B resulted in a rapid and marked translocation of granzyme B from the cytoplasm to the nucleus (Trapani et al. 1998). Furthermore, this nuclear localization correlated precisely with subsequent apoptotic cell death. Similar findings were observed with purified granzyme A, indicating that both granzymes undergo very similar nuclear localization (Jans et al. 1997). The kinetics for granzyme A nuclear localization were considerably slower than for granzyme B, consistent with the less efficient induction of apoptosis seen with this agent, which has tryptase- rather than aspase-activity (Shi et al. 1992b; Shiver et al. 1992). Entry of granzyme B (and granzyme A) into the nucleus preceded the onset of apoptosis in that: (1) the onset of DNA fragmentation and annexin V binding lagged well behind the extremely rapid (<2 min) nuclear targeting of granzyme, and (2) the nuclear membrane remained impermeable to 70-kDa dextran pre-loaded into the cytoplasm, despite the induction of apoptosis. This suggested that penetration of granzymes into the nucleus preceded, and was not simply a consequence of, nuclear membrane disruption during apoptosis.

The mechanism of granzyme nuclear accumulation is unclear. Whilst proteins of less than ~45 kDa can normally enter the nucleus by diffusion, larger nuclear proteins require a specific targeting signal (the nuclear localization signal, NLS) that permits active transport through the nuclear envelope (Jans et al. 1991; Jans 1995; Jans and Hübner 1996). Interestingly, NLS-like sequences, similar to that of the archetypal NLS of the SV40 large tumour-antigen (NLS: Pro-Lys-Lys-Lys-Arg-Lys-Val[132]), can be found in granzyme A and B. For example, human and mouse granzymes contain the basic sequence Lys-Ala-Lys-Ile-Asn-Lys[96] or a closely related sequence which is poorly conserved in the mast cell proteases. Significantly, both granzyme A and B accumulate rapidly in the nucleoplasm and nucleoli (Jans et al. 1996, 1997) in a semi-intact permeabilized cell system which has been used extensively to examine NLS-dependent nuclear protein import (Jans et al. 1991; Ymer and Jans, 1995; Xiao et al. 1996; Efthymiadis et al. 1998). As illustrated in Table 4, conventional NLS-dependent nuclear import is dependent on both cytosolic factors and ATP, and is inhibitable by non-hydrolysable GTP analogues (Jans and Hübner 1996). In contrast, granzyme uptake is energy-independent and not inhibited by non-hydrolysable GTP analogues, although it does require cytosolic factor(s) (Jans et al. 1996, 1997), possibly in the form of a carrier/transporter molecule (see Sect. 5). As shown for granzyme B, nuclear uptake is not dependent on proteolytic activity since it is not inhibited by tripeptide granzyme B inhibitors (Jans et al. 1996).

Permeabilization of the nuclear envelope resulted in accumulation of granzyme A and B, but not of NLS-containing fusion proteins, implying that

Table 4. Comparison of *in vitro* nuclear transport properties of granzymes, CPP32 and NLS-containing proteins

Nuclear import characteristics[a]	Protein		
	Granzyme A/B	NLS-containing (T-ag)	CPP32
Nuclear accumulation	Yes	Yes	No
– dependence on cytosol	Yes	Yes[b]	
– dependence on ATP	No	Yes	
– inhibition by GTPγS[c]	No	Yes[d]	
Binding to nuclear components	Yes	No	No
Nucleolar accumulation	Yes	No	No
Binding to nucleolar components	Yes	No	No
Recognition by NLS-binding proteins (ELISA)	No	Yes	ND[e]

NLS, nuclear localization signal; T-ag, tumour antigen; ND, no data.

[a] Transport results for granzyme A are from Jans et al. (1997); for granzyme B from Jans et al. (1996), Trapani et al. (1996) and Jans et al. (1997); for CPP32 from Jans et al. (1997); and for the NLS-containing simian virus SV40 large tumour-antigen (T-ag) fusion protein from Trapani et al. (1996), Jans et al. (1996) and Efthymiadis et al. (1997). ELISA results for granzyme A and B are from Jans et al. (1997) and for T-ag from Efthymiadis et al. (1997).

[b] Cytosolic factors required for NLS-dependent nuclear import (see Jans and Hübner 1996) are the NLS-binding importin proteins and the GTP-binding protein Ran/TC4.

[c] Non-hydrolysable GTP analogue.

[d] GTPγS blocks Ran/TC4 activity (see Jans and Hübner 1996).

[e] Not done.

the granzymes accumulate in the nucleus through binding to nuclear components (Jans et al. 1997). Interestingly, nuclear accumulation was not dependent on cytosol under these conditions, indicating that the cytosolic factor required for nuclear accumulation in vitro mediates entry through an intact nuclear envelope, rather than binding to nuclear components. The characteristics of nuclear and nucleolar accumulation of granzymes A and B are summarized in Table 4 to highlight the differences between it and NLS-dependent transport, the clear implication being that nuclear/nucleolar accumulation of granzymes is unlikely to be an NLS-dependent phenomenon. Consistent with this, both granzyme A and B fail to react in an ELISA-based binding assay (Efthymiadis et al. 1998) using the conventional NLS-binding importin proteins (Jans et al. 1997), suggesting that the NLS-like sequences alluded to above are not functional as NLSs.

The observation that the nuclear uptake of granzyme A closely resembles that of granzyme B suggests a common mechanism for nuclear entry and accumulation. Both the apoptotic caspase CPP32 (Jans et al. 1997) and the serine protease chymotrypsin (Trapani et al. 1996) do not accumulate in the nucleus, implying that nuclear accumulation is a specific property of granzymes, and not of serine- or apoptotic proteases in general. Nucleolin, which shuttles between the cytoplasm and nucleus/nucleolus, and can bind to NLS-containing proteins, is bound and cleaved by granzyme A (Pasternack et al. 1991), and therefore seemed a possible candidate as a carrier or nuclear binding site for granzyme A and perhaps B. Anti-nucleolin antibodies, however, did not interfere with nuclear access or accumulation of either granzyme (Jans et al. 1997). Neither granzyme A (Jans et al. 1997) nor granzyme B (Jans et al. 1996) can directly disrupt the integrity of the nuclear membrane, in contrast to degradative proteases such as chymotrypsin (D. A. Jans, unpubl. observ.), and neither appears to impair the normal passive or active NLS-dependent nuclear transport system (Jans et al. 1996, 1997).

The above studies indicated that despite the necessity for perforin in apoptosis, the role of this molecule is clearly not as a passive transmembrane conduit. Perforin must therefore exert its effect by alternative means. Firstly, perforin may provide release of granzymes into the cytoplasm from some protective compartment such as an endosome. A model that addresses this likelihood was advanced recently by Froelich et al. (1996b), who showed evidence of a saturable granzyme B cell surface receptor that enabled it to enter the cell, presumably by endocytosis. The identity of the receptor is unclear; however, the same group has shown that granzyme A induces IL-6 and IL-8 production also by binding to the cell surface (Sower et al. 1996a,b). In the experiments of the Froelich laboratory, cells exposed to granzyme B alone or to a non-cytopathic replication deficient adenovirus remained viable, but if granzyme B was introduced with the virus, the cells underwent rapid apoptosis. A key facet of adenovirus pathogenicity is that it can escape from endosomes into the cytosol following its endocytic uptake, and this property has previously been used to introduce foreign proteins into cells (Seth 1994). Escape of granzyme B into the cytoplasm due to its packaging with adenovirus might allow it to access key caspase substrates, mimicking the role normally played by perforin. Perforin's main role may thus be to disrupt endocytic vesicles, in imitation of adenovirus. Alternatively, perforin may generate a membrane signal in its own right which is in some way simu-lated by a kinase or other signalling component present in the cytosol that is essential for nuclear accumulation of granzyme A and B in vitro in the semi-intact cell system (Jans et al. 1996, 1997). No receptor has been found for perforin other than lipid molecules with phosphorylcholine headgroups (Tschopp et al. 1989), however, and although Ca^{2+} is clearly required for its activity, there is no report of phosphorylation or other membrane signalling events being generated following perforin's attachment to the cell membrane.

4.3
Downstream Substrates of Granzymes

The ability of granzyme A to induce apoptosis in several experimental systems independently of granzyme B (e.g. Hayes et al. 1989; Shi et al. 1992a,b; Shiver et al. 1992; Jans et al. 1997) argues strongly for its role as an apoptotic protease that might bolster or substitute the actions of granzyme B, should it be inhibited (e.g. by a CrmA-like molecule) or absent. Granzyme A has ICE-like activity; that is, it can produce active IL-1β by cleaving pro-IL-1 at a basic residue close to the Asp which is usually cleaved by ICE (Irmler et al. 1995). However, granzyme A has not been reported to activate any of the caspases, and the mechanism by which apoptosis is brought about by granzyme A is uncertain. Additionally, granzyme A-deficient mice have no obvious defect of apoptosis induction (Ebnet et al. 1995), although they do suffer from an acute susceptibility to the poxvirus ectromelia (Mullbacher et al. 1996). By contrast, there is ample evidence that granzyme B's aspase specificity can activate many of the pro-caspases. In vitro, granzyme B has been shown to activate pro-CPP32 (Darmon et al. 1995; Fernandes-Alnemri et al. 1996; Martin et al. 1996) and pro-Nedd-2 (Harvey et al. 1996), but it cannot cleave pro-ICE (Darmon et al. 1994). It has not yet been adequately demonstrated whether granzyme B acts directly on pro-CPP32 in vivo. Activation of CPP32 by CL appears to be granzyme B-dependent, but the effect might be indirect, as granzyme B is also capable of activating upstream caspases including pro-FLICE (Mch5) and pro-Mch4, both of which can activate CPP32 (Fernandes-Alnemri et al. 1996). Granzyme B's ability to activate pro-FLICE should be sufficient to activate the whole cascade, but in vitro data suggest that granzyme B can cleave multiple pro-caspases, and thus augments the cascade proximally through FLICE and Mch4 and more distally through CPP32, Mch3 (Fernandes-Alnemri et al. 1996), Mch2 and Mch6 (Srinivasula et al. 1996) and CMH-1/ICE-LAP3 (Gu et al. 1996). The cleavage of pro-Mch6 is at a site distinct from that utilized by CPP32 (Srinivasula et al. 1996). The overall result is the generation of multiple active proteases that can cleave nuclear structures such as PARP and lamins in addition to cytoplasmic substrates (see Chap. 6, this Vol.). Granzyme B does not appear to activate Mch2a directly, although this caspase is thought to act on lamins directly (Fernandes-Alnemri et al. 1995b).

Darmon and colleagues (1995) used specific tetrapeptide inhibitors to dissect the roles of the different caspases in DNA fragmentation and cell membrane damage. The CPP32/Mch3 inhibitor Ac-Asp-Glu-Val-Asp-CHO (Nicholson et al. 1995) had no effect on ^{51}Cr release from target cells killed by a granule-dependent mechanism, but caused a marked reduction in DNA fragmentation from the same cells. In contrast, Ac-Tyr-Val-Ala-Asp-CHO, which inhibits members of the ICE (Thornberry et al. 1992) but not of the CPP32 (Nicholson et al. 1995) caspase subfamily, did not affect DNA fragmentation. Furthermore, mice deficient in ICE show apparently normal nuclear apoptotic morphology (Li et al. 1995), suggesting that the CPP32 family of protease is instrumental in eliciting nuclear damage, while the ICE-related proteases have little effect on this process. This study, like that of Sarin et al. (1997a), suggested caspases play no significant role in cell membrane damage resulting from granule-induced apoptosis.

In vitro evidence suggests that granzyme B is capable of cleaving key nuclear proteins such as PARP (Froelich et al. 1996a) and the catalytic subunit of DNA-dependent protein kinase (Song et al. 1996) directly, but in each case the cleavage sites are different from those conventionally used by the caspases. Whilst there is good evidence that granzyme B is able to reach the nucleus in high concentrations (see Sect. 4.2) and is thus likely to access and cleave nuclear substrates to contribute to nuclear apoptotic morphology (Jans et al. 1996; Pinkoski et al. 1996; Trapani et al. 1996), this is not the case for the caspases. As mentioned above (see Table 4), activated CPP32 does not accumulate within the nucleus (Jans et al. 1997), and nuclear accumulation has not been reported for Mch2 and Mch 3. Indeed, there is recent evidence that CPP32 may activate another cytoplasmic molecule, termed DFF (DNA fragmentation factor), a heterodimeric factor composed of 40- and 45-kDa subunits that is responsible for DNA fragmentation (Liu et al. 1997), suggesting that most of the nuclear events associated with CPP32 proteolysis may be achieved indirectly (see also Krajewski et al. 1997a,b).

5
Nucleolysis Versus Cytolysis

Although the most obvious morphological changes of apoptosis take place in the nucleus, nuclear collapse is clearly not indispensible for apoptotic cell death, and cell death is readily seen in the absence of a nucleus (Jacobson et al. 1994; Schultze-Ostoff et al. 1994; Nakajima et al. 1995). Similarly, DNA fragmentation is not an invariable consequence of apoptosis, as mutant fibroblasts lacking endogenous nucleases undergo apoptosis in response to CTL just as rapidly as their wild-type counterparts (Ucker et al. 1992). It therefore follows that nuclear damage is downstream of "cytoplasmic death", indicating a pressing need to define the cytoplasmic processes which are the basis of the entire mechanism. Some exciting recent findings have begun to address these issues. Two groups of investigators have shown that during apoptosis, the apoptotic protein Ced-4 of C. elegans can interact with Ced-3 and its mammalian homologues, ICE and FLICE (Chinnaiyan et al. 1997; Wu et al. 1997). Overexpression of Ced-4 in mammalian cells induced cell death, and further overexpression of Bcl-X_L in these cells blocked it, as did the addition of caspase inhibitors (Chinnaiyan et al. 1997). The anti-apoptotic protein Ced-9 can negatively regulate apoptosis by binding to Ced-4, and Ced-4 interacts with Ced-3 or the mammalian equivalent proteins such as ICE and FLICE, which have large prodomains, but not with CPP32 or Mch2a, which have small prodomains (see

Chaps. 2 and 11, this vol.). Thus, Ced-4 plays a pivotal role in regulating cell death, and links pro-apoptotic proteases of the ICE family with inhibitory proteins of the Bcl-2 family, by its ability to directly interact with both families of molecules. Normally, Ced-4 is expressed in the cytoplasm, but when Bcl-X_L is coexpressed, the Ced-4/Bcl-X_L complex is relocated to mitochondrial membranes (Wu et al. 1997). In so doing, Ced-9 appears to modify Ced-4, thereby preventing it from activating Ced-3, and thus dislocating upstream signalling events from caspase activation. Conversely, in the absence of Ced-9/Bcl-2, Ced-4 is free to activate the protease cascade.

It is evident from the above discussion that granule exocytosis and the FasL/ Fas pathway function, at least partly, through the same sets of effector proteases, the caspases. Why is it that both mechanisms used by CL should converge at the same biochemical pathway, and why does this not render CL killing vulnerable to viral inhibitors? It is now becoming clear that the two pathways are clearly distinguishable at the biochemical level (Fig. 1). In particular, unlike the Fas pathway, the granule exocytosis mechanism can kill cells independently of caspase-induced proteolysis. Sarin et al. (1997a,b) used synthetic peptide inhibitors to show that all of the manifestations of Fas-induced apoptosis, both nuclear and cytoplasmic, are caspase-dependent; that is, blocking all caspases (or just key caspases such as FLICE) can completely abrogate apoptosis through Fas. In contrast, the *nuclear* apoptotic changes brought about by granule exocytosis were also blocked by caspase inhibitors, but the *non-nuclear* consequences (mitochondrial dysfunction, loss of membrane integrity) still occurred, resulting in cell death. Furthermore, these manifestations were not due simply to the unopposed actions of perforin acting alone on the cell membrane, as proteolysis (presumably by granzymes acting independently of caspases) was also required (Sarin et al. 1997a,b).

Applying these observations hypothetically to models of cells infected with virus raises the following possibilities. If a cell is infected with a virus *in*capable of blocking the caspase cascade, the cell could be killed by either or both granule exocytosis and FasL, and dies a rapid death as a consequence of both nuclear and cytoplasmic damage. In this scenario, nuclear degradation would involve the combined actions of granzyme B and activated caspases acting on different cleavage sites on targeted proteins. Alternatively, the cell might be infected with a virus that completely blocks the caspase pathway (and therefore Fas-mediated apoptosis) to prolong its survival. Such a situation should probably arise relatively infrequently, as it is not an advantage for a virus to kill its natural host, but rather to *delay* apoptosis, thus facilitating its spread to uninfected cells. Nonetheless, when confronting such a potentially serious infection, the CL must be capable of having the "final word" on the survival of an infected cell, and it does so by use of the granule exocytosis mechanism. Thus, if the caspase cascade is blocked, the cell still dies because of the non-nuclear (caspase-independent) consequences of apoptosis. The means by

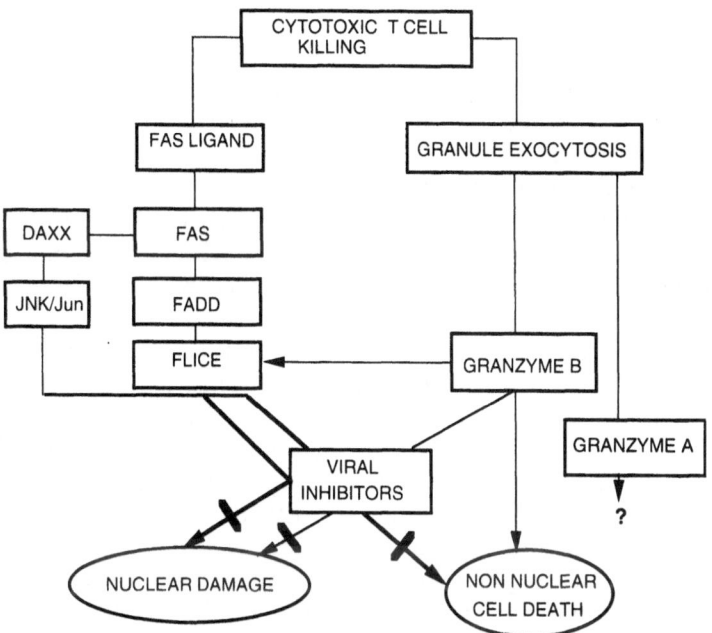

Fig. 1. The signalling pathways involved in granule- and FasL-mediated cytolysis, indicating the relative dependence on caspase activation. Inhibition studies using the broadly active caspase inhibitor p35 and the tripeptide inhibitor val-ala-asp-fmk (Sarin et al. 1997a,b) suggested that both the nuclear and cytoplasmic consequences of triggering through the Fas pathway are caspase-dependent (*thick arrows*), irrespective of whether the membrane signal is carried through the FADD or through the JNK pathway. The nuclear apoptotic changes observed with granzyme B and perforin are also caspase-dependent; however, non-nuclear effects were still seen when these inhibitors were used, indicating the presence of as yet unidentified cytoplasmic substrates for the granzymes (*thin arrows*). The mechanism of action of granzyme A is probably caspase-independent, but is otherwise uncharacterized

which CTL granules achieve caspase-independent apoptosis is an important unresolved issue, one possibility being that granzymes can target their own unique substrates in the cytoplasm. The search for these substrates is one of the key tasks facing T cell biologists.

6
Concluding Remarks

The past several years have brought a realization that very similar processes control cell death brought about by a host of diverse stimuli. It is fascinating that CLs, which have two quite diverse means of inducing target cell apoptosis, utilize pathways that impinge on the same basic processes. Furthermore, various subsets of T and NK cells differentially utilize the granule and FasL pathways, and this may shape the immune responses to viruses and in chronic inflammatory diseases. Although the granule and FasL pathways intersect at the level of caspase activation, the resilience of granule-mediated cell death to caspase inhibitors suggests that this mechanism may have evolved to outflank viruses that have devised ways to thwart the endogenous cell death cascade. Granzymes play an important role in inducing nuclear changes of apoptosis; however, cells can still be killed in the absence of a nucleus. There must therefore be cytoplasmic substrates other than caspases that are crucial for the granzymes' ability to kill cells.

Whilst the granzyme–perforin pathway involves caspase activation, it also embodies a novel mechanism by which the apoptotic signal is transduced, namely translocation of the proteolytic granzymes A and B directly into the nucleus. In this context, future studies will aim to (1) elucidate the membrane or other signalling role of perforin in triggering granzyme translocation from the outside of the cell ultimately to the nucleus; (2) identify the cytosolic factors able to substitute for perforin's role in granzyme nuclear accumulation in vitro; (3) identify the specific physiological substrates of granzymes in the nucleus; and (4) determine the exact relationship of the cytolytic and nucleolytic arms of the apoptotic response. The intriguing question of whether nuclear granzyme activity has any role in triggering cytolysis in the absence of caspase activity may be resolved using reconstituted in vitro models of apoptosis similar to that used to characterize the nuclear import properties of granzymes.

Acknowledgments. The authors wish to thank the staff of their laboratories, especially Mark Smyth, Lyndall Briggs, Patricia Jans, Vivien Sutton, Lisa McDonald and Kylie Browne for their support over many years. We also thank our collaborators, particularly Sharad Kumar, David Vaux, Arnold Greenberg and Philip Bird and their laboratories for sharing their reagents, expertise and unpublished data. At various times, the laboratory of J.A.T. has received generous support from The Wellcome Trust, the National Health and Medical Research Council of Australia and the Anti-Cancer Council of Victoria, and that of D.A.J. from the Clive and Vera Ramaciotti Foundation, the Australian Research Council and International Union against Cancer (UICC). The work has also been supported by the Institute of Advanced Studies Australian Universities Research Collaboration Scheme.

References

Allbritton NL, Verret CR, Wolley RC, Eisen HN (1988) Calcium ion concentrations and DNA fragmentation in target cell destruction by murine cloned cytotoxic T lymphocytes. J Exp Med 167:514–527

Alnemri ES, Livingston DJ, Nicholson DW, Salvesan G, Thornberry NA, Wong WW, Yuan J (1996) Human ICE/CED-3 protease nomenclature. Cell 87:171–174

Banner D, D'arcy A, James W, Gentz R, Shoenfeld H-J, Broger C, Loetscher H, Lesslauer W (1993) Crystal structure of the soluble human 55 kD TNF receptor-human TNF beta complex: implications for TNF receptor activation. Cell 73:431–445

Berke G (1989) The cytolytic T lymphocyte and its mode of action. Immunol Lett 20:169–178

Berke G (1991) Debate: the mechanism of lymphocyte-mediated killing. Lymphocyte-triggered internal target disintegration. Immunol Today 12:396–399

Berke G, Rosen D (1988) Highly lytic in vivo primed cytolytic T lymphocytes devoid of lytic granules and BLT esterase activity acquire these constituents in the presence of T cell growth factors upon blast transformation in vitro. J Immunol 141:1429–1436

Boldin MP, Varfoloeev EE, Pancer Z, Mett IL, Camonis JH, Wallach D (1995) A novel protein that interacts with the death domain of Fas/APO-1 contains a sequence motif related to the death domain. J Biol Chem 270:7795–7798

Boldin MP, Goncharov TM, Golstev YV, Wallach D (1996) Involvement of MACH, a novel MORT1/FADD-interacting protease, in Fas/APO-1- and TNF receptor-induced cell death. Cell 85:803–815

Bykovskaja SN, Rytenko AN, Rauschenbach MO, Bykovsky AF (1978) Ultrastructural alteration of cytolytic T lymphocytes following their interaction with target cells. Cell Immunol 40:175–185

Casciola-Rosen L, Nicholson DW, Chong T, Rowan KR, Thornberry NA, Miller DK, Rosen A (1996) Apopain/cpp32 cleaves proteins that are essential for cellular repair: a fundamental principle of apoptotic death. J Exp Med 183:1957–1964

Chang TW, Eisen HN (1980) Effects of N-tosyl-L-lysyl-chloromethylketone on the activity of cytotoxic T lymphocytes. J Immunol 124:1028–1033

Chinnaiyan AM, O'Rourke Tewari M, Dixit VM (1995a) FADD, a novel death domain-containing protein, interacts with the death domain of Fas and initiates apoptosis. Cell 81:505–512

Chinnaiyan AM, Tepper CG, Seldin MF, O'Rourke K, Kischkel FC, Hellbardt S, Krammer PH, Peter ME, Dixit VM (1995b) FADD/MORT1 is a common mediator of CD95 (Fas/APO-1) and tumor necrosis factor-induced apoptosis. J Biol Chem 271:4961–4965

Chinnaiyan AM, O'Rourke K, Lane BR, Dixit VM (1997) Interaction of CED-4 with CED-3 and CED-9: a molecular framework for cell death. Science 275:1122–1126

Cleveland JL, Ihle JN (1995) Contenders in FasL/TNF death signaling. Cell 81:479–482

Darmon AJ, Ehrman N, Caputo A, Fujinaga J, Bleackley RC (1994) The cytotoxic T cell proteinase granzyme B does not activate interleukin-1β-converting enzyme. J Biol Chem 269:32043–32046

Darmon AJ, Nicholson DW, Bleackley RC (1995) Activation of the apoptotic protease CPP32 by cytotoxic T cell-derived granzyme B Nature 377:446–448

Dennert G, Podack ER (1983) Cytolysis by H-2-specific T killer cells: assembly of tubular complexes on target membranes. J Exp Med 157:1483–1495

Dennert G, Anderson CG, Prochazka G (1987) High activity of N-benzyloxycarbonyl-L-lysin thiobenzylester serine esterase and cytolytic perforin in cloned cell lines is not demonstrable in in vivo-induced cytotoxic effector cells. Proc Natl Acad Sci USA 84:5004–5008

Dourmashkin P, Deteix P, Simone CB, Henkary P (1980) Electron microscopic demonstration of lesions on target cell membranes associated with antibody dependent cellular cytotoxicity. Clin Exp Immunol 43:554–560

Drappa J, Vaishnaw AK, Sullivan KE, Chu J, Elkon KB (1996) Fas gene mutations in the Canale-Smith syndrome, an inherited lymphoproliferative disorder associated with autoimmunity. N Engl J Med 335:1643–1649

Duke RC, Persechini PM, Chang S, Liu CC, Cohen JJ, Young JD (1989) Purified perforin induces target cell lysis but not DNA fragmentation. J Exp Med 170:1451–1456

Ebnet K, Hausmann M, Lehman-Grubbe F, Mullbacher A, Kopf M, Lamers M, Simon MM (1995) Granzyme A-deficient mice retain potent cell-mediated cytotoxicity. EMBO J 14:4230–4239

Efthymiadis A, Shao H, Hübner S, Jans DA (1998) Kinetic characterization of the human retinoblastoma protein bipartite nuclear localization sequence in vivo and in vitro: a comparison with the SV40 large T-antigen NLS. J Biol Chem (in press)

Fernandes-Alnemri T, Takahashi A, Armstrong R, Krebs J, Friz L, Tomaselli KJ, Wang L, Yu Z, Croce CM, Earnshaw WC, Litwack G, Alnemri ES (1995a) Mch3, a novel apoptotic human cysteine protease highly related to CPP32. Cancer Res 55:6045–6052

Fernandes-Alnemri T, Litwack G, Alnemri ES (1995b) Mch2, a new member of the apoptotic Ced-3/ICE cysteine protease gene family. Cancer Res 55:2737–2742

Fernandes-Alnemri T, Armstron RC, Krebs J, Srivasula SM, Wang L, Bullrich F, Fritz LC, Trapani JA, Croce C, Tomaselli KJ, Litwack G, Alnemri ES (1996) In vitro activation of CPP32 and Mch3 by Mch4 a novel human apoptotic cysteine protease and a substrate for granzyme B. Proc Natl Acad Sci USA 93:7464–7469

Froelich CJ, Hanna WL, Poirier GG, Duriez PJ, D'Amours D, Salvesan GS, Alnemri ES, Earnshaw WC, Shah GM (1996a) Granzyme B/perforin mediated apoptosis of Jurkat cells results in cleavage of poly(ADP-ribose) polymerase to the 89-kDa apoptotic fragment and less abundant 64-kDa fragment. Biochem Biophys Res Commun 227:658–667

Froelich CJ, Orth K, Turbov J, Seth P, Gottlieb R, Babior B, Shah G, Bleackley RC, Dixit V, Hanna W (1996b) New paradigm for lymphocyte granule-mediated cytotoxicity. Target cells bind and internalize granzyme B, but an endosmolytic agent is necessary for cytosolic delivery and subsequent apoptosis. J Biol Chem 271:29073–29079

Gagliardini V, Fernandez P-A, Lee R, Drexler HC, Rotello R, Fishman MC, Yuan J (1994) Prevention of vertebrate neuronal death by the crmA gene. Science 263:826–828

Garcia-Sanz JA, MacDonald HR, Jenne DE, Tschopp J, Nabholz M (1990) Cell specificity of granzyme gene expression. J Immunol 145:3111–3118

Goillot E, Raingeaud J, Ranger A, Tepper RI, Davis RJ, Harlow E, Sanchez I (1997) Mitogen-activated protein kinase-mediated Fas apoptotic signaling pathway. Proc Natl Acad Sci USA 94:3302–3307

Gu Y, Sarnecki C, Fleming MA, Lippke JA, Bleackley RC, Su M (1995) Processing and activation of CMH-1 by granzyme B. J Biol Chem 271:10816–10820

Harvey NL, Trapani JA, Fernandes-Alnemri T, Litwack G, Alnemri E, Kumar S (1996) Processing of the Nedd-2 precursor by ICE-like proteases and granzyme B. Genes Cells 1:673–685

Hayes MP, Berrebi GA, Henkart PA (1989) Induction of target cell DNA release by the cytotoxic T lymphocyte granule protease granzyme A. J Exp Med 170:933–946

Henkart, PA (1994) Lymphocyte-mediated cytotoxicity: two pathways and multiple effector molecules. Immunity 1:343–346

Heusel JW, Wesselschmidt RL, Shestra S, Russell JH, Ley TJ (1994) Cytotoxic lymphocytes require granzyme B for the rapid induction of DNA fragmentation and apoptosis in allogeneic target cells. Cell 76:977–987

Hsu H, Xiong J, Goedell DV (1995) The TNF receptor 1-associated protein TRADD signals cell death and NFx-B activation. Cell 81:495–504

Hsu H, Shu H-B, Pan M-P, Goedell DV (1996) TRADD–TRAF2 and TRADD–FADD interactions define two distinct TNF receptor-1 signal transduction pathways. Cell 84:299–308

Hudig D, Allison NJ, Pickett TM, Winkler U, Kam CM, Powers JC (1991) The function of lymphocyte proteases. Inhibition and restoration of granule-mediated lysis with isocoumarin serine protease inhibitors. J Immunol 147:1360–1368

Illum N, Ralfkiger E, Palleson G, Geisler C (1991) Phenotypical and functional characterization of double negative (CD4-CD8-) alpha beta T-cell receptor positive cells from an imunodeficient patient. Scand J Immunol 34:635–640

Irmler M, Hertig S, MacDonald HR, Sadoul R, Becherer JD, Proudfoot A, Solari R, Tschopp J (1995) Granzyme A is an interleukin-1β-converting enzyme. J Exp Med 181:1917–1922

Itoh N, Nagata S (1993) A novel protein domain required for apoptosis. Mutational analysis of human Fas antigen. J Biol Chem 268:10932–10937

Jacobson MD, Burne JF, Raff MC (1994) Programmed cell death and Bcl-2 protection in the absence of a nucleus. EMBO J 13:1899–1908

Janicke RU, Walker PA, Lin XY, Porter AG (1996) Specific cleavage of the retinoblastoma protein by an ICE-like protease in apoptosis. EMBO J 15:6969–6978

Jans DA (1995) Regulation of protein transport to the nucleus by phosphorylation. Biochem J 311:705–716

Jans DA, Hübner S (1996) Regulation of protein transport to the nucleus – the central role of phosphorylation. Physiol Rev 76:651–685

Jans DA, Ackerman M, Bischoff JR, Beach DH, Peters RJ (1991) p34cdc2-mediated phosphorylation at T124 inhibits nuclear import of SV-40 T antigen proteins. J Cell Biol 115:1203–1212

Jans DA, Jans P, Briggs LJ, Sutton V, Trapani JA (1996) Reconstitution of nuclear and nucleolar transport of the natural killer cell serine protease granzyme B – dependence on perforin in vivo and cytosolic factors in vitro. J Biol Chem 271:30781–30789

Jans DA, Briggs LJ, Jans P, Froelich CJ, Parasivam G, Williams EA, Kumar S, Sutton VR, Trapani JA (1997) Nuclear targeting of granzyme A (fragmentin-1): dependence on perforin in intact cells and cytosolic factors in vitro. (submitted)

Ju ST, Cui H, Panka DJ, Ettinger R, Marshak-Rothstein A (1994) Participation of target Fas protein in apoptosis pathway induced by CD4+ Th1 and CD8+ cytotoxic T cells. Proc Natl Acad Sci USA 91:4185–4189

Kagi D, Ledermann B, Burki K, Seiler P, Odermatt B, Olsen KJ, Podack ER, Zinkernagel RM, Hengartner H (1994) Cytotoxicity mediated by T cells and natural killer cells is greatly impaired in perforin-deficient mice. Nature 369:31–37

Kaufmann SH (1989) Induction of endonucleolytic DNA cleavage in human acute myelogenous leukemia cells by etoposide, camptothesin, and other cytotoxic anti cancer drugs: a cautionary note. Cancer Res 52:3976–3985

Kaufmann SH, Desnoyers SM, Ottaviano Y, Davidson NE, Poirier GG (1993) Specific proteolytic cleavage of poly(ADP-ribose) polymerase: an early marker of chemotherapy induced apoptosis. Cancer Res 49:5870–5878

Kayalar C, Ord T, Testa MP, Zhong LT, Bredesen DE (1996) Cleavage of actin by interleukin-1β-converting enzyme to reverse DNase 1 inhibition. Proc Natl Acad Sci USA 93:2234–2238

Kischkel FC, Peter ME, Medema JP, Scaffidi C, Scheuerpflug C, Debatin K, Krammer PH (1997) FLICE is involved in regulation of activation induced cell death (AICD) in peripheral T cells. Proc 6th EMBO Worksh on Cell-mediated cytotoxicity, Kerkrade, The Netherlands, p36 (Abstr)

Kojima H, Shinohara N, Hanaoka S, Someya-Shirota Y, Takagaki Y, Ohno H, Saito T, Katayama T, Yagita H, Okumura K, Shinkai Y, Alt FW, Matsuzawa A, Yonehara A, Takayama H (1994) Two distinct pathways of specific killing revealed by perforin mutant cytotoxic T lymphocytes. Immunity 1:357–364, 1994

Kostura MJ, Tocci MJ, Limjuco G, Chin J, Cameron AG, Hillman NA, Chartrain NA, Schmidt JA (1989) Identification of a monocyte-specific pre-interlerukin-1β convertase activity. Proc Natl Acad Sci USA 86:5527–5531

Krahenbuhl O, Tschopp J (1991) Debate: the mechanism of lymphocyte-mediated killing. Perforin-induced pore formation. Immunol Today 12:399–402

Krajewski S, Gasgoyne RD, Zapata JM, Krajewski M, Kitada S, Chhanabhal M, Horsman D, Berean K, Piro LD, Fugier-Vivier I, Liu YJ, Wang H-C, Reed JC (1997a) Immunolocalization of the ICE/Ced-3-family protease CPP32 (caspase-3), in non-Hodgkin's lymphomas, chronic lymphocytic leukemias, and reactive lymph nodes. Blood 89:3817–3825

Krajewski M, Wang H-C, Krajewski S, Zapata JM, Shabaik A, Gasgoyne R, Reed JC (1997b) Immunohistochemical analysis of in vivo patterns of expression of CPP32 (caspase-3), a cell death protease. Cancer Res 57:1605–1613

Kumar S (1995) Inhibition of apoptosis by the expression of antisense Nedd2. FEBS Lett 368:69–72

Kumar S, Lavin MF (1996) The ICE family of cysteine proteases as effectors of cell death. Cell Death Differ 3:155–267

Lazebnik YA, Kaufmann SH, Desnoyers S, Poirier GG, Earnshaw WC (1994) Cleavage of poly(ADP-ribose) polymerase by a proteinase with properties like ICE. Nature 371:346–347

Lazebnik YA, Takahashi A, Moir RD, Goldman RD, Poirier GG, Kaufmann SH, Earnshaw WC (1995) Studies of the lamin proteinase reveal multiple parallel biochemical pathways during apoptotic execution. Proc Natl Acad Sci USA 92:9042–9046

Li P, Allen H, Bannerjee S, Franklin S, Herzog L, Johnston C, McDowell J, Paskind M, Rodman L, Salfeld J, Towne E, Tracey D, Wardwell S, Wei F-Y, Wong W, Kamen R, Seshadri T (1995) Mice deficient in IL-1β converting enzyme are defective in production of mature IL-1β and resistant to endotoxin shock. Cell 80:401–411

Lindahl LT, Satoh MS, Poirier GG, Klungland A (1995) Post-translational modification of poly(ADP-ribose) polymerase induced by DNA strand breaks. Trends Biochem Sci 20:405–411

Liu X, Zou H, Slaughter C, Wang X (1997) DFF, a heterodimeric protein that functions downstream of caspase-3 to trigger DNA fragmentation during apoptosis. Cell 89:175–184

Lowin B, Beermann F, Schmidt A, Tschopp J (1994) A null mutation in the perforin gene impairs cytolytic T lymphocyte- and natural killer cell-mediated cytotoxicity. Proc Natl Acad Sci USA 91:11571–11575

Martin SJ, Amarante-Mendes GP, Shi L, Chuang TH, Casiano CA, O'Brien GA, Fitzgerald P, Tan EM, Bokoch GM, Greenberg AH, Green DR (1996) The cytotoxic cell protease granzyme B initiates apoptosis in a cell-free system by proteolytic processing and activation of the ICE/CED-3 family protease, CPP32, via a novel two-step mechanism. Embo J 15:2407–2416

Masson D, Tschopp J (1987) A family of serine esterases in lytic granules of cytolytic T lymphocytes. Cell 49:679–685

Miura M, Zho H, Rotello R, Hartwieg EA, Yuan J (1993) Induction of apoptosis in fibroblasts by Il-1β-converting enzyme, a mammalian homolog of the C. elegans cell death gene ced-3. Cell 75:653–660

Miura M, Friedlander RM, Yuan J (1995) Tumor necrosis factor induced apoptosis is mediated by a crmA-sensitive cell death pathway. Proc Natl Acad Sci USA 92:8318–8322

Mullbacher A, Ebnet K, Blanden RV, Hla RT, Stehle T, Museteanu C, Simon MM (1996) Granzyme A is critical for recovery for mice from infection with the natural cytopathic viral pathogen, Ectromelia. Proc Natl Acad Sci USA 93:5783–5787

Munday NA, Vaillancourt JP, Ali A, Casano FJ, Miller DJ, Milineaux SM, Yamin T-T, Yu VL, Nicholson DW (1995) Molecular cloning and pro-apoptotic activity of ICE_relII and ICE_relIII, members of the ICE/CED-3 family of cysteine proteases. J Biol Chem 270:15870–15876

Muzio M, Chinnaiyan AM, Kischkel FC, O'Rourke K, Shevchenko A, Ni J, Scaffidi C, Bretz JD, Zhang M, Gentz R, Mann M, Krammer PH, Dixit VM (1996) FLICE, a novel FADD-homologous ICE-CED-3-like protease, is recruited to the CD95 (Fas/APO-1) death-inducing signaling complex. Cell 85:817–827

Nagata S, Golstein P (1995) The Fas death factor. Science 267:1449–1456

Nakajima H, Henkart PA (1994) Cytotoxic lymphocyte granzymes trigger a target cell internal disintegration pathway leading to cytolysis and DNA breakdown. J Immunol 152:1057–1063

Nakajima H, Golstein P, Henkart PA (1995) The target cell nucleus is not required for cell mediated granzyme- or Fas-based cytotoxicity. J Exp Med 181:1905–1913

Nicholson DW, Ali A, Thornberry NA, Vaillancourt JP, Ding CK, Gallant M, Garaeau Y, Griffin PR, Labelle M, Lazebnik YA, Munday NA, Raju SM, Smulson ME, Yamin T-T, Yu VL, Miller DK (1995) Identification and inhibition of the ICE/CED-3 protease necessary for mammalian apoptosis. Nature 376:37–43

Ostergaard HL, Clark WR (1989) Evidence for multiple lytic pathways used by cytotoxic T lymphocytes. J Immunol 143:2120–2126

Ostergaard HL, Kane KP, Mescher MF, Clark WR (1987) Cytotoxic T lymphocyte mediated lysis without release of serine esterase. Nature 330:71–72

Pasternack MS, Bleier KJ, McInerney TN (1991) Granzyme A binding to target cell proteins. Granzyme A binds to and cleaves nucleolin in vitro. J Biol Chem 266:14703–14708

Pinkoski MJ, Winkler U, Hudig D, Bleackley RC (1996) Binding of granzyme B in the target cell nucleus: recognition of an 80-kilodalton protein. J Biol Chem 271:10225–10229

Podack ER (1995) Functional significance of two cytolytic pathways of cytotoxic T lymphocytes. J Leukoc Biol 57:548–552

Podack ER, Dennert G (1983) Assembly of two types of tubules with putative cytolytic function by cloned natural killer cells. Nature 302:442–445

Poe M, Blake JT, Boulton DA, Gammon M, Sigal NH, Wu JK, Zweerink HJ (1991) Human cytotoxic lymphocyte granzyme B. Its purification from granules and the characterization of substrate and inhibitor specificity. J Biol Chem 266:98–103

Ray CA, Black RA, Kronheim SR, Greenstreet TA, Sleath PR, Salvesan GS, Pickup DJ (1992) Viral inhibition of inflammation: cowpox virus encodes an inhibitor of the interleukin-1β converting enzyme. Cell 69:597–604

Rouvier E, Luciani MF, Golstein P (1993) Fas involvement in Ca^{2+}-independent T cell-mediated cytotoxicity. J Exp Med 177:195–200

Russell JH (1983) Internal disintegration model of cytotoxic lymphocyte-induced target damage. Immunol Rev 72:97–118

Sarin A, Williams MS, Alexander-Miller MA, Berzofsky JA, Zacharchuk CM, Henkart P (1997a) Target cell lysis by CTL granule exocytosis is independent of ICE-ced-3 family proteases. Immunity 6:209–215

Sarin A, Williams MS, Alexander-Miller MA, Berzofsky JA, Zacharchuk CM, Henkart PA (1997b) CTL lysis via the two CTL pathways differs with respect to utilization of target caspases: lysis via Fas requires caspases while granule exocytosis does not. Proc 6th EMBO Worksh on Cell-mediated cytotoxicity, Kerkrade, The Netherlands, p 28 (Abstr)

Sayers TJ, Wiltrout TA, Sowder R, Munger WL, Smyth MJ, Henderson LE (1992) Purification of a factor from the granules of a rat natural killer cell line (RNK) that reduces tumor cell growth and changes tumor morphology. Molecular identity with a granule serine protease (RNKP-1). J Immunol 148:292–300

Schmid DS, Tite JP, Ruddle NH (1986) DNA fragmentation: manifestation of target cell destruction mediated by cytotoxic T-cell lines, lymphotoxin-secreting helper T-cell clones, and cell-free lymphotoxin-containing supernatant. Proc Natl Acad Sci USA 83:1881–1885

Schultze-Ostoff K, Walczak KH, Droge W, Krammer PH (1994) Cell nucleus and DNA fragmentation are not required for apoptosis. J Cell Biol 127:15–22

Seth P (1994) A simple and efficient method of protein delivery into cells using adenovirus. Biochem Biophys Res Commun 203:582–587

Shi L, Kraut RP, Aebersold R, Greenberg AH (1992a) A natural killer cell granule protein that induces DNA fragmentation and apoptosis. J Exp Med 175:553–566

Shi L, Kam CM, Powers JC, Aebersold R, Greenberg AH (1992b) Purification of three cytotoxic lymphocyte granule serine proteases that induce apoptosis through distinct substrate and target cell interactions. J Exp Med 176:1521–1529

Shi L, Mai S, Israels S, Browne K, Trapani JA, Greenberg AH (1997) Granzyme B autonomously crosses the cell membrane and perforin initiates apoptosis and granzyme B nuclear accumulation. J Exp Med 185:855–866

Shiver JW, Henkart PA (1991) A noncytotoxic mast cell tumor line exhibits potent IgE-dependent cytotoxicity after transfection with the cytolysin/perforin gene. Cell 64:1175–1181

Shiver JW, Su L, Henkart PA (1992) Cytotoxicity with target DNA breakdown by rat basophilic leukemia cells expressing both cytolysin and granzyme A. Cell 71:315–322

Simon MM, Kramer MD (1994) Granzyme A. Methods Enzymol 244:68–79

Simon MM, Hoschutzky H, Fruth U, Simon HG, Kramer MD (1986) Purification and characterization of a T cell specific serine proteinase (TSP-1) from cloned cytolytic T lymphocytes. Embo J 5:3267–3274

Simon MM, Simon HG, Fruth U, Epplen J, Muller-Hermelink HK, Kramer MD (1987) Cloned cytolytic T-effector cells and their malignant variants produce an extracellular matrix degrading trypsin-like serine proteinase. Immunology 60:219–230

Simon MM, Kramer MD, Prester M, Gay S (1991) Mouse T-cell associated serine proteinase 1 degrades collagen type IV: a structural basis for the migration of lymphocytes through vascular basement membranes. Immunology 73:117–119

Smyth MJ, Trapani JA (1995) Granzymes: exogenous proteases that induce target cell apoptosis. Immunol Today 16:202–206

Smyth MJ, O'Connor MD, Trapani JA (1996) Granzymes: a variety of serine protease specificities encoded by genetically distinct subfamilies. J Leukoc Biol 60:555–562

Sneller MC, Straus SE, Jaffe ES et al. (1992) A novel lymphoproliferative/autoimmune syndrome resembling murine lpr/gld disease. J Clin Invest 90:334–341

Song Q, Lees-Miller SP, Kumar S, Zhang N, Smith GC, Jackson SP, Alnemri ES, Litwack G, Lavin MF (1996) DNA-dependent protein kinase catalytic subunit: a target for the ICE-like protease CPP32 in apoptosis. EMBO J 15:3238–3246

Sower L, Froelich CJ, Allegretto N, Rose PM, Hanna WD, Klimpel DR (1996a) Extracellular activation of granzyme A. Monocyte activation by granzyme A versus α-thrombin. J Immunol 156:2585–2590

Sower L, Klimpel GR, Hanna W, Froelich CJ (1996b) Extracellular activities of human granzymes. I. Granzyme A induces IL6 and IL8 production in fibroblast and epithelial cell lines. Cell Immunol 171:159–163

Srinivasula SM, Fernandes-Alnemri T, Zangrilli J, Robertson N, Armstrong RC, Wang L, Trapani JA, Tomaselli KJ, Litwack G, Alnemri ES (1996) The ced-3/interleukin-1β converting enzyme-like homolog Mch6 and the lamin-cleaving enzyme Mch2α are substrates for the apoptotic mediator CPP32. J Biol Chem 271:27099–27106

Stalder T, Hahn S, Erb P (1994) Fas antigen is the major target molecule for CD4+ T cell-mediated cytotoxicity. J Immunol 152:1127–1133

Stanger BZ, Leder P, Lee TH, Kim E, Seed B (1995) RIP: a novel protein containing a death domain that interacts with Fas/APO-1 (CD95) in yeast and causes cell death. Cell 81:513–523

Strasser A, Harris AW, Huang DC, Krammer PH, Cory S (1995) Bcl-2 and Fas/APO-1 regulate distinct pathways to lymphocyte apoptosis. EMBO J 14:6136–6147

Suda T, Nagata S (1994) Purification and characterization of the Fas-ligand that induces apoptosis. J Exp Med 179:873–879

Suidan HS, Bouvier J, Schaerer E, Stone SR, Monard D, Tschopp J (1994) Granzyme A released upon stimulation of T lymphocytes activates the thrombin receptor on neuronal cells and astrocytes. Proc Natl Acad Sci USA 91:8112–8116

Suidan HS, Clemetson KJ, Brown-Luedi M, Niclou SP, Clemetson JM, Tschopp J, Monard D (1996) The serine protease granzyme A does not induce platelet aggregation but inhibits responses triggered by thrombin. Biochem J 315:939–945

Sun J, Bird CH, Sutton V, McDonald L, Coughlin PB, De Jong TA, Trapani JA, Bird PI (1996) A cytosolic granzyme B inhibitor related to the viral apoptotic regulator cytokine response modifier A is present in cytotoxic lymphocytes. J Biol Chem 271:27802–27809

Sutton VR, Vaux DL, Trapani JA (1997) Bcl-2 prevents apoptosis induced by perforin and granzyme B but not that mediated by whole cytotoxic lymphocytes. J Immunol 158:5783–5790

Talento A, Nguyen M, Law S, Wu JK, Poe M, Blake JT, Patel M, Wu TJ, Manyak CL, Silberklang M (1992) Transfection of mouse cytotoxic T lymphocyte with an antisense granzyme A vector reduces lytic activity. J Immunol 149:4009–4015

Tewari M, Dixit VM (1995) Fas- and tumor necrosis factor-induced apoptosis is inhibited by the cowpox crmA gene product. J Biol Chem 270:18738–18741

Tewari M, Quan LT, O'Rourke K, Desnoyers S, Zeng Z, Beidler DR, Poirier GG, Salvesan GS, Dixit VM (1995) Yama/CPP32b, a mammalian homologue of CED-3, is a crmA inhibitable protease that cleaves the death substrate poly(ADP-ribose) polymerase. Cell 81:801–809

Thornberry NA, Bull HG, Calaycay JR, Chapman KT, Howard AD, Kostura MJ, Miller DK, Molineaux SM, Weidner JR, Aunins J et al. (1992) A novel heterodimeric cysteine protease is required for interleukin-1β processing in monocytes. Nature 356:768–774

Trapani JA, Browne KA, Smyth MJ, Jans DA (1996) Localization of granzyme B in the nucleus: a putative role in the mechanism of lymphocyte-mediated apoptosis. J Biol Chem 271:4127–4133

Trapani JA, Jans P, Froelich CJ, Smyth MJ, Sutton VR, Jans D (1998) Perforin-dependent nuclear entry of granzyme B precedes apoptosis, and is not a consequence of nuclear membrane dysfunction (in press)

Trauth BC, Klas C, Peters AM, Matzku S, Moller P, Falk W, Debatin KM, Krammer PH (1989) Monoclonal antibody-mediated tumor regression by induction of apoptosis. Science 245:301–305

Trenn G, Takayama H, Sitkovsky MV (1987) Exocytosis of cytolytic granules may not be required for target cell lysis by cytotoxic T-lymphocytes. Nature 330:72–74

Tschopp J, Schafer S, Masson D, Peitsch MC, Heusser C (1989) Phosphorylcholine acts as a calcium dependent receptor for lymphocyte perforin. Nature 337:272–274

Ucker DS, Obermiller PS, Eckhart W, Apgar JR, Berger NA, Meyers J (1992) Genome digestion is a disposable consequence of physiological cell death mediated by cytotoxic T lymphocytes. Mol Cell Biol 12:3060–3069

Varfolomeev EE, Boldin MP, Goncharov TM, Wallach D (1996) A potential mechanism of "cross-talk" between the p55 tumor necrosis factor (TNF) receptor and Fas/APO1: proteins binding to the death domains of the two receptors bind also to each other. J Exp Med 183:1272–1275

Vaux DL, Weissman IL, Kim SK (1992) Prevention of programmed cell death in Caenorhabditis elegans by human bcl-2. Science 258:1955–1957

Vaux DL, Haecker G, Strasser A (1994) An evolutionary perspective on apoptosis. Cell 76:777–779

Walsh CM, Matloubian M, Liu C-C et al. (1994) Immune function in mice lacking the perforin gene. Proc Natl Acad Sci USA 91:10854–10858

Wang Z-Q, Auer B, Sting L, Berghammer H, Haidacher D, Schweiger M, Wagner EF (1995) Mice lacking ADPRT and poly(ADP-ribosyl)ation develop normally but are susceptible to skin disease. Genes Dev 9:509–520

Wang X, Zelenski NG, Yang G, Sakai J, Brown MS, Goldstein JL (1996) Cleavage of sterol regulatory element binding proteins (SREBPs) by CPP32 during apoptosis. EMBO J 15:1012–1020

White K, Grether ME, Abrams JM, Young L, Farrell K, Steller H (1994) Genetic control of programmed cell death in drosophila. Science 264:677–683

Wilson KP, Black JF, Thomson JA, Kim EE, Griffith JP, Navia MA, Murcko MA, Chambers SP, Aldape RA, Raybuck SA, Livingston DJ (1994) Structure and mechanism of interleukin-1β converting enzyme. Nature 370:270–275

Wu D, Wallen HD, Nunez G (1997) Interaction and regulation of subcellular localization of CED-4 by CED-9. Science 275:1126–1128

Xiao C-Y, Hübner S, Elliot RM, Caon M, Jans DA (1996) A consensus PK-A site in place of the CcN motif casein kinase II site of SV40 large T-antigen confers PK-A-mediated regulation of nuclear import. J Biol Chem 271:6451–6457

Yang X, Khosravi-Far R, Chang HY, Baltimore D (1997) Daxx, a novel Fas-binding protein that activates JNK and apoptosis. Cell 89:1067–1076

Ymer S, Jans DA (1995) In vivo chromatin structure of the murine interleukin-5 gene region: a new intact cell system. Biotechniques 20:834–837

Yonehara S, Ishii A, Yonehara M (1989) A cell-killing monoclonal antibody (anti-Fas) to a cell surface antigen co-downregulated with the receptor of tumor necrosis factor. J Exp Med 169:1747–1756

Yuan J, Horwitz HR (1992) The Caenorhabditis elegans cell death gene ced-4 encodes a novel protein and is expressed during the period of extensive programmed cell death. Development 116:309–320

Yuan J, Shaham S, Ledoux S, Ellis H, Horwitz HR (1993) The C. elegans cell death gene ced-3 encodes a protein similar to mammalian interleukin-1β-converting enzyme. Cell 75:641–652

Zagury D, Bernard J, Thierness N, Feldman M, Berke G (1975) Isolation and characterization of individual functionally reactive cytotoxic T lymphocytes: conjugation killing and recycling at the single cell level. Eur J Immunol 5:818–822

Zhitovitsky B, Gahm A, Ankarcrona M, Nicotera P, Orrenius S (1995) Multiple proteases are involved in thymocyte apoptosis. Exp Cell Res 221:404–412

Granule-Mediated Cytotoxicity

Alison J. Darmon[1], Michael J. Pinkoski[2] and R. Chris Bleackley[2]

1
Introduction

Cytotoxic T lymphocytes (CTLs) represent the body's primary line of defence against virus-infected and tumourigenic cells, and may play a role in auto-immune diseases and organ transplant rejection. CTLs are able to carry out their essential functions due to their ability to specifically recognize and destroy "foreign" cells. In this case, foreign may be defined as self cells which have mutated or become infected, or, alternatively, if specificity is somehow disrupted, they may be normal self cells. Whatever the case, CTLs may use one of two mechanisms to destroy the recognized target cell – one which is granule based, and one mediated by cell surface receptors on the target cell, for which the ligand is expressed on the CTL (see Chap. by Trapani and Jans in this Volume). This chapter focuses on mechanisms of target cell destruction during granule-mediated cytotoxicity.

2
Granule-Mediated Cytotoxicity

CTL recognition of a target cell results in the generation of intracellular signals within the CTL, including the release of intracellular calcium and protein phosphorylation (Cantrell 1996), which culminates in the transcription of function-related genes. Many of the protein products of these genes are stored in the electron dense granules which subsequently appear in the CTL cyto-plasm. Abundant evidence indicates that the "lethal hit" inflicted by these cells on their targets involves the contents of these granules. The granule exocytosis model of cytotoxicity states that, following conjugate formation with a target cell bearing a foreign antigen, the lytic granule is vectorally secreted in a calcium-dependent manner into the intercellular space between target cell and

[1] Medical Research Council Laboratory for Molecular Cell Biology, University College London, Gower Street, London WC1E 6BT, UK
[2] Department of Biochemistry, Medical Sciences Building, University of Alberta, Edmonton, Alberta T6G 2H7, Canada

effector cell. The granule contents then cause target cell death by inflicting a "lethal hit" (Henkart 1985). Often the lethal hit involves the formation of membrane lesions on the target cell which are visible by electron microscopy (Dourmashkin et al. 1980; Dennert and Podack 1983). Much effort has gone into determining the nature of the lytic granule contents and the mechanism of death induction.

2.1
Nature of the Lytic Granule and Degranulation

The lytic granule is a "secretory lysosome" having characteristics of both regulated secretory granules and lysosomes (reviewed by Griffiths 1995). In most cells, lysosomal and secretory granule components are separated in the *trans*-Golgi and packaged into distinct organelles (Kelly 1985; Burgoyne and Morgan 1993). In the CTL, however, lysosomal and secretory proteins are sorted and packaged into the same organelle (Burkhardt et al. 1989, 1990; Peters et al. 1991). This is perfectly demonstrated by a summary of proteins found in the lytic granule, presented in Table 1. While the granule contains function-related proteins such as the granzymes and perforin, it also contains lysosomal enzymes such as β-glucoronidase. Some proteins, such as granzymes A and B and the lysosomal proteins, are targeted to the lytic granule through the mannose-6-phosphate receptor (Griffiths and Isaaz 1993) which was originally defined as a lysosomal trafficking molecule. However, at least one other mechanism mediates this trafficking since perforin is correctly targeted in CTLs from patients with I-cell disease, in which the mannose-6-phosphate modification cannot be made (Griffiths 1995).

Like lysosomes, the lytic granule is an acidic vesicle with an internal pH of 5.5 (Henkart et al. 1987; Masson et al. 1990). This acidity is maintained by the presence of a proton pump in granule membranes, and is an essential part of the granule: inhibition of its activity with concanamycin A results in granule breakdown and loss of cytotoxicity of the CTL (Kataoka et al. 1994). The acidic pH may also play a role in protecting the CTL from lysis due to its own granule contents (see Sect. 2.2.3).

Following target cell recognition by the CTL, the granules, microtubule-organizing centres and the Golgi apparatus of the CTL all reorient towards the point of contact with the target cell (Geiger et al. 1982; Kupfer and Dennert 1984; Kupfer et al. 1985; Yanelli et al. 1986). This reorientation helps to ensure that the exocytosed granule contents are directed toward the target cell, and that bystander killing (that is, killing of cells which have not been specifically recognized by the CTL) is minimized. The granules move along the CTL's microtubules in a kinesin-dependent manner (Burkhardt et al. 1993). Kinesin activity in turn can be regulated by phosphorylation of kinesin-associated proteins (McIlvain et al. 1994); thus the phosphorylation events initiated by T cell receptor activation are directly linked to the exocytosis of granule con-

Table 1. Summary of granule contents

Protein	Function	Reference
Perforin (cytolysin)	Pore-former	Podack et al. (1985)
		Young et al. (1986a)
		Groscurth et al. (1987)
Granzymes	Proteolysis	See Table 2
Dipeptidyl peptidase I (cathepsin C)	Granzyme activation	McGuire et al. (1993)
		Smyth et al. (1995)
Chondroitin sulphate	Complexes with perforin and the granzymes	Tschopp and Masson (1987)
		Masson et al. (1990)
		Peters et al. (1991)
Calreticulin	Binds perforin Ca^{2+} binding	Burns et al. (1992)
		Dupuis et al. (1993)
TIA-1	RNA binding	Anderson et al. (1990)
		Tian et al. (1991)
Leukalexin	TNF-like cytokine	Liu et al. (1987)
Leukophysin	Granule mobility	Abdelhaleem et al. (1991)
		Abdelhaleem et al. (1996)
Mannose-6-phosphate receptor	Protein targeting	Burkhardt et al. (1990)
		Peters et al. (1991)
		Griffiths and Isaaz (1993)
H^+-ATPase	Acidification	Kataoka et al. (1994)
Cathepsin D	Lysosomal enzyme	Tschopp and Nabholz (1990)
Arylsulphatase	Lysosomal enzyme	Tschopp and Nabholz (1990)
β-Glucoronidase	Lysosomal enzyme	Tschopp and Nabholz (1990)
β-Hexosamidase	Lysosomal enzyme	Tschopp and Nabholz (1990)
Lamp-1	Lysosomal protein	Peters et al. (1991)
Lamp-2	Lysosomal protein	Peters et al. (1991)
CD63	Lysosomal protein	Peters et al. (1991)

TNF, tumour necrosis factor.

tents, consistent with a requirement for phosphorylation for degranulation (Anel et al. 1994).

The cytoskeletal rearrangements involved in granule exocytosis may be mediated by a family of small GTPases. Two GTPases, Rac and Rho, have been shown to regulate mast cell secretion (Price et al. 1994) and have been identified in CTLs (Lang et al. 1992), along with the haematopoietic cell-specific GTPase Rac2 which is upregulated following T cell activation (Reibel et al. 1991). Rac and Rho mediate the cytoskeletal rearrangements which precede degranulation (Norman et al. 1994), suggesting that CTL degranulation may be regulated by the same GTPases involved in mast cell secretion.

2.2
Granule Proteins

A summary of known granule proteins is shown in Table 1. The roles that many of these proteins play within the granule and/or during CTL-mediated cytotoxicity are unknown. Some of these proteins are lysosomal proteins and likely play no role in CTL-mediated cytotoxicity; however, others clearly have functional roles in target cell killing.

2.2.1
Perforin

Perforin (cytolysin, pore-forming protein, C9-related protein) is the granule protein responsible for the calcium-dependent lytic activity of the CTL. Perforin was originally isolated from the lytic granules of natural killer (NK) cells and CTLs, and was subsequently shown to be capable of inducing target cell lysis in the presence of calcium (Masson and Tschopp 1985; Podack et al. 1985; Liu et al. 1986; Zalman et al. 1987). In the lytic granules, perforin is in monomeric form and is found in association with proteoglycans and calreticulin (see Sect. 2.2.3). Granule exocytosis releases perforin into the extracellular space where it is exposed to calcium and neutral pH. This neutral pH causes perforin to be released from the proteoglycans (Persechini et al. 1989) and the perforin monomers bind the target cell, possibly by recognizing phosphorylcholine molecules on the cell surface (Tschopp et al. 1989), and insert into the target cell lipid bilayer in a calcium-dependent manner (Blumenthal et al. 1984; Young et al. 1987; Yue et al. 1987; Tschopp et al. 1989; Ishiura et al. 1990). The perforin monomers then aggregate in the target cell membrane to form pores which can be visualized by electron microscopy (EM) (Dennert and Podack 1983; Dourmashkin et al. 1980). Young has reported that while only 3–4 perforin monomers are required to form a functional channel, it takes 10–20 monomers to form an EM-visible channel. It is likely that these smaller pores which retain activity are actually more physiologically relevant than the larger ones which could perturb membrane permeability and result in cell death by osmolysis (Liu et al. 1995). In support of this notion, Rochel and Cowan have shown that polypeptides of the cytolytic N-terminal 22 residues of perforin exhibit negative cooperativity that limits the number of perforin monomers to $4+/-1$ (Rochel and Cowan 1996).

Perforin itself is able to induce lysis in a number of cell types, a finding which led to the suggestion that perforin alone accounts for CTL-induced cytolysis. Indeed, the requirement for perforin during granule-mediated cytotoxicity has been demonstrated by the generation of perforin "knockout" mice (Kagi et al. 1994a,b; Kojima et al. 1994; Lowin et al. 1994). However, perforin cannot induce target cell DNA fragmentation (Duke et al. 1989), an event which precedes membrane damage during CTL attack (Duke et al. 1983; Cohen et al. 1985), suggesting that perforin alone cannot mediate all of the events involved in CTL-mediated cytotoxicity and that other proteins, probably contained in the lytic granules, are involved (Munger et al. 1988). It is now believed that the prime role of perforin is not in cytolysis but rather to allow other cytotoxic mediators to enter the target cell and deliver the lethal hit (Liu et al. 1995) although recent evidence suggests a more active role for perforin.

2.2.2
The Granzymes

The granzymes are a family of CTL-specific serine proteases (reviewed by Smyth et al. 1996) which colocalize with perforin to the cytolytic granules (Redmond et al. 1987; Ojcius et al. 1991). Table 2 shows a summary of the known murine granzymes and their properties. These proteases are synthesized as inactive precursors with an activation dipeptide at the amino terminus

Table 2. Summary of granzymes[a]

Protease	Synonym	Specificity (residue at P_1)	References
Granzyme A	HF	Arg/Lys	Pasternack and Eisen (1985) Gershenfeld and Weissman (1986) Masson et al. (1986a,b)
	SE1		Young et al. (1986b)
	CTLA-3		Brunet et al. (1986)
	TSP-1		Simon et al. (1986)
Granzyme B	CCP1	Asp/Glu	Lobe et al. (1986a,b, 1988) Masson and Tschopp (1987)
	SE2		Young et al. (1986b)
	CTLA-1		Brunet et al. (1986)
Granzyme C	CCP2	Asn/Ser	Lobe et al. (1986a,b, 1988) Jenne et al. (1988b)
Granzyme D	CCP5	Phe/Leu	Bleackley et al. (1988a,b) Jenne et al. (1988a) Prendergast et al. (1991)
Granzyme E	CCP3	Phe/Leu	Bleackley et al. (1988a,b) Jenne et al. (1988a) Prendergast et al. (1991)
	MCSP-2		Kwon et al. (1988)
Granzyme F	CCP4	Phe/Leu	Bleackley et al. (1988a,b) Jenne et al. (1988a) Prendergast et al. (1991)
	MCSP-3		Kwon et al. (1988)
Granzyme G	CCP6	Phe/Leu	Masson and Tschopp (1987) Jenne et al. (1989)

[a] Only mouse granzymes are shown here. In addition, fragmentin-1 and fragmentin-2 have been isolated from rat natural killer cells and have homology to granzyme A and granzyme B respectively (Shi et al. 1992a,b). Human homologues of granzymes A and B have been identified (Schmid and Weissman 1987; Caputo et al. 1988, 1990; Gershenfeld et al. 1988; Trapani et al. 1988) as well as granzyme H (HuCCPX) (Meier et al. 1990; Haddad et al. 1991) and granzyme 3 (Hameed et al. 1988).

and require removal of this dipeptide for enzyme activation (Caputo et al. 1993), a process believed to be involved in protecting the CTL from its own lytic proteins (see Sect. 2.2.3).

Although evidence is only just beginning to accumulate regarding the biological role of the granzymes in CTL-mediated killing it has been known for some time that protease inhibitors can protect cells from cell-mediated lysis (Brogan and Targan 1986; Hudig et al. 1991; Kaiser and Hoskin 1992; Helgason et al. 1995). However, it was not known whether the granzymes or intracellular protease(s) were a target for these inhibitors. Loading of cells with chymotrypsin, proteinase K or trypsin has been found to cause cell lysis accompanied in most cases by DNA fragmentation and nuclear damage (Williams and Henkart 1994) suggesting that proteases are involved in the induction of target cell death and that the granzymes likely exert their effect inside the cell. Multiple mechanisms have been proposed by which the granzymes may induce target cell death, including the activation of endogenous endonucleases (Smyth et al. 1994).

The first two granzymes to be identified were granzyme A and granzyme B. These proteases have received the most attention thus far, probably because human homologues have been isolated. Additionally, in murine CTL stimulated ex-vivo only granzymes A and B, and maybe minor amounts of C, are expressed (Garcia-Sanz et al. 1990; Ebnet et al. 1991) suggesting that these granzymes are key mediators of CTL-mediated cytotoxicity. Quantitative polymerase chain reaction (PCR) of granzyme transcript levels in CTLs activated by mitogen, allogeneic cells or anti-CD3 revealed that only granzyme B transcripts correlated with cytotoxicity for all modes of stimulation (Prendergast et al. 1992) suggesting that granzyme B may be a direct effector in the lytic process. Unfortunately, granzyme A levels were not examined in this study and it is not known whether granzyme A expression also correlates with cytotoxicity in response to these different activators.

While granzyme A has substrate specificity resembling trypsin (cleavage after Arg or Lys), granzyme B has a substrate specificity which is unique among eukaryotic serine proteases. Molecular modelling of the murine homologue revealed that the side chain of murine granzyme B residue Arg^{208} partially fills the specificity pocket of the protease, predicting a requirement for acidic residues (Asp or Glu) at the P_1 site (Murphy et al. 1988). Subsite mapping and inhibitor studies confirmed that granzyme B cleaves following aspartic acid residues (Odake et al. 1991; Poe et al. 1991). Finally, replacement of Arg^{208} with a glycine residue converted the substrate specificity of granzyme B from Asp to cleavage following hydrophobic residues, demonstrating that Arg^{208} is indeed responsible for determining substrate specificity (Caputo et al. 1994).

Interestingly, granzyme A exists as a homodimer, whereas granzyme B is monomeric. When combined with their differing substrate specificities, it seems likely that granzyme A and granzyme B have separate cellular substrates, a fact supported by the findings of Irmler et al. (1995) who found that granzyme A cleaves and activates pro-interleukin-1β (pro-IL-1β) while granzyme B does not. Granzyme A has also been shown to cleave various intracellular and extracellular proteins, including nucleolin (Pasternack et al. 1991), collagen type IV (Simon et al. 1991), and the thrombin receptor (Suidan et al. 1994) although these studies were performed using in vitro systems.

Both granzymes are thought to play a role in the induction of target cell DNA fragmentation. Two groups (Hayes et al. 1989; Shi et al. 1992a) have shown that granzyme A can induce target cell DNA fragmentation in a perforin-dependent manner. Similar studies using granzyme B purified from rat NK cells (called fragmentin-2 but having homology to granzyme B) showed a similar result with this protease (Shi et al. 1992a,b). Other groups have confirmed these results using a complementary technique. The non-cytotoxic rat mast cell tumour line RBL, which can be triggered to degranulate by the immunoglobulin E (IgE)-specific Fc receptor, was transfected with perforin or the granzymes either alone or in combination and then transfectants were tested for their ability to lyse IgE-coated target cells. Cells expressing granzyme A alone were found to be non-cytolytic while those transfected with perforin alone were cytolytic but failed to induce target cell DNA fragmentation. However, the combination of granzyme A and perforin could induce both target cell lysis and DNA fragmentation (Shiver and Henkart 1991; Shiver et al. 1992). Studies using granzyme B yielded similar results. Again, coexpression of granzyme B with perforin could enhance cytolytic and nucleolytic activity of these cells compared to cells expressing perforin alone (Nakajima et al. 1995). Interestingly, perforin/granzyme A- or perforin/granzyme B-expressing RBL were not as effective against tumour cell targets as RBL expressing all three cytolytic proteins, suggesting that there is a synergism between the two granzymes (Nakajima et al. 1995).

Besides DNA fragmentation, granzyme A may also play a role in other aspects of target cell killing. Talento et al. (1992) found that expression of an antisense granzyme A construct in a cloned CTL line interfered not only with target cell DNA fragmentation but also with release of ^{51}Cr-labelled proteins (as a measure of cytolytic activity), suggesting granzyme A plays either a direct or indirect role in the induction of target cell membrane damage. These results were confirmed by Nakajima and Henkart (1994) who showed that pre-loading of target cells with aprotinin (a granzyme A inhibitor) suppressed both cytolysis and DNA fragmentation in target cells exposed to RBL expressing both granzyme A and perforin. In contrast to this, Ebnet et al. (1995) have shown that both the CTL and NK cells derived from granzyme A-deficient mice are indistinguishable from wild-type cells in causing target cell membrane damage, death and DNA fragmentation, seeming to suggest that granzyme A is not essential for cell-mediated cytotoxicity. Therefore the role of granzyme A during CTL-mediated killing is still unresolved.

Similar studies have been more successful at determining the role of granzyme B. Besides the above-mentioned work, other experiments have accumulated data to suggest that granzyme B is involved in the induction of target cell DNA fragmentation, and may play a secondary role in regulating membrane damage. Bochan et al. (1995) have demonstrated that stable transfection of an NK cell which contains no granzyme A (Su et al. 1994) with an antisense granzyme B construct inhibits the lytic ability of these cells (measured as ^{51}Cr release) by >95%. However, their results may be a consequence of looking at NK cell-mediated killing. Using CTL, NK and lymphokine-activated killers (LAKs) isolated from mice homologous for a null mutation in the granzyme B gene, other workers have shown that granzyme B plays a critical and non-redundant role in the rapid induction of target cell DNA fragmentation and cell death (Heusel et al. 1994; Shresta et al. 1995) and in NK cells (but not in CTL or LAK cells). The reduced DNA fragmentation in target cells treated with these effectors is due to reduced kinetics since longer incubation times resulted in target cell DNA fragmentation. Therefore, other granzymes, possibly granzyme A, may be able to induce DNA fragmentation but granzyme B is involved in its rapid induction. Consistent with a role for granzyme B in inducing DNA damage, two groups (Pinkoski et al. 1996; Trapani et al. 1996) have shown that granzyme B localizes to the nucleus of a target cell.

Although the evidence is confusing and contradictory at times, there definitely seems to be a role for both granzymes A and B in the induction of target cell death (see also Chap. 5). The requirement for multiple proteases during CTL-mediated killing, each with different substrate specificities, is not surprising in light of the fact that CTLs are primarily involved in the removal of viral-infected and tumourigenic cells – both of these mutant cell types being potentially lethal. Therefore, the CTL has developed multiple mechanisms to ensure the death of these cells.

Besides their roles in the elimination of tumourigenic and virus-infected cells, granzymes may also mediate immunomodulatory functions by interacting with cell surface receptors. Sower et al. (1996) have shown that catalytically active, but not inactive, granzyme A can stimulate IL-6, IL-8 and tumour necrosis factor α (TNFα) production by human peripheral blood monocytes and purified monocytes. Interestingly, although these cells possess a cell surface receptor for the serine protease thrombin, and granzyme A has previously been shown to be capable of cleaving and activating the thrombin receptor (Suidan et al. 1994), this activity of granzyme A against monocytes was found to be separate from that mediated by thrombin. This implies the existence of a separate cell surface receptor for granzyme A and further suggests that once activated, CTLs may modulate the immune response by constitutive secretion of their granule contents (as demonstrated by Isaaz et al. 1995; Sower et al. 1996).

2.2.3
Other Granule Proteins

Besides perforin and the granzymes, additional granule proteins have been identified, although in many cases the roles of these proteins are unknown.

Dipeptidyl peptidase I (DPPI, cathepsin C), a cysteine protease with specificity for cleaving dipeptides from the amino terminus of proteins, has been found in cytolytic granules. DPPI has been shown to activate both granzyme A (Kummer et al. 1996) and granzyme B (McGuire et al. 1993; Smyth et al. 1995) from their zymogen forms, suggesting a role for DPPI in granzyme activation once the zymogens have reached the granules, thereby ensuring protection of the CTL from its own lytic proteins.

A second mechanism for protecting CTLs is due to the presence of proteoglycans in the lytic granule (Schmidt et al. 1985; Stevens et al. 1987). The granule, as a secretory vesicle, has an acidic pH of 5.5 (Henkart et al. 1987; Masson et al. 1990). At this pH both perforin (Tschopp and Masson 1987) and the granzymes (Peters et al. 1991) are bound to chondroitin sulphate and are maintained in an inactive state ensuring CTL protection from the action of these proteins. Following granule exocytosis, these complexes are exposed to neutral pH which releases the chondroitin sulphate from the lytic proteins, allowing them to act on the target cell.

It has long been known that perforin polymerizes in the presence of calcium and it seems that the lytic granules may possess a mechanism, besides association with proteoglycans, which prevents perforin polymerization within the cell. Lytic granules have been reported to contain the calcium-binding protein calreticulin (Dupuis et al. 1993), a protein whose expression is induced following T cell activation (Burns et al. 1992). When first isolated, this protein was believed to be localized only to the endoplasmic reticulum (ER) since it possesses the carboxy terminal ER retention signal KDEL (Michalak et al. 1992). Although it is not known how calreticulin escapes the ER, one possibility is that calreticulin "escorts" perforin out of the ER and to the granules (Dupuis et al. 1993; Burns et al. 1994). While bound to perforin, the KDEL retention signal of calreticulin may be masked, allowing it to exit the ER. It has been shown that although granzymes A and B are targeted to the granules by the mannose-6-phosphate receptor (Griffiths and Isaaz 1993) another mechanism must account for perforin targeting (Griffiths 1995), a mechanism which may involve calreticulin. The perforin–calreticulin complex would then be packaged into granules and would then be disrupted upon exocytosis and exposure to high calcium levels. Calreticulin may also serve to sequester calcium away from perforin in the granule and maintain it in a monomeric state. Alternatively, if calreticulin enters the target cell it could contribute to the calcium flux seen in target cells following CTL attack (Allbritton et al. 1988).

The use of monoclonal antibodies directed against lytic granule contents allowed Anderson et al. (1990) to identify TIA-1, a 15-kDa RNA-binding protein whose expression is restricted to NK cells and CTLs, and is upregulated following CTL activation (Anderson et al. 1990). Interestingly, TIA-1 has been found to induce DNA fragmentation in digitonin-permeabilized cells (Tian et al. 1991), suggesting that this protein may play a role in inducing target cell DNA fragmentation during CTL attack.

Leukalexin is a TNF-like molecule with an undetermined role in CTL-mediated cytotoxicity (Liu et al. 1987). One possibility is that it binds an as yet unidentified target cell surface receptor and may be able to transduce a death signal in much the same way as TNF and the ligand of Fas following granule exocytosis.

2.3
CTL Protection from Lysis

One of the questions arising in the study of CTL-mediated cytotoxicity is how the T cell is able to induce target cell lysis but remain intact itself. Studies have shown that CTLs and NK cells are much more resistant to lytic granule contents than other cells (Golstein 1974; Kranz and Eisen 1987; Nagler-Anderson et al. 1988) and that this resistance correlates with cytotoxicity (Liu et al. 1989). Certain mechanisms are in place to protect the CTL from its granule contents, such as the fact that granzymes are not activated from the zymogen form until they reach the granules, and the binding of proteoglycans to perforin and the granzymes to maintain them in an inactive state. However, it has been proposed that the CTL possesses certain membrane features which protect it from its lytic proteins once they are exocytosed (Blakely et al. 1987; Martin et al. 1988; Zalman et al. 1988; Jiang et al. 1989; Muller and Tschopp 1994). Another possibility is that degranulation results in the exposure of a protective layer of proteoglycan on the CTL surface, which was on the inside surface of the lytic granule and is exposed by exocytosis, and protects the CTL from its lytic proteins. Perforin is also inhibited by lipoproteins found in the serum, a mechanism believed to limit bystander lysis once perforin has been released from CTLs, and which may protect the CTL from its lytic proteins (Tschopp et al. 1986). Finally, CTLs express a serine protease inhibitor in their cytoplasm which binds to granzyme B. This is believed to inhibit any granzyme B which may escape the granule, or manage to re-enter the CTL after degranulation, thereby protecting the CTL from its own lytic proteins (Sun et al. 1996).

3
Induction of Cell Death

3.1
Entry of Granzyme into the Target Cell

The macromolecular pore model in which polyperforin facilitates entry of cytolytic effector molecules has come under closer scrutiny recently. The pore size of perforin channels in physiological conditions has long been suggested to be large enough for the passage of small molecules and there is evidence of ion channel activity attributed to perforin. It is not currently known how many perforin subunits constitute a biological active complex, but the self-limiting polymer size of the membrane spanning polypeptide suggest that the maximum number of perforin subunits would be limited to four or five (Rochel and Cowan 1996). If these data are a reflection of the physiological channels then it is unlikely that polyperforin channels could form pores large enough for the passage of the proteins in the cytolytic granule.

There is no doubt that, in the presence of high concentrations of perforin, cells are lysed rapidly. However, with targets that are under attack by CTL it takes approximately 4 h before membrane disruption is revealed through the leakage of chromium labelled intracellular proteins. We know that the apoptotic signal has been delivered because DNA fragmentation is detected much earlier. If perforin channels allow passage of granzymes into the cell, then why don't macromolecules come out? Finally, to our knowledge, perforin has not been successfully used to introduce any macromolecule into a cell and all attempts to replace perforin, in in vitro killing assays, with a variety of channel formers have failed.

An interesting variation on the mechanism of perforin-mediated entry of granule proteases is the suggestion that perforin may stimulate endocytic uptake of granzymes (see Liu et al. 1995). Perforin may recapitulate the complement system which causes membrane damage that results in the stimulation of repair processes that involve enhanced endocytosis (Morgan et al. 1987). Inhibition of killing by the microtubule inhibitor cytochalasin B suggests that endocytosis by target cells may be involved in some aspect of uptake of granule proteins (Shi et al. 1992b).

One of the key effector molecules that is believed to pass through the perforin channel is the serine protease granzyme B. However, Froelich et al. (1996) demonstrated that target cells can be treated with soluble granzyme B and after washing will die only when perforin is subsequently added to the media. Radioiodinated granzyme B was used to demonstrate that binding of granzyme B to the target cell occurred in a saturable manner, suggesting a receptor-ligand binding arrangement. The possible existence of a granzyme receptor raises serious questions regarding the mechanism of granzyme entry through the perforin pore.

Support of the notion that perforin acts in some capacity other than as a macromolecular pore has come from the observation that granzyme B can enter the target independently of perforin (Pinkoski et al. 1998; Shi et al. 1997). Although granzyme B enters the target, apoptosis is not initiated until perforin is added to the granzyme-containing target, thus suggesting a role for perforin downstream of granzyme entry. In contrast, direct injection of granzyme B into the cytoplasm does induce apoptosis. Thus there is a fundamental difference between granzyme B internalized from the medium and "free" cytoplasmic enzyme, and perforin is needed to convert the former to the latter. Recent immunocytochemical data suggests that the internalized granzyme is sequestered within intracellular vesicles, and thus would not have access to its substrates (Pinkoski et al. 1998).

The apoptosis inducing of activity of perforin can be replaced by a replication deficient strain of type 2 adenovirus (Ad2) (Froelich et al. 1996). It appears that Ad2 acts to release granzyme B into the cytoplasm from an endocytic vesicle (Pinkoski et al. 1998). Since the cytoplasmic penetration of Ad2 occurs by disruption of endosomes by the viral hexon protein, it is possible that perforin may also act in this manner. Alternatively perforin may act in a more pleotrophic manner to bring about a change in intracellular membrane permeability either directly or through an intermediary set of signals.

Based on the currently available results we suggest a model of granule-mediated cytotoxicity that depends on both perforin and granzyme B. After degranulation, the latter is internalized by receptor-mediated endocytosis into early endosomes and may pass further into the cell through retrograde transport. Perforin then mediates the release of the granzyme from its intracellular locale by a mechanism that is still unclear. However, the end result is that granzyme B is freed into the cytoplasm where it can cleave and activate its substrates, including the caspases, to bring about initiation of the apoptotic pathway. In addition, the granzyme rapidly translocates into the nucleus where it can mediate other effects on the doomed cell (see Chap. by Trapani and Jans in this Vol.).

3.2
Caspases and Granule-Mediated Cytotoxicity

The discovery of the caspases as mediators of apoptosis also had implications in CTL-mediated cytotoxicity. Since the requirement for caspase activation (cleavage after Asp) coincides with the substrate specificity of granzyme B, it was hypothesized that granzyme B could activate a target cell's endogenous ICE (caspase-1) activity and thereby induce cell death (Vaux et al. 1994). However, early studies indicated that the caspase-1 precursor does not act as a granzyme B substrate (Darmon et al. 1994).

This did not preclude other caspase precursors from acting as substrates, however. In particular, caspase-3 (CPP32) was a promising candidate that was identified as the proteolytic activity responsible for cleaving the nuclear protein poly(ADP-ribose) polymerase (PARP) during apoptosis (Nicholson et al. 1995; Tewari et al. 1995). PARP had previously been implicated in the negative regulation of a Ca^{2+}/Mg^{2+}-dependent endonuclease thought to be involved in DNA fragmentation (Yoshihara et al. 1974; Tanaka et al. 1984). Thus, by inactivating PARP activity, caspase-3 could lead to DNA fragmentation (Nelipovich et al. 1988), and, furthermore, caspase-3 could link granzyme B to DNA fragmentation if it acted as a cellular substrate for granzyme B. Additionally, caspase-3 had been shown to be activated by caspase-1 in vitro (Tewari et al. 1995).

In earlier studies it had been shown that granzyme B could hydrolyse a synthetic caspase-1 substrate (Darmon et al. 1994), suggesting that granzyme B and caspase-1 could share substrates. This turns out to be the case since granzyme B can cleave the caspase-3 precursor both in vitro (Darmon et al. 1995; Martin et al. 1996a; Quan et al. 1996) and in vivo (Darmon et al. 1995, 1996). Furthermore, the ability to cleave and activate caspase-3 is linked to the induction of DNA fragmentation. Using CTLs from mice deficient in granzyme B it has been shown that those CTLs are unable to induce rapid DNA fragmentation and unable to activate target cell caspase-3. Also, use of an inhibitor of caspase-3-like enzymes demonstrated that DNA fragmentation is substantially reduced in the presence of this inhibitor, further confirming the link between granzyme B activation of caspase-3 and induction of DNA fragmentation. However, in both the absence of granzyme B and the presence of the caspase-3 inhibitor, low levels of fragmentation were still seen, suggesting a granzyme B/caspase-3-independent pathway to fragmentation (Darmon et al. 1996). This pathway may utilize granzyme A which, as outlined above (see Sect. 2.2.2), has previously been implicated in the induction of target cell DNA fragmentation and is known to induce fragmentation with slower kinetics than granzyme B. However, it appears that the pathway to DNA fragmentation that is normally observed in a 2–4h CTL assay is likely primarily due to granzyme B and directly involves caspase-3. The mechanism that results in other apoptotic events is presumably mediated by other substrates for granzyme B.

Subsequent studies have shown that granzyme B is able to cleave and activate all members of the caspase-3 subfamily (see Table 3). Thus it is difficult to know what the main target of granzyme B is. Additional complications arise from the fact that autocatalysis and proteolytic cascades are possible within the caspase-3-like subfamily. Although in vitro data suggest that caspase-7 is cleaved more efficiently by granzyme B than caspase-3 (Gu et al. 1996) and therefore may make a better target, other data suggest that caspases-7, -3 and -6 are activated concurrently when targets are treated with granzyme B and adenovirus (Froelich et al. 1996). The latter results suggest that granzyme B activates a caspase upstream of these three; however, we have recently obtained evidence that activation of caspase-3 is mainly, but not exclusively, direct (E. A. Atkinson et al., in press). Many of the apparently discrepant results are likely due to cell-specific differences, and it is most likely that granzyme B can activate at multiple caspases in the pathway. Thus, although catalytic cascades are possible, it is likely that granzyme B can bypass the cascade by directly activating the key proteases that hydrolyse the "apoptotic substrates" responsible for the various morphologic and biochemical events observed. This is obviously advantageous to the immune system when destroying tumourigenic or virus-infected cells in which individual caspase genes may be mutated or their activity suppressed. A multispecific activator, such as

Table 3. CPP32-like proteases cleaved by granzyme B

Protease	Cleavage site	References
Caspase-3/CPP32/Yama/apopain	IETD-S	Darmon et al. (1995)
		Martin et al. (1996a)
		Quan et al. (1996)
Caspase-6/Mch2	TEVD-A	Orth et al. (1996)
		Srinivasula et al. (1996)
Caspase-7/CMH1/ICE-LAP3	IQAD-S	Fernandes-Alnemri et al. (1996)
		Gu et al. (1996)
Caspase-8/MACH/FLICE/Mch5	VETD-S	Fernandes-Alnemri et al. (1996)
		Muzio et al. (1996)
Caspase-9/ICE-LAP6/Mch6	PEPD-A	Duan et al. (1996b)
		Srinivasula et al. (1996)
Caspase-10/Mch4	IEAD-A	Fernandes-Alnemri et al. (1996)

granzyme B, should be able to deal with many of these potentially pathogenic roadblocks.

4
Concluding Remarks

Although 10 years have passed since the first identification of lytic molecules found in CTLs, it is only recently that we have begun to understand the mechanisms used by these cells to destroy target cells. It is now clear that these killers use a target cell's endogenous death molecules to induce target cell death. This clearly involves the CTL-proteases known as granzymes and their activation of caspases in the target. The mechanism of transfer of the granzymes is now being reconsidered and appears to involve internalization through a receptor. Thus this novel model would suggest that susceptibility to killing may be a reflection of the expression of this molecule. After internalization, the promiscuity of granzyme B (i.e. the ability to activate multiple caspases), and the redundancy in granzymes, ensures that even in cells in which caspases have become mutated or suppressed, the CTL is able to destroy the target. Thanks to these findings, it should now be possible to identify novel therapeutic targets for the treatment of autoimmune diseases and organ transplant rejection.

Acknowledgments. M.J.P. was the recipient of an Alberta Heritage Foundation for Medical Research (AHFMR) Studentship. R.C.B. is an AHFMR Medical Scientist, Distinguished Scientist of the Medical Research Council (MRC) of Canada, and International Research Scholar of the Howard Hughes Medical Institute. Research in R.C.B.'s laboratory is supported by the MRC and the National Cancer Institute (NCI) of Canada.

References

Abdelhaleem MM, Hatskelzon L, Dalal BI, Gerrard JM, Greenberg AH (1991) Leukophysin: a 28 kD granule membrane protein of leukocytes. J Immunol 147:3053–3059

Abdelhaleem MM, Hameed S, Klassen D, Greenberg AH (1996) Leukophysin: an RNA helicase A-related molecule identified in cytotoxic T cell granules and vesicles. J Immunol 156:2026–2035

Allbritton NL, Verret CR, Wolley RC, Eisen HN (1988) Calcium ion concentrations and DNA fragmentation in target cell destruction by murine cloned cytotoxic T lymphocytes. J Exp Med 167:514–527

Anderson P, Nagler-Anderson C, O'Brien C, Levine H, Watkins S, Slayter HS, Blue M-L, Schlossman SF (1990) A monoclonal antibody reactive with a 15-kDa cytoplasmic granule-associated protein defines a subpopulation of CD8[+] T lymphocytes. J Immunol 144:574–582

Anel A, Richieri GV, Kleinfeld AM (1994) A tyrosine phosphorylation requirement for cytotoxic T lymphocyte degranulation. J Biol Chem 269:9506–9513

Atkinson EA, Barry M, Darmon AJ, Shostak I, Turner PC, Moyer RW, Bleackley RC (1998) Cytotoxic T lymphocyte assisted suicide: Caspase-3 activation is primarily the result of the direct action of granzyme B. (in press) J. Biol. Chem.

Blakely A, Gorman K, Ostergaard H, Svoboda K, Liu C-C, Young JD-E, Clark WR (1987) Resistance of cloned cytotoxic T lymphocytes to cell-mediated cytotoxicity. J Exp Med 166:1070–1083

Bleackley RC, Duggan B, Ehrman N, Lobe CG (1988a) Isolation of two cDNA sequences which encode cytotoxic cell proteases. FEBS Lett 234:153–159

Bleackley RC, Lobe CG, Havele C, Shaw J, Pohajdak B, Redmond M, Letellier M, Paetkau VH (1988b) A molecular-genetic analysis of cytotoxic T lymphocyte function. Ann NY Acad Sci 532:359–366

Blumenthal R, Millard PJ, Henkart MP, Reynolds CW, Henkart PA (1984) Liposomes as targets for granule cytolysin from cytotoxic large granular lymphocyte tumors. Proc Natl Acad Sci USA 81:5551–5555

Bochan MR, Goebel WS, Brahmi Z (1995) Stably transfected antisense granzyme B and perforin constructs inhibit human granule-mediated lytic ability. Cell Immunol 164:234–239

Boldin MP, Goncharov TM, Goltsev YV, Wallach D (1996) Involvement of MACH, a novel MORT1/FADD-interacting protease, in Fas/APO-1- and TNF receptor-induced cell death. Cell 85:803–815

Brogan M, Targan S (1986) Evidence for involvement of serine proteases in the late stages of the natural killer cell lytic reaction. Cell Immunol 103:426–433

Brunet JF, Dosseto M, Denizot F, Mattei M-G, Clark WR, Haqqi TM, Ferrier P, Nabholz M, Schmitt-Verhulst A-M, Luciani M-F, Golstein P (1986) The inducible cytotoxic T-lymphocyte-associated gene transcript CTLA-1 sequence and gene localization to mouse chromosome 14. Nature 322:268–271

Burgoyne RD, Morgan A (1993) Regulated exocytosis. Biochem J 293:305–316

Burkhardt JK, Hester S, Argon Y (1989) Two proteins targeted to the same lytic granule compartment undergo very different posttranslational processing. Proc Natl Acad Sci USA 86:7128–7132

Burkhardt JK, Hester S, Lapham CK, Argon Y (1990) The lytic granules of natural killer cells are dual-function organelles combining secretory and prelysosomal compartments. J Cell Biol 111:2327–2340

Burkhardt JK, McIlvain J Jr., Sheetz MP, Argon Y (1993) Lytic granules from cytotoxic T cells exhibit kinesin-dependent motility on microtubules in vitro. J Cell Sci 104:151–162

Burns K, Helgason CD, Bleackley RC, Michalak M (1992) Calreticulin in T-lymphocytes: identification of calreticulin in T-lymphocytes and demonstration that activation of T cells correlates with increased levels of calreticulin mRNA and protein. J Biol Chem 267:19039–19042

Burns K, Duggan B, Atkinson EA, Famulski KS, Nemer M, Bleackley RC, Michalak M (1994) Modulation of gene expression by calreticulin binding to the glucocorticoid receptor. Nature 367:476–480

Cantrell D (1996) T cell antigen receptor signal transduction pathways. Annu Rev Immunol 14:259–274

Caputo A, Fahey D, Lloyd C, Vozab R, McCairns E, Rowe PB (1988) Structure and differential mechanisms of regulation of expression of a serine protease gene in activated human T lymphocytes. J Biol Chem 263:6363–6369

Caputo A, Sauer DEF, Rowe PB (1990) Nucleotide sequence and genomic organization of a human T lymphocyte serine protease gene. J Immunol 145:737–744

Caputo A, Garner RS, Winkler U, Hudig D, Bleackley RC (1993) Activation of recombinant murine cytotoxic cell proteinase-1 requires deletion of an amino-terminal dipeptide. J Biol Chem 268:17672–17675

Caputo A, James MNG, Powers JC, Hudig D, Bleackley RC (1994) Conversion of the substrate specificity of mouse proteinase granzyme B. Nat Struct Biol 1:364–367

Cohen JJ, Duke RC, Chervenak R, Sellins KS, Olson LK (1985) DNA fragmentation in targets of CTL: an example of programmed cell death in the immune system. Adv Exp Med Biol 184:493–508

Darmon AJ, Ehrman N, Caputo A, Fujinaga J, Bleackley RC (1994) The cytotoxic T cell proteinase granzyme B does not activate interleukin-1β-converting enzyme. J Biol Chem 269:32043–32046

Darmon AJ, Nicholson DW, Bleackley RC (1995) Activation of the apoptotic protease CPP32 by cytotoxic T-cell-derived granzyme B. Nature 377:446–448

Darmon AJ, Ley TJ, Nicholson DW, Bleackley RC (1996) Cleavage of CPP32 by granzyme B represents a critical role for granzyme B in the induction of target cell DNA fragmentation. J Biol Chem 271:21709–21712

Dennert G, Podack ER (1983) Cytolysis by H-2-specific T killer cells: assembly of tubular complexes on target membranes. J Exp Med 157:1483–1495

Dourmashkin RR, Deteix P, Simone CB, Henkart P (1980) Electron microscopic demonstration of lesions in target cell membranes associated with antibody-dependent cellular cytotoxicity. Clin Exp Immunol 42:554–560

Duan H, Chinnaiyan AM, Hudson PL, Wing JP, He W-W, Dixit VM (1996a) ICE-LAP3, a novel mammalian homologue of the *Caenorhabditis elegans* cell death protein Ced-3 is activated during Fas- and tumor necrosis factor-induced apoptosis. J Biol Chem 271:1621–1625

Duan H, Orth K, Chinnaiyan AM, Poirier GG, Froelich CJ, He W-W, Dixit VM (1996b) ICE-LAP6, a novel member of the ICE/Ced-3 gene family, is activated by the cytotoxic T cell protease granzyme B. J Biol Chem 271:16720–16724

Duke RC, Chervenak R, Cohen JJ (1983) Endogenous endonuclease-induced DNA fragmentation: an early event in cell-mediated cytolysis. Proc Natl Acad Sci USA 80:6361–6365

Duke RC, Persechini PM, Chang S, Liu C-C, Cohen JJ, Young JD-E (1989) Purified perforin induces target cell lysis but not DNA fragmentation. J Exp Med 170:1451–1456

Dupuis M, Schaerer E, Krause KH, Tschopp J (1993) The calcium-binding protein calreticulin is a major constituent of lytic granules in cytolytic T lymphocytes. J Exp Med 177:1–7

Ebnet K, Chluba de Tapia J, Hurtenbach U, Kramern MD, Simon MM (1991) In vivo primed mouse T cells selectively express T cell-specific serine proteinase-1 and the proteinase-like molecules granzyme B and C. Int Immunol 3:9–19

Ebnet K, Hausmann M, Lehmann-Grube F, Mullbacher A, Kopf M, Lamers M, Simon MM (1995) Granzyme A-deficient mice retain potent cell-mediated cytotoxicity. EMBO J 14:4230–4239

Faucheu C, Diu A, Chan AW, Blanchet AM, Miossec C, Herve F, Collard Dutilleul V, Gu Y, Aldape RA, Lippke JA, Rocher C, Su MS-S, Livingston DJ, Hercend T, Lalanne J-L (1995) A novel human protease similar to the interleukin-1β converting enzyme induces apoptosis in transfected cells. EMBO J 14:1914–1922

Faucheu C, Blanchet AM, Collard-Dutilleul V, Lalanne J-L, Diu-Hercend A (1996) Identification of a cysteine protease closely related to interleukin-1β-converting enzyme. Eur J Biochem 236:207–213

Fernandes-Alnemri T, Litwack G, Alnemri ES (1994) CPP32, a novel human apoptotic protein with homology to *Caenorhabditis elegans* cell death protein Ced-3 and mammalian interleukin-1β-converting enzyme. J Biol Chem 269:30761–30764

Fernandes-Alnemri T, Armstrong RC, Krebs J, Srinivasula SM, Wang L, Bullrich F, Fritz LC, Trapani JA, Tomaselli KJ, Litwack G, Alnemri ES (1996) In vitro activation of CPP32 and Mch3 by Mch4, a novel human apoptotic cysteine protease containing two FADD-like domains. Proc Natl Acad Sci USA 93:7464–7469

Froelich CJ, Orth K, Turbov J, Seth P, Gottleib R, Babior B, Shah GM, Bleackley RC, Dixit VM, Hanna W (1996) New paradigm for lymphocyte granule mediated cytotoxicity: target cells bind and internalize granzyme B but an endosomolytic agent is necessary for cytosolic delivery and subsequent apoptosis. J Biol Chem 271:29073–29079

Garcia-Sanz JA, MacDonald HR, Jenne DE, Tschopp J, Nabholz M (1990) Cell specificity of granzyme gene expression. J Immunol 145:3111–3118

Geiger B, Rosen D, Berke G (1982) Spatial relationships of microtubule-organizing centers and the contact area of cytotoxic T lymphocytes. J Cell Biol 95:137–143

Gershenfeld HK, Weissman IL (1986) Cloning of a cDNA for a T cell-specific serine protease from a cytotoxic T lymphocyte. Science 232:854–858

Gershenfeld HK, Hershberger RJ, Shows TB, Weissman IL (1988) Cloning and chromosomal assignment of a human cDNA encoding a T cell- and natural killer cell-specific trypsin-like serine protease. Proc Natl Acad Sci USA 85:1184–1188

Golstein P (1974) Sensitivity of cytotoxic T cells to T cell-mediated cytotoxicity. Nature 252:81–86

Griffiths GM (1995) The cell biology of CTL killing. Curr Opin Immunol 7:343–348

Griffiths GM, Isaaz S (1993) Granzymes A and B are targeted to the lytic granules of lymphocytes by the mannose-6-phosphate receptor. J Cell Biol 120:885–896

Groscurth P, Qiao BY, Podack ER, Hengartner H (1987) Cellular localization of perforin 1 in murine cloned cytotoxic T lymphocytes. J Immunol 138:2749–2752

Gu Y, Sarnecki C, Fleming MA, Lippke JA, Bleackley RC, Su MS-S (1996) Processing and activation of CMH-1 by granzyme B. J Biol Chem 271:10816–10820

Haddad P, Jenne D, Tschopp J, Clement MV, Mathieu Mahul D, Sasportes M (1991) Structure and evolutionary origin of the human granzyme H gene. Int Immunol 3:57–66

Hameed A, Lowrey DM, Lichtenheld M, Podack ER (1988) Characterization of three serine esterases isolated from human IL-2 activated killer cells. J Immunol 141:3142–3147

Hayes MP, Berrebi GA, Henkart PA (1989) Induction of target cell DNA release by the cytotoxic T lymphocyte granule protease granzyme A. J Exp Med 170:933–946

Helgason CD, Atkinson EA, Pinkoski MJ, Bleackley RC (1995) Proteinases are involved in both DNA fragmentation and membrane damage during CTL-mediated target cell killing. Exp Cell Res 218:50–56

Henkart PA (1985) Mechanism of lymphocyte mediated cytotoxicity. Annu Rev Immunol 3:31–58

Henkart PA, Berrebi GA, Takayama H, Munger WE, Sitkovsky MV (1987) Biochemical and functional properties of serine esterases in acidic cytoplasmic granules of cytotoxic T lymphocytes. J Immunol 139:2398–2405

Heusel JW, Wesselschmidt RL, Shresta S, Russell JH, Ley TJ (1994) Cytotoxic lymphocytes require granzyme B for the rapid induction of DNA fragmentation and apoptosis in allogeneic target cells. Cell 76:977–987

Hudig D, Allison NJ, Pickett TM, Winkler U, Kam C-M, Powers JC (1991) The function of lymphocyte proteases: inhibition and restoration of granule-mediated lysis with isocoumarin serine protease inhibitors. J Immunol 147:1360–1368

Irmler M, Hertig S, MacDonald HR, Sadoul R, Becherer JD, Proudfoot A, Solari R, Tschopp J (1995) Granzyme A is an interleukin 1β-converting enzyme. J Exp Med 181:1917–1922

Isaaz S, Baetz K, Olsen K, Podack E, Griffiths GM (1995) Serial killing by cytotoxic T lymphocytes: T cell receptor triggers degranulation, re-filling of the lytic granules and secretion of lytic proteins via a non-granule pathway. Eur J Immunol 25:1071–1079

Ishiura S, Matsuda K, Koizumi H, Tsukahara T, Arahata K, Sugita H (1990) Calcium is essential for both the membrane binding and lytic activity of pore-forming protein (perforin) from cytotoxic T-lymphocytes. Mol Immunol 27:803–807

Jenne D, Rey C, Haefliger JA, Qiao BY, Groscurth P, Tschopp J (1988a) Identification and sequencing of cDNA clones encoding the granule-associated serine proteases granzymes D, E, and F of cytolytic T lymphocytes. Proc Natl Acad Sci USA 85:4814–4818

Jenne D, Rey C, Masson D, Stanley KK, Herz J, Plaetinck G, Tschopp J (1988b) cDNA cloning of granzyme C, a granule-associated serine protease of cytolytic T lymphocytes. J Immunol 140:318–323

Jenne DE, Masson D, Zimmer M, Haefliger JA, Li WH, Tschopp J (1989) Isolation and complete structure of the lymphocyte serine protease granzyme G, a novel member of the granzyme multigene family in murine cytolytic T lymphocytes: evolutionary origin of lymphocyte proteases. Biochemistry 28:7953–7961

Jiang S, Persechini PM, Perussia B, Young JD-E (1989) Resistance of cytolytic lymphocytes to perforin-mediated killing: murine cytotoxic T lymphocytes and human natural killer cells do not contain functional soluble homologous restriction factor or other specific soluble protective factors. J Immunol 143:1453–1460

Kagi D, Ledermann B, Burki K, Seiler P, Odermatt B, Olsen KJ, Podack ER, Zinkernagel RM, Hengartner H (1994a) Cytotoxicity mediated by T cells and natural killer cells is greatly impaired in perforin-deficient mice. Nature 369:31–37

Kagi D, Vignaux F, Ledermann B, Burki K, Depraetere V, Nagata S, Hengartner H, Golstein P (1994b) Fas and perforin pathways as major mechanisms of T cell-mediated cytotoxicity. Science 265:528–530

Kaiser M, Hoskin DW (1992) Expression and utilization of chymotrypsin-like but not trypsin-like serine protease enzymes by nonspecific T killer cells activated by anti-CD3 monoclonal antibody. Cell Immunol 141:84–98

Kamens J, Paskind M, Hugunin M, Talanian RV, Allen H, Banach D, Bump N, Hackett M, Johnston CG, Li P, Mankovich JA, Terranova M, Ghayur T (1995) Identification and characterization of ICH-2, a novel member of the interleukin-1β-converting enzyme family of cysteine proteases. J Biol Chem 270:15250–15256

Kataoka T, Takaku K, Magae J, Shinohara N, Takayama H, Kondo S, Nagai K (1994) Acidification is essential for maintaining the structure and function of lytic granules of CTL: effect of concanamycin A, an inhibitor of vacuolar type H⁺-ATPase, on CTL-mediated cytotoxicity. J Immunol 153:3938–3947

Kelly RB (1985) Pathways of protein secretion in eukaryotes. Science 230:25–32

Kojima H, Shinohara N, Hanaoka S, Someya-Shirota Y, Takagaki Y, Ohno H, Saito T, Katayama T, Yagita H, Okumura K, Shinkai Y, Alt FW, Matsuzawa A, Yonehara S, Takayama H (1994) Two distinct pathways of specific killing revealed by perforin mutant cytotoxic T lymphocytes. Immunity 1:357–364

Kranz DM, Eisen HN (1987) Resistance of cytotoxic T lymphocytes to lysis by a clone of cytotoxic T lymphocytes. Proc Natl Acad Sci USA 84:3375–3379

Kumar S, Kinoshita M, Noda M, Copeland NG, Jenkins NA (1994) Induction of apoptosis by the mouse Nedd2 gene, which encodes a protein similar to the product of the *Caenorhabditis elegans* cell death gene ced-3 and the mammalian IL-1β-converting enzyme. Genes Dev 8:1613–1626

Kummer JA, Kamp AM, Citarella F, Horrevoets AJG, Hack CE (1996) Expression of human recombinant granzyme A zymogen and its activation by the cysteine protease cathepsin C. J Biol Chem 271:9281–9286

Kupfer A, Dennert G (1984) Reorientation of the microtubule-organizing center and the Golgi apparatus in cloned cytotoxic lymphocytes triggered by binding to lysable target cells. J Immunol 133:2762-2766

Kupfer A, Dennert G, Singer SJ (1985) The reorientation of the Golgi apparatus and the microtubule-organizing center in the cytotoxic effector cell is a prerequisite in the lysis of bound target cells. J Mol Cell Immunol 2:37-49

Kwon BS, Kestler D, Lee E, Wakulchik M, Young JD-E (1988) Isolation and sequence analysis of serine protease cDNAs from mouse cytolytic T lymphocytes. J Exp Med 168:1839-1854

Lahti JM, Xiang J, Heath LS, Campana D, Kid VJ (1995) PITSLRE protein kinase activity is associated with apoptosis. Mol Cell Biol 15:1-11

Lang P, Guizani L, Vitte-Mony I, Stancou R, Dorseuil O, Gacon G, Bertoglio J (1992) ADP-ribosylation of the ras-related, GTP-binding protein RhoA inhibits lymphocyte-mediated cytotoxicity. J Biol Chem 267:11677-11680

Liu C-C, Perussia B, Cohn ZA, Young JD-E (1986) Identification and characterization of a pore-forming protein of human peripheral blood natural killer cells. J Exp Med 164:2061-2076

Liu C-C, Steffen M, King F, Young JD-E (1987) Identification, isolation, and characterization of a novel cytotoxin in murine cytolytic lymphocytes. Cell 51:393-403

Liu C-C, Jiang S, Persechini PM, Zychlinsky A, Kaufman Y, Young JD-E (1989) Resistance of cytolytic lymphocytes to perforin-mediated killing: induction of resistance correlates with increase in cytotoxicity. J Exp Med 169:2211-2225

Liu C-C, Walsh CM, Young JD-E (1995) Perforin: structure and function. Immunol Today 16:194-201

Lobe CG, Finlay BB, Paranchych W, Paetkau VH, Bleackley RC (1986a) Novel serine proteases encoded by two cytotoxic T lymphocyte-specific genes. Science 232:858-861

Lobe CG, Havele C, Bleackley RC (1986b) Cloning of two genes that are specifically expressed in activated cytotoxic T lymphocytes. Proc Natl Acad Sci USA 83:1448-1452

Lobe CG, Upton C, Duggan B, Ehrman N, Letellier M, Bell J, McFadden G, Bleackley RC (1988) Organization of two genes encoding cytotoxic T lymphocyte-specific serine proteases CCPI and CCPII. Biochemistry 27:6941-6946

Lowin B, Hahne M, Mattmann C, Tschopp J (1994) Cytolytic T-cell cytotoxicity is mediated through perforin and Fas lytic pathways. Nature 370:650-652

Martin DE, Zalman LS, Muller-Eberhard HJ (1988) Induction of expression of a cell-surface homologous restriction factor upon anti-CD3 stimulation of human peripheral lymphocytes. Proc Natl Acad Sci USA 85:213-217

Martin SJ, Amarante-Mendes GP, Shi L, Chuang TH, Casiano CA, O'Brien GA, Fitzgerald P, Tan EM, Bokoch GM, Greenberg AH, Green DR (1996a) The cytotoxic cell protease granzyme B initiates apoptosis in a cell-free system by proteolytic processing and activation of the ICE/CED-3 family protease, CPP32, via a novel two-step mechanism. EMBO J 15:2407-2416

Masson D, Tschopp J (1985) Isolation of a lytic, pore-forming protein (perforin) from cytolytic T-lymphocytes. J Biol Chem 260:9069-9072

Masson D, Tschopp J (1987) A family of serine esterases in lytic granules of cytolytic T lymphocytes. Cell 49:679-685

Masson D, Nabholz M, Estrade C, Tschopp J (1986a) Granules of cytolytic T-lymphocytes contain two serine esterases. EMBO J 5:1595-1600

Masson D, Zamai M, Tschopp J (1986b) Identification of granzyme A isolated from cytotoxic T-lymphocyte-granules as one of the proteases encoded by CTL-specific genes. FEBS Lett 208:84-88

Masson D, Peters PJ, Geuze HJ, Borst J, Tschopp J (1990) Interaction of chondroitin sulfate with perforin and granzymes of cytolytic T-cells is dependent on pH. Biochemistry 29:11229-11235

McGuire MJ, Lipsky PE, Thiele DL (1993) Generation of active myeloid and lymphoid granule serine proteases requires processing by the granule thiol protease dipeptidyl peptidase I. J Biol Chem 268:2458-2467

McIlvain JM, Burkhardt J, Hamm-Alvarez S, Argon Y, Sheetz M (1994) Regulation of kinesin activity by phosphorylation of kinesin-associated proteins. J Biol Chem 269:19176–19182

Meier M, Kwong PC, Fregeau CJ, Atkinson EA, Burrington M, Ehrman N, Sorenson O, Lin CC, Wilkins J, Bleackley RC (1990) Cloning of a gene that encodes a new member of the human cytotoxic cell protease family. Biochemistry 29:4042–4049

Michalak M, Milner RE, Burns K, Opas M (1992) Calreticulin. Biochem J 285:681–692

Morgan BP, Dankert JR, Esser AF (1987) Recovery of human neutrophils from complement attack: removal of the membrane attack complex by endocytosis and exocytosis. J Immunol 138:246–253

Muller C, Tschopp J (1994) Resistance of CTL to perforin-mediated lysis: evidence for a lymphocyte membrane protein interacting with perforin. J Immunol 153:2470–2478

Munday NA, Vaillancourt JP, Ali A, Casano FJ, Miller DK, Molineaux SM, Yamin TT, Yu VL, Nicholson DW (1995) Molecular cloning and pro-apoptotic activity of ICE$_{rel}$II and ICE$_{rel}$III, members of the ICE/CED-3 family of cysteine proteases. J Biol Chem 270:15870–15876

Munger WE, Berrebi GA, Henkart PA (1988) Possible involvement of CTL granule proteases in target cell DNA breakdown. Immunol Rev 103:99–109

Murphy ME, Moult J, Bleackley RC, Gershenfeld H, Weissman IL, James MNG (1988) Comparative molecular model building of two serine proteinases from cytotoxic T lymphocytes. Proteins 4:190–204

Muzio M, Chinnaiyan AM, Kischkel FC, O'Rourke K, Shevchenko A, Ni J, Scaffidi C, Bretz JD, Zhang M, Gentz R, Mann M, Krammer PH, Peter ME, Dixit VM (1996) FLICE, a novel FADD-homologous ICE/CED-3-like protease, is recruited to the CD95 (Fas/APO-1) death-inducing signaling complex. Cell 85:817–827

Nagler-Anderson C, Verret CR, Firmenich AA, Berne N, Eisen HR (1988) Resistance of primary CD8 cytotoxic T lymphocytes to lysis by cytotoxic granules from cloned T cell lines. J Immunol 141:3299–3305

Nakajima H, Henkart PA (1994) Cytotoxic lymphocyte granzymes trigger a target cell internal disintegration pathway leading to cytolysis and DNA breakdown. J Immunol 152:1057–1063

Nakajima H, Park HL, Henkart PA (1995) Synergistic roles of granzymes A and B in mediating target cell death by rat basophilic leukemia mast cell tumors also expressing cytolysin/perforin. J Exp Med 181:1037–1046

Nelipovich PA, Nikonova LV, Umansky SR (1988) Inhibition of poly(ADP-ribose) polymerase as a possible reason for activation of Ca^{2+}/Mg^{2+}-dependent endonuclease in thymocytes of irradiated rats. Int J Radiat Biol 53:749–765

Nicholson DW, Ali A, Thornberry NA, Vaillancourt JP, Ding CK, Gallant M, Gareau Y, Griffin PR, Labelle M, Lazebnik YA, Munday NA, Raju SM, Smulson ME, Yamin T-T, Yu VL, Miller DK (1995) Identification and inhibition of the ICE/CED-3 protease necessary for mammalian apoptosis. Nature 376:37–43

Norman JC, Price LS, Ridley AJ, Hall A, Koffer A (1994) Actin filament organization in activated mast cells is regulated by heterotrimeric and small GTP-binding proteins. J Cell Biol 126:1005–1015

Odake S, Kam CM, Narasimhan L, Poe M, Blake JT, Krahenbuhl O, Tschopp J, Powers JC (1991) Human and murine cytotoxic T lymphocyte serine proteases: subsite mapping with peptide thioester substrates and inhibition of enzyme activity and cytolysis by isocoumarins. Biochemistry 30:2217–2227

Ojcius DM, Zheng LM, Sphicas EC, Zychlinsky A, Young JD-E (1991) Subcellular localization of perforin and serine esterase in lymphokine-activated killer cells and cytotoxic T cells by immunogold labeling. J Immunol 146:4427–4432

Orth K, Chinnaiyan AM, Garg M, Froelich CJ, Dixit VM (1996) The CED-3/ICE-like protease Mch2 is activated during apoptosis and cleaves the death substrate lamin A. J Biol Chem 271:16443–16446

Pasternack MS, Eisen HN (1985) A novel serine esterase expressed by cytotoxic T lymphocytes. Nature 314:743–745

Pasternack MS, Bleier KJ, McInerney TN (1991) Granzyme A binding to target cell proteins: granzyme A binds to and cleaves nucleolin in vitro. J Biol Chem 266:14703–14708

Persechini PM, Liu C-C, Jiang S, Young JD-E (1989) The lymphocyte pore-forming protein perforin is associated with granules by a pH-dependent mechanism. Immunol Lett 22:23–27

Peters PJ, Borst J, Oorschot V, Fukuda M, Krahenbuhl O, Tschopp J, Slot JW, Geuze HJ (1991) Cytotoxic T lymphocyte granules are secretory lysosomes, containing both perforin and granzymes. J Exp Med 173:1099–1109

Pinkoski MJ, Winkler U, Hudig D, Bleackley RC (1996) Binding of granzyme B in the nucleus of target cells: recognition of an 80–kilodalton protein. J Biol Chem 271:10225–10229

Pinkoski MJ, Hobman M, Heibein JA, Tomaselli K, Li F, Seth P, Froelich CJ, Bleackley RC (1998) Entry and trafficking of granzyme B in target cells during granzyme B-perforin mediated apoptosis. Blood 92:1–12

Podack ER, Young JD-E, Cohn ZA (1985) Isolation and biochemical and functional characterization of perforin 1 from cytolytic T-cell granules. Proc Natl Acad Sci USA 82:8629–8633

Poe M, Blake JT, Boulton DA, Gammon M, Sigal NH, Wu JK, Zweerink HJ (1991) Human cytotoxic lymphocyte granzyme B: its purification from granules and the characterization of substrate and inhibitor specificity. J Biol Chem 266:98–103

Prendergast JA, Pinkoski M, Wolfenden A, Bleackley RC (1991) Structure and evolution of the cytotoxic cell proteinase genes CCP3, CCP4 and CCP5. J Mol Biol 220:867–875

Prendergast JA, Helgason CD, Bleackley RC (1992) Quantitative polymerase chain reaction analysis of cytotoxic cell proteinase gene transcripts in T cells: pattern of expression is dependent on the nature of the stimulus. J Biol Chem 267:5090–5095

Price LS, Norman JC, Ridley AJ, Koffer A (1994) The small GTPases Rac and Rho as regulators of secretion in mast cells. Curr Biol 5:68–73

Quan LT, Tewari M, O'Rourke K, Dixit V, Snipas SJ, Poirier GG, Ray C, Pickup DJ, Salvesen GS (1996) Proteolytic activation of the cell death protease Yama/CPP32 by granzyme B. Proc Natl Acad Sci USA 93:1972–1976

Redmond MJ, Letellier M, Parker JM, Lobe C, Havele C, Paetkau V, Bleackley RC (1987) A serine protease (CCP1) is sequestered in the cytoplasmic granules of cytotoxic T lymphocytes. J Immunol 139:3184–3188

Reibel L, Dorseuil O, Stancou R, Bertoglio T, Gacon G (1991) A hemopoietic specific gene encoding a small GTP binding protein is overexpressed during T cell activation. Biochem Biophys Res Commun 175:451–458

Rochel N, Cowan JA (1996) Negative cooperativity exhibited by the lytic amino-terminal domain of human perforin: implications for perforin-mediated cell lysis. Chem Biol 3:31–36

Schmid J, Weissman C (1987) Induction of mRNA for a serine protease and a B-thromboglobulin-like protein in mitogen-stimulated human leukocytes. J Immunol 139:250–256

Schmidt RE, MacDermott RP, Bartley G, Bertovich M, Amato DA, Austen KF, Schlossman SF, Stevens RL, Ritz J (1985) Specific release of proteoglycans from human natural killer cells during target lysis. Nature 318:289–291

Shi L, Kam CM, Powers JC, Aebersold R, Greenberg AH (1992a) Purification of three cytotoxic lymphocyte granule serine proteases that induce apoptosis through distinct substrate and target cell interactions. J Exp Med 176:1521–1529

Shi L, Kraut RP, Aebersold R, Greenberg AH (1992b) A natural killer cell granule protein that induces DNA fragmentation and apoptosis. J Exp Med 175:553–566

Shi L, Mai S, Israels S, Browne K, Trapani JA, Greenberg AH (1997) Granzyme B (GraB) autonomously crosses the cell membrane and perforin initiates apoptosis and GraB nuclear localization. J Exp Med 185:855–866

Shiver JW, Henkart PA (1991) A noncytotoxic mast cell tumor line exhibits potent IgE-dependent cytotoxicity after transfection with the cytolysin/perforin gene. Cell 64:1175–1181

Shiver JW, Su L, Henkart PA (1992) Cytotoxicity with target DNA breakdown by rat basophilic leukemia cells expressing both cytolysin and granzyme A. Cell 71:315–322

Shresta S, Heusel JW, MacIvor DM, Wesselschmidt RL, Russell JH, Ley TJ (1995) Granzyme B plays a critical role in cytotoxic lymphocyte-induced apoptosis. Immunol Rev 146:211–221

Simon MM, Hoschutzky H, Fruth U, Simon HG, Kemaer MD (1986) Purification and characterization of a T cell specific serine proteinase (TSP-1) from cloned cytolytic T lymphocytes. EMBO J 5:3267–3274

Simon MM, Kramer MD, Prester M, Gay S (1991) Mouse T-cell associated serine proteinase 1 degrades collagen type IV: a structural basis for the migration of lymphocytes through vascular basement membranes. Immunology 73:117–119

Smyth MJ, Browne KA, Thia KY, Apostolidis VA, Kershaw MH, Trapani JA (1994) Hypothesis: cytotoxic lymphocyte granule serine proteases activate target cell endonucleases to trigger apoptosis. Clin Exp Pharmacol Physiol 21:67–70

Smyth MJ, McGuire MJ, Thia KY (1995) Expression of recombinant human granzyme B: a processing and activation role for dipeptidyl peptidase I. J Immunol 154:6299–6305

Smyth MJ, O'Connor MD, Trapani JA (1996) Granzymes: a variety of serine-protease specificities encoded by genetically distinct subfamilies. J Leuk Biol 60:555–562

Sower LE, Froelich CJ, Allegretto N, Rose PM, Hanna WD, Klimpel GR (1996) Extracellular activities of human granzyme A: monocyte activation by granzyme A versus a-thrombin. J Immunol 156:2585–2590

Strinivasula SM, Fernandes-Alnemri T, Zangrilli J, Robertson N, Armstrong RC, Wang L, Trapani JA, Tomaselli KJ, Litwack G, Alnemri ES (1996) The Ced-3/interleukin 1β converting enzyme-like homolog Mch6 and the lamin-cleaving enzyme Mch2α are substrates for the apoptotic mediator CPP32. J Biol Chem 271:27099–27106

Stevens RL, Otsu K, Weis JH, Tantravahi RV, Austen KF, Henkart PA, Galli MC, Reynolds CW (1987) Co-sedimentation of chondroitin sulfate A glycosaminoglycans and proteoglycans with the cytolytic secretory granules of rat large granular lymphocyte (LGL) tumor cells, and identification of a mRNA in normal and transformed LGL that encodes proteoglycans. J Immunol 139:863–868

Su B, Bochan MR, Hanna WL, Froelich CJ, Brahmi Z (1994) Human granzyme B is essential for DNA fragmentation of susceptible target cells. Eur J Immunol 24:2073–2080

Suidan HS, Bouvier J, Schaerer E, Stone SR, Monard D, Tschopp J (1994) Granzyme A released upon stimulation of cytotoxic T lymphocytes activates the thrombin receptor on neuronal cells and astrocytes. Proc Natl Acad Sci USA 91:8112–8116

Sun J, Bird CH, Sutton V, McDonald L, Coughlin PB, De Jong TA, Trapani JA, Bird PI (1996) A cytosolic granzyme B inhibitor related to the viral apoptotic regulator cytokine response modifier A is present in cytotoxic lymphocytes. J Biol Chem 271:27802–27809

Talento A, Nguyen M, Law S, Wu JK, Poe M, Blake JT, Patel M, Wu TJ, Manyak CL, Silberklang M, Mark G, Springer M, Sigal NH, Weissman IL, Bleackley RC, Podack ER, Tykocinski ML, Koo GC (1992) Transfection of mouse cytotoxic T lymphocyte with an antisense granzyme A vector reduces lytic activity. J Immunol 149:4009–4015

Tanaka Y, Yoshihara K, Itaya A, Kamiya T, Koide SS (1984) Mechanism of the inhibition of Ca^{2+}, Mg^{2+}-dependent endonuclease of bull seminal plasma induced by ADP-ribosylation. J Biol Chem 259:6579–6585

Teraoka H, Yumoto Y, Watanabe F, Tsukada K, Suwa A, Enari M, Nagata S (1996) CPP32/Yama/apopain cleaves the catalytic component of DNA-dependent protein-kinase in the holoenzyme. FEBS Lett 393:1–6

Tewari M, Quan LT, O'Rourke K, Desnoyers S, Zeng Z, Beidler RD, Poirier GG, Salvesen GS, Dixit VM (1995) Yama/CPP32β, a mammalian homolog of CED-3, is a CrmA-inhibitable protease that cleaves the death substrate poly(ADP-ribose) polymerase. Cell 81:801–809

Tian Q, Streuli M, Saito H, Schlossman SF, Anderson P (1991) A polyadenylate binding protein localized to the granules of cytolytic lymphocytes induces DNA fragmentation in target cells. Cell 67:629–639

Trapani JA, Klein JL, White PC, Dupont B (1988) Molecular cloning of an inducible serine esterase gene from human cytotoxic lymphocytes. Proc Natl Acad Sci USA 85:6924–6928

Trapani JA, Browne KA, Smyth MJ, Jans DA (1996) Localization of granzyme B in the nucleus: a putative role in the mechanism of cytotoxic lymphocyte-mediated apoptosis. J Biol Chem 271:4127–4133

Tschopp J, Masson D (1987) Inhibition of the lytic activity of perforin (cytolysin) and of late complement components by proteoglycans. Mol Immunol 24:907–913

Tschopp J, Nabholz M (1990) Perforin-mediated target cell lysis by cytolytic T lymphocytes. Annu Rev Immunol 8:279–302

Tschopp J, Masson D, Schafer S (1986) Inhibition of the lytic activity of perforin by lipoproteins. J Immunol 137:1950–1953

Tschopp J, Schafer S, Masson D, Peitsch MC, Heusser C (1989) Phosphorylcholine acts as a Ca^{2+}-dependent receptor molecule for lymphocyte perforin. Nature 337:272–274

Vaux DL, Haecker G, Strasser A (1994) An evolutionary perspective on apoptosis. Cell 76:777–779

Wang L, Miura M, Bergeron L, Zhu H, Yuan J (1994) Ich-1, an Ice/ced-3-related gene, encodes both positive and negative regulators of programmed cell death. Cell 78:739–750

Wang S, Miura M, Jung Y-k, Zhu H, Gagliardini V, Shi L, Greenberg AH, Yuan J (1996a) Identification and characterization of Ich-3, a member of the interleukin-1β converting enzyme (ICE)/Ced-3 family and an upstream regulator of ICE. J Biol Chem 271:20580–20587

Williams MS, Henkart PA (1994) Apoptotic cell death induced by intracellular proteolysis. J Immunol 153:4247–4255

Yanelli JR, Sullivan JA, Mandell GL, Engelhard VH (1986) Reorientation and fusion of cytotoxic T lymphocyte granules after interaction with target cells as determined by high resolution cinemicrography. J Immunol 136:377–382

Yoshihara K, Tanigawa Y, Koide SS (1974) Inhibition of rat liver Ca^{2+}, Mg^{2+}-dependent endonuclease activity by nicotinamide adenine dinucleotide and poly (adenosine diphosphate ribose) synthetase. Biochem Biophys Res Commun 59:658–665

Young JD-E, Hengartner H, Podack ER, Cohn ZA (1986a) Purification and characterization of a cytolytic pore-forming protein from granules of cloned lymphocytes with natural killer activity. Cell 44:849–859

Young JD-E, Leong LG, Liu C-C, Damiano A, Wall DA, Cohn ZA (1986b) Isolation and characterization of a serine esterase from cytolytic T cell granules. Cell 47:183–194

Young JD-E, Clark WR, Liu C-C, Cohn ZA (1987) A calcium- and perforin-independent pathway of killing mediated by murine cytolytic lymphocytes. J Exp Med 166:1894–1899

Yue CC, Reynolds CW, Henkart PA (1987) Inhibition of cytolysin activity in large granular lymphocyte granules by lipids: evidence for a membrane insertion mechanism of lysis. Mol Immunol 24:647–653

Zalman LS, Martin DE, Jung G, Muller-Eberhard HJ (1987) The cytolytic protein of human lymphocytes related to the ninth component (C9) of human complement: isolation from anti-CD3-activated peripheral blood mononuclear cells. Proc Natl Acad Sci USA 84:2426–2429

Zalman LS, Brothers MA, Muller-Eberhard HJ (1988) Self-protection of cytotoxic lymphocytes: a soluble form of homologous restriction factor in cytoplasmic granules. Proc Natl Acad Sci USA 85:4827–4831

The Cell Cycle and Apoptosis

Hugh J. M. Brady[1] and Gabriel Gil-Gómez[2]

1
Introduction

Which decisions can be more fundamental to the cell of a living organism than to divide or to die? Hence the tremendous efforts by biologists to understand both these processes. All somatic cells proliferate using the common mechanism of mitosis following progression through the cell cycle. Apoptosis, or programmed cell death, also appears to be a ubiquitous process in multicellular organisms where it serves to eliminate unwanted or damaged cells. In this chapter, we shall address the question of whether these two most fundamental of all biological pathways are interlinked. We shall discuss the differences between cell death in cycling cells from resting or quiescent cells. We shall examine the evidence that cell cycle regulatory molecules influence apoptosis and, that similarly, some of the molecules known to regulate apoptosis can impinge on cell cycle control. Finally we shall discuss models that attempt to interconnect the two processes.

2
A Connection Between Proliferation and Apoptosis in Vivo

The link between cell death and cell cycle is suggested by the numerous studies that have found large numbers of dying cells in proliferating tissues in vivo. At any site in the body where cells are dividing there is accompanying apoptosis. This is seen in areas as diverse as the intestinal crypt cells (Merritt et al. 1995) to the spermatogonia (Heiskanen et al. 1996) and the germinal centres of lymph nodes (Liu et al. 1991). Apoptosis is most clearly seen after periods of rapid cell proliferation such as in mammary tissue following weaning (Walker et al. 1989) or in the uterine endometrium during the menstrual cycle (Otsuki et al. 1994).

[1] Division of Molecular Immunology, National Institute for Medical Research, The Ridgeway, Mill Hill, London, NW7 1AA, UK
[2] Division of Molecular Genetics, The Netherlands Cancer Institute, Plesmanlaan 121, 1066 CX Amsterdam, The Netherlands

Studies on the nervous system have also associated proliferation with apoptosis. In mice lacking the retinoblastoma protein (Rb) dividing cells are found well outside of the normal neurogenic regions in both the central and peripheral nervous systems (Lee et al. 1994). Many of these ectopically dividing cells die by apoptosis shortly after their entrance into S phase, as judged by bromodeoxyuridine (BrdU) incorporation. Another interesting study focused on detecting and quantifying dying cells during pre- and early-postnatal development of the rat cerebral cortex using in situ end-labelling of DNA fragmentation (TUNEL) and electron microscopy (Thomaidou et al. 1997). The proliferative zones that give rise to the neuronal and glial cell types of the cortex, the ventricular and, to a larger extent, the subventricular zones showed higher incidence of cell death than other regions of the developing cortex during neurogenesis. In this same study, cumulative labelling with BrdU showed that 71% of TUNEL-labelled (i.e. apoptotic) cells in the subventricular zone of the cerebral cortex of neonate rats had also taken up the marker for cells in S phase of the cell cycle, thus relating proliferation to a predisposition to cell death in the developing rat brain.

3
Mitotic Catastrophe

On account of the observed connections between dying cells and cycling cells the question arises if apoptosis is simply an aberrant form of mitosis. When normal progression through the cell cycle is disrupted then abnormal mitosis can occur and this can lead to death of the cell. This process has been termed "mitotic catastrophe" (King and Cidlowski 1995). Cells undergoing mitotic catastrophe display many of the morphological changes seen in cells during normal mitosis. Cultured mammalian cells undergoing mitotic catastrophe become non-adherent as they do during mitosis (Heald et al. 1993). In addition, cell volume is reduced and chromatin aggregation is observed. While a mitotic spindle forms in the cells the chromosomes fail to orientate properly.

As well as sharing features with normal mitosis, cells undergoing mitotic catastrophe also share features with apoptotic cells (King and Cidlowski 1995). For instance, in all three situations the cells are non-adherent, rounded and have reduced volume. They also display chromatin condensation and disassemble the nuclear lamina (Heald et al. 1993; Lazebnik et al. 1993).

Mitotic catastrophe arises when components of the cell cycle machinery are expressed inappropriately. In yeast, the Wee1 kinase can inhibit mitosis by blocking the activity of the cell cycle regulatory protein Cdc2 (Lundgren et al. 1991). Cultured mammalian cells co-transfected with Cdc2 and either Cyclin A or Cyclin B1 go into mitotic catastrophe (Heald et al. 1993). This is presumed to follow from the hyperactivation of Cdc2 at an inappropriate point during the cell cycle since the same study shows that Wee1 blocks this Cdc2-Cyclin induced mitotic catastrophe. A separate study has suggested a requirement for Cdc2 activity in cell death induced by Granzyme B, a protease component of the lytic granules released by cytotoxic T cells during destruction of their targets. Premature activation of Cdc2 was found to be required for Granzyme B induced apoptosis (Shi et al. 1994).

Thus, it could be argued that mitotic catastrophe and apoptosis are interrelated. However, mitotic catastrophe is certainly not a general mechanism of programmed cell death since many of the characteristics intrinsic to apoptosis are not present in cells undergoing mitotic catastrophe. Unlike the latter situation, apoptosing cells activate specific sets of gene products and proteases, show cytoplasmic blebbing and characteristic DNA fragmentation into nucleosomal ladders, but do not form mitotic spindles with incorrectly orientated chromosomes (King and Cidlowski 1995). The endpoint of both processes may be death of the cell, but it would appear that the journey to get there is rather different.

4
Cell Cycle Arrest and Apoptosis

Two ways to directly examine the relationship between cell cycle and apoptosis are either to observe the effect of apoptotic stimuli on cycling cells or to arrest cells in specific parts of their cycle and then observe their response to apoptotic stimuli. For example, treatment of the mature B cell lymphoma cell line, WEHI-231, with anti-IgM results in apoptosis after 24–48 h (Benhamou et al. 1990; Hasbold and Klaus 1990). These cells are phenotypically immature B cells and anti-IgM treatment mimics antigen induced surface IgM crosslinking. As a consequence of anti-IgM treatment, the cells were arrested in G1, apoptosed and did not show characteristics of S phase entry such as Rb phosphorylation (Maheswaran et al. 1991). T cell hybridomas are also sensitive to activation induced cell death following ligation of their antigen receptor (Green and Scott 1994). After stimulation with antigen, the cells rapidly accumulate in G1 prior to the onset of apoptosis (Ashwell et al. 1987). Furthermore, lymphoid cell lines treated with the glucocorticoid, dexamethasone, arrest in G1 prior to cell death (Harmon et al. 1979).

Another system in which this phenomenon has been studied is the chicken bursa of Fabricius which is the primary site of B cell development in birds. Disruption of cell-to-cell contact leads to apoptosis within an hour and the cells continue to progress into S phase where the bulk of apoptosis occurs (Neiman et al. 1994). The authors detect the presence of Cyclin A (only present during S phase) in the nuclei of cells which also manifest DNA breakage.

A further extension is to exploit the availability of a large number of agents which can arrest cells in a specific stage of the cell cycle (Halicka et al. 1997). An elegant example of this is found in the study of cycling mature T lymphocytes which undergo apoptosis following cross-linking of their antigen receptor

(Boehme and Lenardo 1993). Mature non-transformed CD4+ and CD8+ T lymphocytes were made susceptible to T cell receptor mediated apoptosis by pretreatment with IL-4 or IL-2. The degree of susceptibilty to death could be correlated with the level of cell cycling as measured by thymidine or BrdU incorporation during S phase. The cells were incubated with pharmacological agents leading to cell cycle arrest prior to T cell antigen receptor (TCR) ligation. When the cells were blocked in G1 by mimosine, deferoxamine or dibutyrl cAMP they were resistant to TCR mediated apoptosis. However, cells arrested at G1/S by aphidicolin or in S phase by excess thymidine were susceptible to TCR mediated cell death. The authors thus implicate molecules acting in late G1/early S as being involved in the induction of apoptosis.

5
Resting Cells also Undergo Apoptosis

The above discussion of the relationship between cycling cells and programmed cell death should not overwhelm the obvious point that resting and even "post-mitotic" cells also readily undergo apoptosis. The best studied example of these cells are thymocytes. In the thymus at any given point about 90% of thymocytes are in the G0/G1 phase of cell cycle (Brady et al. 1996b). These thymocytes can all undergo apoptosis in response to various stimuli such as dexamethasone (Wyllie 1980) or γ-radiation (Sellins and Cohen 1987). There is no evidence to suggest that during apoptosis the thymocytes progress any further in the cell cycle. For example, although DNA fragmentation and ultimately apoptosis in irradiated thymocytes require transcription and protein synthesis, however, DNA synthesis is not required (Sellins and Cohen 1987; Yamada and Ohyama 1988). It has also been shown that those thymocytes in other phases of the cell cycle undergo apoptosis with the same frequency as those in G0/G1 (Pellicciari et al. 1996).

Neuronal cells are also characterized as being resting or even post-mitotic in that after differentiation they do not enter the cell cycle. Both sympathetic neurons (Martin et al. 1992) and a neuronal cell line, PC12, derived from a neuronal phaeochromocytoma (Mesner et al. 1992) will undergo apoptosis following withdrawal of nerve growth factor (NGF). There is no evidence that this process is associated with DNA synthesis. However, it has been postulated that it is the result of an abortive attempt by the neurons to somehow re-enter the cell cycle (Rubin et al. 1993).

The withdrawal of serum induces apoptosis in density arrested quiescent murine 3T3 cells which are dependent on growth factors for their survival. The upregulation of certain proteins associated with proliferation and cell cycle entry such as c-Myc, c-Jun, c-Fos and Cdc2 as well as phosphorylation of Rb was detected (Pandey and Wang 1995). The induction of apoptosis in quiescent cells was also correlated with the appearance of proliferating cell nuclear antigen (PCNA), as well as some incorporation of BrdU and thymidine. However, this incorporation is very likely due to DNA repair rather than DNA synthesis. The authors suggest that quiescent cells undergoing apoptosis also manifest events typical of G1 traverse.

Thus far, we have surveyed numerous physiological and in vitro models to imply that apoptosis and progression through the cell cycle are overlapping processes. However, equally cells arrested in G0/G1 can undergo apoptosis and in fact programmed cell death can happen at any point in the cell cycle. So we have arrived at a seemingly paradoxical situation. It is instructive at this stage to outline some of the experimental data relating, firstly, proteins established as regulators of cell cycle to specific effects upon apoptosis and, secondly, regulators of apoptosis to specific effects on cell cycle.

6
Cell Cycle Related Proteins and Apoptosis

6.1
C-myc

The c-myc proto-oncogene has been implicated in the control of normal cell proliferation and the induction of neoplasia (reviewed in Desbarats et al. 1996). The c-myc gene product is a transcription factor that acts in conjunction with its partner protein Max to form a heterodimeric protein complex. C-myc expression is maintained throughout the cell cycle and is rapidly downregulated following mitogen withdrawal, leading to cell cycle arrest in G1 (Waters et al. 1991). However, constitutive deregulated c-myc expression removes cell cycle arrest in Rat-1 fibroblasts following serum deprivation and leads to greatly enhanced apoptosis (Evan et al. 1992). A similar effect was also shown in an IL-3 dependent cell line, 32D. Constitutive c-myc overexpression prevented G1 arrest following IL-3 withdrawal and greatly accelerated cell death (Askew et al. 1991). The c-Myc induced acceleration in apoptosis of fibroblasts in low serum can be inhibited by Bcl-2 (Fanidi et al. 1992) as well as the addition of platelet derived growth factor or insulin-like growth factor-1 (Harrington et al. 1994). This led the authors of the latter study to propose that apoptosis is a normal physiological aspect of c-Myc function whose execution is regulated by the availability of survival factors.

Although c-Myc is a transcription factor, very few of its target genes have been identified, though recently one attractive candidate has been. This is the cell cycle regulated phosphatase, Cdc25A, a known proto-oncogene, whose transcription is regulated by the Myc/Max heterodimer (Galaktionov et al. 1996). Furthermore, the same authors show that Cdc25A is required for Myc-induced cell death in fibroblasts in low serum conditions. Intriguingly, Cdc25A is known to be an activator of cyclin dependent kinases (Cdks) that regulate cell cycle entry; in particular, it can regulate activation of Cdk2 (Hoffmann et al. 1994).

6.2
p53

The p53 tumour suppressor gene is of singular importance in the regulation of apoptosis and its various roles and properties are fully discussed in a Chapter by Choisy-Rossi et al. in this Volume. p53 is required for G1 arrest and/or apoptosis following DNA damage (reviewed in Cox and Lane 1995). p53 activates the transcription of several genes, the best studied of which is $p21^{Cip1}$ which has a role in p53's ability to cause G1 arrest (Harper et al. 1993). However, this effect on G1 arrest is not a prerequisite for p53 mediated apoptosis. Cells from $p21^{Cip1}(-/-)$ mice exhibit a defect in p53 induced G1 arrest but show a normal response to p53 mediated apoptosis (Brugarolas et al. 1995; Deng et al. 1995). p53 dependent apoptosis does appear to occur in G1 in some cell types. Expression of p53 induced growth arrest followed by cell death in murine erythroleukaemia cells transfected with a temperature sensitive mutant p53 and this was specific to cycling cells in G1 (Ryan et al. 1993). Irradiation of factor dependent murine haematopoietic cell lines following growth factor withdrawal leads to cell death in S phase (Canman et al. 1995). In the presence of the growth factor these cells exhibit G1 arrest following irradiation. p53 induced apoptosis in mouse T lymphoma cells is preceded by G1 arrest (Wang et al. 1995).

Although p53 has a role in the G2/M checkpoint (Cross et al. 1995), p53 independent death appears to occur in G2/M. Transformed B cells containing a mutant p53 do not undergo G1 arrest but accumulate in G2/M and then die (Allday et al. 1995). Apoptosis induced by DNA damage in proliferating T cells from $p53(-/-)$ mice is accompanied by cell cycle arrest in both G1 and G2 (Strasser et al. 1994). The outcome of p53 activation, whether leading to G1 arrest or apoptosis, is not straightforward. It depends on several factors including cell type or growth conditions. Thymocytes undergo p53 dependent apoptosis upon irradiation (Clarke et al. 1993; Lowe et al. 1993). However, fibroblasts irradiated with similar doses undergo G1 arrest with $p21^{Cip1}$ accumulation (Di Leonardo et al. 1994). Similarly, the presence or absence of growth factor after irradiation of haematopoietic cell lines leads to G1 arrest or apoptosis respectively (Canman et al. 1995).

6.3
Rb

The retinoblastoma (Rb) gene product functions as a tumour suppressor gene. Although the level of Rb protein remains relatively unchanged during the cell cycle, its state of phosphorylation does vary. Rb is unphosphorylated or hypophosphorylated in G0 and early G1. It first becomes phosphorylated in the middle of G1 and further phosphorylated (hyperphosphorylated) in late G1, S and G2/M by various Cyclin-Cdk complexes (reviewed in Hunter and Pines 1994; Sherr and Roberts 1995). Rb becomes dephosphorylated when cells re-enter G0/G1 from G2/M (Ludlow et al. 1993). The best characterized target of Rb is the E2F transcription factor family. It has been shown that unphosphorylated (hypophosphorylated) Rb binds E2F-1 and inhibits its activity (reviewed in Nevins 1992). When Rb is phosphorylated it no longer binds E2F-1; free E2F-1 is involved in the transcriptional control of genes responsible for cell growth control.

When the osteosarcoma cell line SAOS-2, which contains mutant Rb, was irradiated then apoptosis occurred in a time and dose dependent manner (Haas-Kogan et al. 1995). Transfection of wild type Rb into these cells inhibited this apoptosis. Transfection of a mutant Rb that did not bind E2F did not protect the cells. This suggests that inhibition of apoptosis by Rb is through inhibition of E2F mediated gene transcription (discussed in Sect. 6.4). Further evidence relating inactivation of Rb function to apoptosis comes from studies of Rb(−/−) mice. The Rb(−/−) mice are embryonic lethal after 13–15 days of gestation. Histological analysis of the embryos revealed widespread cell death in both the peripheral and central nervous systems (Clarke et al. 1992; Jacks et al. 1992; Lee et al. 1992) due to apoptosis as detailed by TUNEL staining. When embryos deficient in both Rb and p53 were examined, apoptosis was inhibited in the ocular lens that had previously been seen to be substantially elevated in embryos only deficient in Rb (Morgenbesser et al. 1994). Consistent with these findings was the observation that treatment of mouse embryonic fibroblasts with chemotherapeutic agents led to p53 accumulation regardless of Rb status (Almasan et al. 1995). However, the induction of p53 leads not only to apoptosis in Rb(−/−) cells but also to growth arrest in both Rb(+/−) and Rb(+/+) cells. Therefore, loss of Rb produces a constitutive DNA damage signal capable of activating a p53 dependent apoptosis pathway.

6.4
E2F

As mentioned above (see Sect. 6.4), the regulation of the activity of the E2F transcription factor family is modulated by its interaction with the Rb gene product. The best characterized member of the E2F family is E2F-1 which has been hypothesized to play a role in activating the expression of genes important for the execution of S phase (reviewed in Weinberg 1995). However, it has also been shown that overexpression of E2F-1 can trigger apoptosis in fibroblasts (Shan and Lee 1994). This apoptotic effect is p53 dependent (Qin et al. 1994; Wu and Levine et al. 1994). It is also seen in both serum starved and hence quiescent or actively growing fibroblasts. Thus, the possibility is raised that E2F-1 may regulate genes which are important in programmed cell death. As a means to address this further, mice have been generated that lack E2F-1. Thymocytes from E2F-1(−/−) mice show increased viability over wild type thymocytes after in vitro culture (Field et al. 1996). The E2F-1(−/−) mice have also an increased number of both mature thymocytes and peripheral T cells in the lymph nodes. The authors suggest this may be due to a defect in the selection process whereby immature thymocytes are deleted. This was partially addressed by injecting anti-CD3 antibodies into mice and the effect on thymic apoptosis was analysed. Anti-CD3 injection to some extent may mimic the selection process whereby thymocytes that recognize self-antigen are elimi-nated. The results showed a decreased response to anti-CD3 in E2F-1(−/−) as opposed to wild type mice as assayed by TUNEL staining on apoptosis. This gives a hint that E2F-1 may exert some effect on negative selection but more rigorous experiments are required before this assertion can be made. It has also not been ruled out that the effect of E2F-1 on T cell maturation could be related to its role in cell cycle progression since it has been shown that the dividing or non-dividing status of a thymocyte can affect its selection (Huesmann et al. 1991).

6.5
C-Myb

C-Myb is a transcription factor expressed almost exclusively in haematopoietic lineages (Thompson and Ramsay 1995). C-Myb has been suggested to be a regulator of the cell cycle. In T cells, for example, the levels of c-myb mRNA are higher at the G1/S transition following mitogenic stimulation of peripheral T cells (Stern and Smith 1986). The same study shows that c-myb is expressed at high levels in immature thymocytes, then switched off as they mature, but can be activated in mature T cells during a proliferative response to antigen.

Transgenic mice were generated expressing two dominant negative Myb mutants exclusively in T cells. One mutant comprised the Myb DNA binding domain and the other (referred to as MEnT) acted as an active repressor in that it was made up of the DNA binding domain of Myb fused to the *Drosophila* engrailed repressor domain (Badiani et al. 1994). Expression of either mutant causes a block in thymocyte development and those T cells that reached matu-rity were unable to proliferate properly. Further studies using an inducible version of the MEnT construction in a murine thymoma cell line, EL4, show that inhibition of Myb activity did not affect cell cycle progression whereas it did trigger apoptosis (Taylor et al. 1996). Similarly, analysis of thymocytes from the transgenic mice showed accelerated apoptosis after in vitro culture. Furthermore, induction of MEnT expression in EL4 cells was seen to repress bcl-2 expression and the results of a nuclear run-on assay suggest that inhibition of Myb might decrease transcription of the bcl-2 gene. In agreement with this the v-myb gene product of the E26 leukaemia virus upregulates bcl-2 expression and suppresses apoptosis in myeloid cells (Frampton et al. 1996).

By contrast, other data suggest that c-Myb rather than being an inhibitor of apoptosis is in fact pro-apoptotic. C-Myb was shown to stimulate apoptosis in

32D and SAOS-2 cells when co-expressed with p53 and this is accompanied by upregulation of bax gene expression (Sala et al. 1996). Therefore, the relationship between c-Myb and apoptosis remains somewhat disputed.

6.6
Cyclins and Cyclin Dependent Kinases

The cyclin dependent kinases (Cdks) and their regulatory subunits, the cyclins, are key regulators of cell progression, catalysing the transitions both into S phase and into mitosis (reviewed in Sherr 1996). For example, in middle to late G1 the cyclin D-Cdk4 complex is activated leading to the hyperphosphorylation of Rb. This in turn leads to the release of the E2F transcription factor family, as discussed above (see Sect. 6.4), permitting transcription of cyclin E and other genes required for the initiation of S phase. Following the synthesis of Cyclin E, Cyclin E-Cdk2 is activated and, with the phosphorylation of other critical substrates, DNA replication begins (reviewed in Nigg 1995). During the S and G2 phases, cells accumulate Cyclin A and Cyclin B which in complex with Cdks trigger the entry into mitosis. Thus, the Cyclins are regulators that control Cdk activity and this control is further enhanced by the fact that most of the activating cyclins are only synthesized at specific times during the cell cycle.

Several lines of evidence exist which point to a requirement for cyclin-Cdk activity in apoptosis. As mentioned above (see Sect. 4), cell cycle arrest prior to S phase entry is often able to protect cells from apoptosis (e.g. Boehme and Lenardo 1993; Meikrantz et al. 1994). This is not merely due to a requirement for p53 induction as the protection also extends to p53 independent apoptotic stimuli such as TCR cross-linking or tumour necrosis factor α (TNFα). The fact that cyclin A synthesis only begins in S phase suggests that this may be a good candidate for involvement in cell death. Another interesting point is that c-Myc which, as described above (see Sect. 6.1), induces apoptosis under restrictive growth conditions in low serum is a transcriptional activator of the cyclin A gene (Jansen-Dürr et al. 1993).

Apoptosis can be induced in S phase arrested cells by compounds such as caffeine and staurosporine, known to elicit premature mitosis (Meikrantz et al. 1994). This led to the activation of Cyclin A dependent kinase within a few hours of the treatment prior to the onset of apoptosis. Furthermore, treatment with TNFα also produced the Cyclin A dependent kinase activation in S phase arrested HeLa cells whereas non-arrested cycling HeLa cells were insensitive. Another example of the involvement of a Cyclin-Cdk complex in cell death is found in Myc-induced apoptosis of serum starved Rat-1a fibroblasts (Hoang et al. 1994). Firstly, analysis of cyclin transcription following serum withdrawal from cells constitutively overexpressing c-myc show elevated cyclin A transcription but no change in cyclins B, C, D1 and E. Secondly, Rat-1 lines were produced expressing cyclin A under the control of the Zn inducible metallothionein promoter. Serum withdrawal plus the addition of Zn led to apoptosis. The cell lines expressed differing levels of Cyclin A whereby the higher levels of expression correlated with apoptosis in a dose dependent manner. Therefore, in these circumstances Cyclin A expression alone was sufficient to accelerate apoptosis in restrictive growth conditions. As mentioned above (see Sect. 4), studies on apoptotic lymphoblasts from the chicken bursa of Fabricius show a co-localization of Cyclin A expression and DNA fragmentation (Neiman et al. 1994).

The requirement for Cdk activity in apoptosis has also been documented from studies of cytolysis induced by Granzyme B. Premature activation of Cdc2 was found to be required for Granzyme B induced apoptosis (Shi et al. 1994) in a process termed "mitotic catastrophe", as discussed in Section 3. Subsequently, Jurkat cells separated into fractions representing different stages of the cell cycle were shown to be susceptible to Granzyme B induced apoptosis regardless of cell cycle stage (Shi et al. 1996). The same study showed that although Cdc2 activity is associated with Cyclin A and Cyclin B, that induced by granzyme B is primarily cyclin A associated. A concomitant activation of Cdk2-cyclin A was also detected.

Dominant negative mutants of Cdc2 and Cdk2, Cdk3 and Cdk5 were transiently expressed in HeLa cells prior to exposure to the apoptosis inducing agents, TNFα or staurosporine (Meikrantz and Schlegel 1996). The dominant negative mutants of Cdc2, Cdk2 and Cdk3 each suppressed apoptosis induction whereas that of Cdk5 did not. Furthermore, overexpression of wild type Cdc2, Cdk2 and Cdk3 as well as cyclin A increased the level of TNFα and staurosporine in bcl-2 transfected HeLa cells which otherwise were reasonably resistant to apoptosis by these stimuli.

While Cdk activity has been implicated in apoptosis in a number of different situations the particular Cyclin-Cdk complex involved seems to vary with the cell type, differentiation state and the particular apoptotic stimulus. For example, overexpression of Cyclin D1 is sufficient to induce apoptosis in differentiated neuronal cells (Kranenburg et al. 1996) while Cyclin B has been described as being required for activation induced cell death in T cells (Fotedar et al. 1995). In addition, overexpression of a distantly related member of the Cdc2-like gene family, PITSLREβ1, has been shown to induce apoptosis in Chinese hamster ovary cells (Lahti et al. 1995). Induction of PITSLREβ1 was also seen in response to dexamethasone and anti-Fas antibody in T cell lines.

6.7
Cyclin Dependent Kinase Inhibitors

Two families of cyclin dependent kinase inhibitors (CKIs) have been described: the p21^{Cip1} family which includes p27^{Kip1} and p57^{Kip2} and the p16^{Ink4} family which includes p15^{Ink4}, p18^{Ink4} and p19^{Ink4} (reviewed in Harper and Elledge 1996). The Ink4 family are selective for complexes of Cyclin D with Cdk4 or Cdk6 whereas the $p21^{Cip1}$ family binds and inactivates several Cdks.

As mentioned above (see Sect. 6.2), $p21^{Cip1}$ is transcriptionally regulated by p53, but the study of $p21^{Cip1}(-/-)$ mice suggests that although its absence abrogates G1 arrest in response to DNA damage it does not have an essential role in the regulation of apoptosis (Deng et al. 1995). This is also suggested from studies on the induction of $p21^{Cip1}$ by a p53 temperature sensitive mutant in K562 erythroleukaemia cells. The induction of $p21^{Cip1}$ did not promote or sensitize the cells to radiation induced apoptosis (Kobayashi et al. 1995). A role for CKIs in inhibiting apoptosis has been suggested from studies of murine myoblasts. Proliferating murine C2C12 myoblasts can undergo either terminal differentiation or apoptosis when deprived of mitogen. whereas differentiated myotubes were resistant to apoptosis (Wang and Walsh, 1996). During myogenesis the resistance to apoptosis was correlated with the induction of $p21^{Cip1}$ but not with the appearance of myogenin, a marker expressed earlier in differentiation. Deregulated expression of $p21^{Cip1}$ or $p16^{Ink4}$ blocked apoptosis during myocyte differentiation. The authors postulate that induction of CKIs may serve to protect differentiating myocytes from programmed cell death as well as playing a role in establishing the post-mitotic state. $p57^{Kip2}$ is found highly expressed in post-mitotic lens fibre cells with lower level expression in the anterior epithelial layer (Zhang et al. 1997). These authors have generated $p57^{Kip2}(-/-)$ mice. They observed inappropriate S phase entry in the lens fibre cells as well as increased apoptosis. Similarly, they saw a ten-fold increase in apoptosis in the anterior epithelial compartment. Thus, evidence has been accumulating to suggest an important regulatory role for the CKIs in apoptosis and, in particular, in quiescent or post-mitotic cells. The role of $p27^{Kip1}$ in apoptosis is discussed below.

7
Apoptosis Related Proteins and the Cell Cycle

Having discussed how the major proteins which regulate the cell cycle can impinge on apoptosis, we can now discuss the logical corollary which is how some of the proteins recognized as regulating apoptosis can influence the cell cycle. This is a somewhat smaller body of evidence concerning as it does only the Bcl-2 family members, in particular Bcl-2 and Bax.

The functions and properties of the Bcl-2 family members are discussed in detail elsewhere in this volume. Their relationship to cell cycle regulation was primarily established from studies on T cells from transgenic mice. One approach was from the point of view of mice carrying a bax transgene overexpressed exclusively in the T cell compartment (Brady et al. 1996a). By pulsing these mice with BrdU and then assessing the BrdU content of thymocytes as well as their DNA content by propidium iodide counterstaining, the number of cells in the various phases was established. About twice as many thymocytes were in the S phase of the cell cycle in bax transgenic mice as in non-transgenic littermates (Brady et al. 1996b). To determine whether this effect was cell autonomous or due to some undefined homeostatic mechanism operating in the thymus, the situation was examined where a synchronized population of primary T cells was allowed to re-enter cell cycle following an exogenous stimulus, namely IL-2 re-addition. As judged by BrdU incorporation the presence of the bax transgene accelerated the entry into S phase of cycling T cells. The same experiment on T cells from bcl-2 transgenic mice showed that Bcl-2 delayed their entry into S phase. Similar data on Bcl-2 was also obtained by others (Linette et al. 1996; Mazel et al. 1996; O'Reilly et al. 1996). Bcl-2 overexpression was also shown to reduce the number of thymocytes in S phase (O'Reilly et al. 1996), whereas bcl-2($-/-$) mice have an increased number of thymocytes over wild type (Linette et al. 1996). A further interesting point regarding the action of Bax and Bcl-2 on the cell cycle entry is that they both seem to influence the level of p27^{Kip1} which in turn regulates the activity of the Cdks. Bax overexpression accelerates p27^{Kip1} degradation whereas Bcl-2 impairs it (Brady et al. 1996b; Linette et al. 1996). This effect is not seen for another CKI examined, p21^{Cip1}. Whether this effect on p27^{Kip1} is direct or indirect remains to be established, but it is a promising point for the apoptotic regulatory proteins to feed into the cell cycle regulatory system.

8
Conclusions

There are few iron clad statements that can be made regarding the cell cycle and apoptosis at this stage in our knowledge. However, from the evidence reviewed here, one or two assertions may be made. Apoptosis is not restricted to a particular part of the cell cycle. However, in any given set of circumstances all the cells do not die at all phases of the cell cycle. Depending on a particular stimulus or stage of growth or cell type or even stage of differentiation cells tend to accumulate at a specific point in the cell cycle before dying by apoptosis. This is very often a consequence of arrest caused by the stimulus to the cycling cells that are its target. Cells in G0/G1 or even post-mitotic cells also readily undergo apoptosis. However, although it can be demonstrated that these cells show some indications of traversing G1 there is no evidence they progress further through the cell cycle before death.

It has previously been put forward that apoptosis in resting or post-mitotic cells is a result of these cells trying to re-enter the cell cycle after receiving an apoptotic stimulus (Rubin et al. 1993). Subsequently, others have suggested that apoptosis may correspond to the cyclin A dependent part of mitosis (Meikrantz et al. 1994). The same authors have also suggested that the G1/S boundary is a crucial point for the cell cycle–apoptosis interface where a choice is made between DNA fragmentation or DNA synthesis (Meikrantz and Schlegel 1995). As an example, they state that while staurosporine and similar agents can induce mitotic catastrophe in G2 arrested cells, they induce apoptosis in G1/S arrested cells. They surmise that apoptosis resembles mitotic catastrophe in that both are characterized by inappropriate Cdk activation and by uncoupling the normal interdependence of cell cycle events.

We suggest a slightly different theory as to how the cell cycle and apoptotic mechanisms overlap. As discussed in Section 7, Bax and Bcl-2, as well as being regulators of apoptosis, can clearly influence cell cycle entry (Brady et al. 1996b; Linette et al. 1996; Mazel et al. 1996; O'Reilly et al. 1996). They can influence the rate of degradation of p27^{Kip1} (Brady et al. 1996b; Linette et al. 1996). We have recently found that this effect extends to thymocytes. The level of p27^{Kip1} degradation in thymocytes reflects the level of apoptosis and this is accelerated by Bax or slowed down by Bcl-2 (G. Gil-Gómez, A. J. M. Berns and H. J. M. Brady, submitted). Furthermore, we also find that this leads to a concomitant activation of Cdk2. When this activity is inhibited by specific chemical inhibitors then apoptosis is also inhibited.

Our model (Fig. 1) predicts that the cross-talk between apoptosis and cell cycle regulation would involve the sharing of common biochemical machinery rather than apoptosis being in any way an abortive entry into cell cycle. During

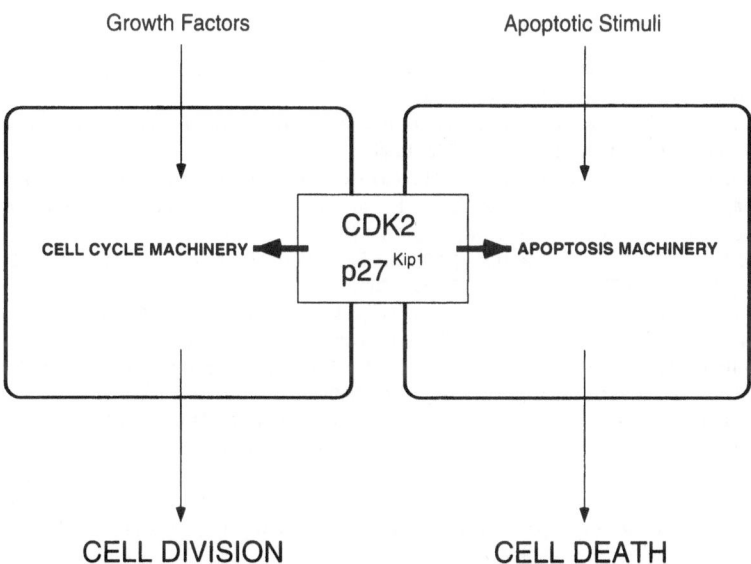

Fig. 1. Our model showing how apoptosis and the cell cycle are interlinked. The model predicts that crosstalk between apoptosis and cell cycle regulation involves sharing of common biochemical machinery rather than apoptosis being an abortive entry into cell cycle. We postulate a role for p27^{Kip1} degradation and subsequently Cdk2 activity in T cell apoptosis which is not a consequence of the ordered progression normally associated with cell cycle

normal cell cycle progression a highly ordered succession of rise and fall in Cdk activity is observed as a consequence of the modulation of synthesis, activation and degradation of cyclins, Cdks and CKIs. During apoptosis, although we postulate a role for Cdk2 activity in T cells, this activity is not a consequence of the ordered progression normally associated with cell cycle. It represents rather the use of some borrowed machinery for a different purpose. Further experiments are underway to test this hypothesis and probe this crucial interface between life and death.

References

Allday MJ, Inman GJ, Crawford DH, Farrell PJ (1995) DNA damage in human B cells can induce apoptosis, proceeding from G1/S when p53 is transactivation competent and G2/M when it is transactivation defective. EMBO J 14:4994–5005

Almasan A, Yin Y, Kelly RE, Lee EY, Bradley A, Li W, Bertino JR, Wahl GM (1995) Deficiency of retinoblastoma protein leads to inappropriate S-phase entry, activation of E2F-responsive genes, and apoptosis. Proc Natl Acad Sci USA 92:5436–5440

Ashwell JD, Cunningham RE, Noguchi PD, Hernandez D (1987) Cell growth cycle block of T cell hybridomas upon activation with antigen. J Exp Med 165:173–194

Askew DS, Ashmun RA, Simmons BC, Cleveland JL (1991) Constitutive c-myc expression in an IL-3-dependent myeloid cell line suppresses cell cycle arrest and accelerates apoptosis. Oncogene 6:1915–1922

Badiani P, Corbella P, Kioussis D, Marvel J, Weston K (1994) Dominant interfering alleles define a role for c-Myb in T-cell development. Genes Dev 8:770–782

Benhamou LE, Cazenave PA, Sarthou P (1990) Anti-immunoglobulins induce death by apoptosis in WEHI-231 B lymphoma cells. Eur J Immunol 20:1405–1407

Boehme SA, Lenardo MJ (1993) Propriocidal apoptosis of mature T lymphocytes occurs at S phase of the cell cycle. Eur J Immunol 23:1552–1560

Brady HJM, Salomons GS, Bobeldijk RC, Berns J (1996a) T cells from baxα transgenic mice show accelerated apoptosis in response to stimuli but do not show restored DNA damage-induced cell death in the absence of p53 gene product. EMBO J 15:1221–1230

Brady HJM, Gil-Gómez G, Kirberg J, Berns AJ (1996b) Baxα perturbs T cell development and affects cell cycle entry of T cells. EMBO J 15:6991–7001

Brugarolas J, Chandrasekaran C, Gordon JI, Beach D, Jacks T, Hannon GJ (1995) Radiation-induced cell cycle arrest compromised by p21 deficiency. Nature 377:552–557

Canman CE, Gilmer TM, Coutts SB, Kastan MB (1995) Growth factor modulation of p53-mediated growth arrest versus apoptosis. Genes Dev 9:600–611

Clarke AR, Maandag ER, van Roon M, van der Lugt N, van der Valk M, Hooper ML, Berns A, te Riele H (1992) Requirement for a functional Rb-1 gene in murine development. Nature 359:328–330

Clarke AR, Purdie CA, Harrison DJ, Morris RG, Bird CC, Hooper ML, Wyllie AH (1993) Thymocyte apoptosis induced by p53-dependent and independent pathways. Nature 362:849–852

Cox LS, Lane DP (1995) Tumour suppressors, kinases and clamps: how p53 regulates the cell cycle in response to DNA damage. Bioessays 17:501–508

Cross SM, Sanchez CA, Morgan CA, Schimke MK, Ramel S, Idzerda RL, Raskind WH, Reid BJ (1995) A p53-dependent mouse spindle checkpoint. Science 267:1353–1356

Deng C, Zhang P, Harper JW, Elledge SJ, Leder P (1995) Mice lacking p21CIP1/WAF1 undergo normal development, but are defective in G1 checkpoint control. Cell 82:675–684

Desbarats L, Schneider A, Muller D, Burgin A, Eilers M (1996) Myc: a single gene controls both proliferation and apoptosis in mammalian cells. Experientia 52:1123–1129

Di Leonardo A, Linke SP, Clarkin K, Wahl GM (1994) DNA damage triggers a prolonged p53-dependent G1 arrest and long-term induction of Cip1 in normal human fibroblasts. Genes Dev 8:2540–2551

Evan GI, Wyllie AH, Gilbert CS, Littlewood TD, Land H, Brooks M, Waters CM, Penn LZ, Hancock DC (1992) Induction of apoptosis in fibroblasts by c-Myc protein. Cell 69:119–128

Fanidi A, Harrington EA, Evan GI (1992) Cooperative interaction between c-myc and bcl-2 proto-oncogenes. Nature 359:554–556

Field SJ, Tsai FY, Kuo F, Zubiaga AM, Kaelin WJ, Livingston DM, Orkin SH, Greenberg ME (1996) E2F-1 functions in mice to promote apoptosis and suppress proliferation. Cell 85:549–561

Fotedar R, Flatt J, Gupta S, Margolis RL, Fitzgerald P, Messier H, Fotedar A (1995) Activation-induced T-cell death is cell cycle dependent and regulated by cyclin B. Mol Cell Biol 15:932–942

Frampton J, Ramqvist T, Graf T (1996) V-Myb of E26 leukemia virus up-regulates bcl-2 and suppresses apoptosis in myeloid cells. Genes Dev 10:2720–2731

Galaktionov K, Chen X, Beach D (1996) Cdc25 cell-cycle phosphatase as a target of c-myc. Nature 382:511–517

Green DR, Scott DW (1994) Activation-induced apoptosis in lymphocytes. Curr Opin Immunol 6:476–487

Haas-Kogan D, Kogan SC, Levi D, Dazin P, T'Ang A, Fung YK, Israel MA (1995) Inhibition of apoptosis by the retinoblastoma gene product. EMBO J 14:461–472

Halicka HD, Seiter K, Feldman EJ, Traganos F, Mittelman A, Ahmed T, Darzynkiewicz Z (1997) Cell cycle specificity of apoptosis during treatment of leukaemias. Apoptosis 2:25–39

Hardie DL, Johnson GD, Khan M, MacLennan IC (1993) Quantitative analysis of molecules which distinguish functional compartments within germinal centers. Eur J Immunol 23:997–1004

Harmon JM, Norman MR, Fowlkes BJ, Thompson EB (1979) Dexamethasone induces irreversible G1 arrest and death of a human lymphoid cell line. J Cell Physiol 98:267–278

Harper JW, Elledge SJ (1996) Cdk inhibitors in development and cancer. Curr Opin Genet Dev 6:56–64

Harper JW, Adami JR, Wei N, Keyomarsi K, Elledge SJ (1993) The p21 Cdk-interacting protein Cip1 is a potent inhibitor of G1 cyclin-dependent kinases. Cell 75:805–816

Harrington EA, Bennett MR, Fanidi A, Evan GI (1994) C-Myc-induced apoptosis in fibroblasts is inhibited by specific cytokines. EMBO J 13:3286–3295

Hasbold J, Klaus GG (1990) Anti-immunoglobulin antibodies induce apoptosis in immature B cell lymphomas. Eur J Immunol 20:1685–1690

Heald R, McLoughlin M, McKeon F (1993) Human wee1 maintains mitotic timing by protecting the nucleus from cytoplasmically activated Cdc2 kinase. Cell 74:463–474

Heiskanen P, Billig H, Toppari J, Kaleva M, Arsalo A, Rapola J, Dunkel L (1996) Apoptotic cell death in the normal and cryptorchid human testis: the effect of human chorionic gonadotropin on testicular cell survival. Pediatr Res 40:351–356

Hoang AT, Cohen KJ, Barrett JF, Bergstrom DA, Dang CV (1994) Participation of cyclin A in Myc-induced apoptosis. Proc Natl Acad Sci USA 91:6875–6879

Hoffmann I, Draetta G, Karsenti E (1994) Activation of the phosphatase activity of human cdc25A by a cdk2-cyclin E dependent phosphorylation at the G1/S transition. EMBO J 13:4302–4310

Huesmann M, Scott B, Kisielow P, von Boehmer H (1991) Kinetics and efficacy of positive selection in the thymus of normal and T cell receptor transgenic mice. Cell 66:533–540

Hunter T, Pines J (1994) Cyclins and cancer. II: cyclin D and CDK inhibitors come of age. Cell 79:573–582

Jacks T, Fazeli A, Schmitt EM, Bronson RT, Goodell MA, Weinberg RA (1992) Effects of an Rb mutation in the mouse. Nature 359:295–300

Jansen-Dürr P, Meichle A, Steiner P, Pagano M, Finke K, Botz J, Wessbecher J, Draetta G, Eilers M (1993) Differential modulation of cyclin gene expression by MYC. Proc Natl Acad Sci USA 90:3685–3689

King KL, Cidlowski JA (1995) Cell cycle and apoptosis: common pathways to life and death. J Cell Biochem 58:175–180

Kobayashi T, Consoli U, Andreeff M, Shiku H, Deisseroth AB, Zhang W (1995) Activation of p21WAF1/Cip1 expression by a temperature-sensitive mutant of human p53 does not lead to apoptosis. Oncogene 11:2311–2316

Kranenburg O, van der Eb A, Zantema A (1996) Cyclin D1 is an essential mediator of apoptotic neuronal cell death. EMBO J 15:46–54

Lahti JM, Xiang J, Heath LS, Campana D, Kidd VJ (1995) PITSLRE protein kinase activity is associated with apoptosis. Mol Cell Biol 15:1–11

Lazebnik YA, Cole S, Cooke CA, Nelson WG, Earnshaw WC (1993) Nuclear events of apoptosis in vitro in cell-free mitotic extracts: a model system for analysis of the active phase of apoptosis. J Cell Biol 123:7–22

Lee EY, Chang CY, Hu N, Wang YC, Lai CC, Herrup K, Lee WH, Bradley A (1992) Mice deficient for Rb are nonviable and show defects in neurogenesis and haematopoiesis. Nature 359:288–294

Lee EY, Hu N, Yuan SS, Cox LA, Bradley A, Lee WH, Herrup K (1994) Dual roles of the retinoblastoma protein in cell cycle regulation and neuron differentiation. Genes Dev 8:2008–2021

Linette GP, Li Y, Roth K, Korsmeyer SJ (1996) Cross talk between cell death and cell cycle progression: BCL-2 regulates NFAT-mediated activation. Proc Natl Acad Sci USA 93:9545–9552

Liu YJ, Mason DY, Johnson GD, Abbot S, Gregory CD, Hardie DL, Gordon J, MacLennan IC (1991) Germinal center cells express bcl-2 protein after activation by signals which prevent their entry into apoptosis. Eur J Immunol 21:1905–1910

Lowe SW, Schmitt EM, Smith SW, Osborne BA, Jacks T (1993) p53 is required for radiation-induced apoptosis in mouse thymocytes. Nature 362:847–849

Ludlow JW, Glendening CL, Livingston DM, DeCaprio JA (1993) Specific enzymatic dephosphorylation of the retinoblastoma protein. Mol Cell Biol 13:367–372

Lundgren K, Walworth N, Booher R, Dembski M, Kirschner M, Beach D (1991) Mik1 and wee1 cooperate in the inhibitory tyrosine phosphorylation of cdc2. Cell 64:1111–1122

Maheswaran S, McCormack JE, Sonenshein GE (1991) Changes in phosphorylation of myc oncogene and RB antioncogene protein products during growth arrest of the murine lymphoma WEHI 231 cell line. Oncogene 6:1965–1971

Martin DP, Ito A, Horigome K, Lampe PA, Johnson EJ (1992) Biochemical characterization of programmed cell death in NGF-deprived sympathetic neurons. J Neurobiol 23:1205–1220

Mazel S, Burtrum D, Petrie HT (1996) Regulation of cell division cycle progression by bcl-2 expression: a potential mechanism for inhibition of programmed cell death. J Exp Med 183:2219–2226

Meikrantz W, Schlegel R (1995) Apoptosis and the cell cycle. J Cell Biochem 58:160–174

Meikrantz W, Schlegel R (1996) Suppression of apoptosis by dominant negative mutants of cyclin-dependent protein kinases. J Biol Chem 271:10205–10209

Meikrantz W, Gisselbrecht S, Tam SW, Schlegel R (1994) Activation of cyclin A-dependent protein kinases during apoptosis. Proc Natl Acad Sci USA 91:3754–8

Morgenbesser SD, Williams BO, Jacks T, DePinho RA (1994) p53-dependent apoptosis produced by Rb-deficiency in the developing mouse lens. Nature 371:72–74

Merritt AJ, Potten CS, Watson AJ, Loh DY, Nakayama K, Nakayama K, Hickman JA (1995) Differential expression of bcl-2 in intestinal epithelia. Correlation with attenuation of apoptosis in colonic crypts and the incidence of colonic neoplasia. J Cell Sci 108:2261–2271

Mesner PW, Winters TR, Green SH (1992) Nerve growth factor withdrawal-induced cell death in neuronal PC12 cells resembles that in sympathetic neurons. J Cell Biol 119:1669–1680

Morgan DO (1995) Principles of CDK regulation. Nature 374:131-134

Neiman PE, Blish C, Heydt C, Loring G, Thomas SJ (1994) Loss of cell cycle controls in apoptotic lymphoblasts of the bursa of Fabricius. Mol Biol Cell 5:763-772

Nevins JR (1992) E2F: a link between the Rb tumor suppressor protein and viral oncoproteins. Science 258:424-429

Nigg EA (1995) Cyclin-dependent protein kinases: key regulators of the eukaryotic cell cycle. Bioessays 17:471-480

Otsuki Y, Misaki O, Sugimoto O, Ito Y, Tsujimoto Y, Akao Y (1994) Cyclic bcl-2 gene expression in human uterine endometrium during menstrual cycle. Lancet 344:28-29

Pandey S, Wang E (1995) Cells en route to apoptosis are characterized by the upregulation of c-fos, c-myc, c-jun, cdc2, and RB phosphorylation, resembling events of early cell-cycle traverse. J Cell Biochem 58:135-150

Pellicciari C, Bottone MG, Schaack V, Barni S, Manfredi AA (1996) Spontaneous apoptosis of thymocytes is uncoupled with progression through the cell cycle. Exp Cell Res 229:370-377

Qin XQ, Livingston DM, Kaelin WJ, Adams PD (1994) Deregulated transcription factor E2F-1 expression leads to S-phase entry and p53-mediated apoptosis. Proc Natl Acad Sci USA 91:10918-10922

Rubin LL, Philpott KL, Brooks SF (1993) Apoptosis - the cell cycle and cell death. Curr Biol 3:391-394

O'Reilly LA, Huang DC, Strasser A (1996) The cell death inhibitor Bcl-2 and its homologues influence control of cell cycle entry. EMBO J 15:6979-6990

Ryan JJ, Danish R, Gottlieb CA, Clarke MF (1993) Cell cycle analysis of p53-induced cell death in murine erythroleukemia cells. Mol Cell Biol 13:711-719

Sala A, Casella I, Grasso L, Bellon T, Reed JC, Miyashita T, Peschle C (1996) Apoptotic response to oncogenic stimuli: cooperative and antagonistic interactions between c-myb and the growth suppressor p53. Cancer Res 56:1991-1996

Sellins KS, Cohen JJ (1987) Gene induction by gamma-irradiation leads to DNA fragmentation in lymphocytes. J Immunol 139:3199-3206

Shan B, Lee WH (1994) Deregulated expression of E2F-1 induces S-phase entry and leads to apoptosis. Mol Cell Biol 14:8166-173

Sherr CJ (1996) Cancer cell cycles. Science 274:1672-1677

Sherr CJ, Roberts JM (1995) Inhibitors of mammalian G1 cyclin-dependent kinases. Genes Dev 9:1149-1163

Shi L, Nishioka WK, Th'ng J, Bradbury EM, Litchfield DW, Greenberg AH (1994) Premature p34cdc2 activation required for apoptosis. Science 263:1143-1145

Shi L, Chen G, He D, Bosc DG, Litchfield DW, Greenberg AH (1996) Granzyme B induces apoptosis and cyclin A-associated cyclin-dependent kinase activity in all stages of the cell cycle. J Immunol 157:2381-2385

Strasser A, Harris AW, Jacks T, Cory S (1994) DNA damage can induce apoptosis in proliferating lymphoid cells via p53-independent mechanisms inhibitable by Bcl-2. Cell 79:329-339

Taylor D, Badiani P, Weston K (1996) A dominant interfering Myb mutant causes apoptosis in T cells. Genes Dev 10:2732-2744

Thomaidou D, Mione MC, Cavanagh JF, Parnavelas JG (1997) Apoptosis and its relation to the cell cycle in the developing cerebral cortex. J Neurosci 17:1075-1085

Stern JB, Smith KA (1986) Interleukin-2 induction of T-cell G1 progression and c-myb expression. Science 233:203-206

Thompson MA, Ramsay RG (1995) Myb: an old oncoprotein with new roles. Bioessays 17:341-350

Walker NI, Bennett RE, Kerr JFR (1989) Cell death by apoptosis during involution of the lactating breast in mice and rats. Am J Anat 185:19-32

Wang J, Walsh K (1996) Resistance to apoptosis conferred by Cdk inhibitors during myocyte differentiation. Science 273:359-361

Waters CM, Littlewood TD, Hancock DC, Moore JP, Evan GI (1991) C-myc protein expression in untransformed fibroblasts. Oncogene 6:797-805

Weinberg RA (1995) The retinoblastoma protein and cell cycle control. Cell 81:323–330

Wu X, Levine AJ (1994) p53 and E2F-1 cooperate to mediate apoptosis. Proc Natl Acad Sci USA 91:3602–3606

Wyllie AH (1980) Glucocorticoid-induced thymocyte apoptosis is associated with endogenous endonuclease activation. Nature 284:555–556

Yamada T, Ohyama H (1988) Radiation-induced interphase death of rat thymocytes is internally programmed (apoptosis). Int J Radiat Biol 53:65–75

Zhang P, Liegeois NJ, Wong C, Finegold M, Hou H, Thompson JC, Silverman A, Harper JW, DePinho RA, Elledge SJ (1997) Altered cell differentiation and proliferation in mice lacking p57KIP2 indicates a role in Beckwith–Wiedemann syndrome. Nature 387:151–158

Wang Y, Okan I, Szekely L, Klein G, Wiman KG (1995) bcl-2 inhibits wild type p53-triggered apoptosis but not G1 cell cycle arrest and transactivation of WAF1 and bax. Cell Growth Differ 6:1071–1075

The p53 Tumor Suppressor Gene: Structure, Function and Mechanism of Action

Caroline Choisy-Rossi, Philippe Reisdorf and Elisheva Yonish-Rouach

1
Introduction

The p53 protein was first identified through its ability to co-precipitate with the large T antigen of simian virus 40 (SV40) (DeLeo et al. 1979; Kress et al. 1979; Lane and Crawford 1979; Linzer and Levine 1979). The gene was subsequently cloned from several species (reviewed by Soussi et al. 1990), and was first suggested to function as an oncogene since overexpression of p53 was reported to participate in cell transformation (Eliyahu et at. 1984; Parada et al. 1984). However, in the late 1980s it became clear that the oncogenic activities of p53 were exhibited by mutant forms of the protein (Eliyahu et al. 1988; Finlay et al. 1988; Hinds et al. 1989). Genetic evidence suggested in fact that the wild-type (wt) form of the p53 gene could be defined as a tumor suppressor gene. Mutations in p53 were found to occur in high frequency in colon cancer (Baker et al. 1989), and this observation was later extended to most of the common types of human cancer (Hollstein et al. 1994).

A significant biological role of p53 as a tumor suppressor gene is also implied by the fact that p53 null mice are susceptible to spontaneous tumors, and over 70% develop tumors by the age of 6 months (Donehower et al. 1992; Jacks et al. 1994; Williams et al. 1994; Harvey et al. 1995). A similar susceptibility is observed in humans: families with the Li-Fraumeni syndrome, which is a rare inherited autosomal condition, have been shown to transmit a mutated form of p53 in their germ line, and have a 50% probability of developing cancer by the age of 30 (reviewed by Donehower and Bradley 1993).

The sequencing of p53 cDNAs from a variety of species revealed five evolutionary-conserved regions, associating these regions with an important role in p53 activity (Soussi et al. 1990). Indeed, four of the conserved regions are located within the central core domain of p53 which was defined as the sequence-specific binding domain through which p53 binds to target genes (Fig. 1). The majority of p53 missense mutations found in human cancers are clustered within this central domain, particularly within the four conserved

Laboratoire de Cancérogenèse Moléculaire, UMR 217 du CNRS/CEA, DRR-DSV, CEA, 92265 Fontenay-aux-Roses Cedex, France

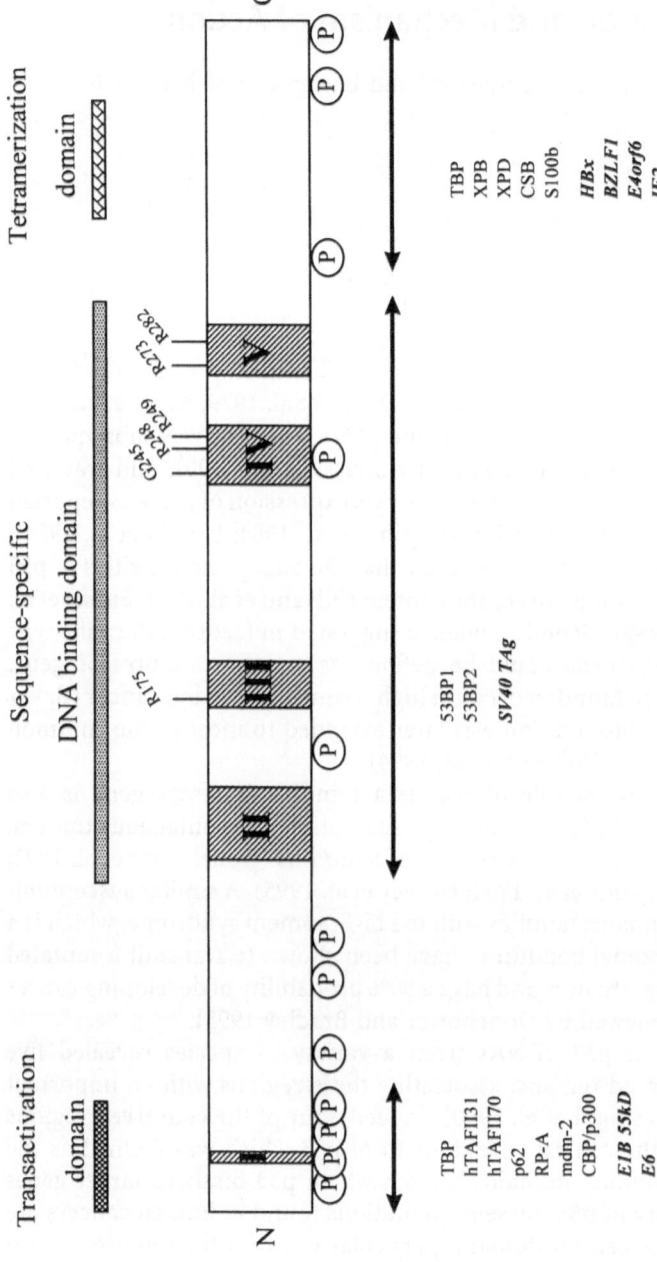

Fig. 1. The p53 protein. The five evolutionary conserved regions are shown by *hatched boxes*. Vertical lines indicate mutation hotspots with the corresponding residue number. *Circled Ps* represent phosphorylation sites, some of which correspond to mouse p53 (reviewed by Meek 1997). Functional domains of p53 are indicated at the *top*, and p53-interacting proteins are listed at the *bottom*. Viral proteins are written in *bold italic letters*

regions. Thus, inactivation of normal p53 function appears to be an important step in cancer development, and the study of p53 function has been the focus of intensive research for several years already. Since it was designated "molecule of the year" in 1993 (Culotta and Koshland 1993) it remained the center of interest for many laboratories, and the increasing amount of data, although sometimes confusing, may add up to contribute to our understanding of this fascinating "molecule of the last twentieth century decade".

2
Structure and Function of p53 Domains

p53 was shown to be a sequence-specific transcription transactivator (Farmer et al. 1992). It can interact with DNA containing the sequence 5'-PuPuPuC(A/T)(T/A)GPyPyPy-3', usually in the context of two such sequence motifs separated by 0–13 bp (Kern et al. 1991; El-Deiry et al. 1992; Funk et al. 1992; Bourdon et al. 1997). Biochemical and functional studies of p53 have defined the domain structure of p53, and have revealed that the central "core" of the protein (amino acid residues 102 to 292; Fig. 1) is responsible for the sequence-specific binding activity (Bargonetti et al. 1993; Pavletich et al. 1993; Wang et al. 1993). The crystal structure of the complex containing the p53 core domain and DNA target sequences was determined at a resolution of 2.2 Å using X-ray diffraction, and was refined to a crystallographic R factor of 20.5% (Cho et al. 1994). The core domain structure consists of a β sandwich, comprising two antiparallel β sheets, which serves as a scaffold for the structural elements at the DNA–protein interface: a loop-sheet-helix (LSH) motif, that binds in the major groove and is involved in contacts with the bases, and two large loops, which are encoded by conserved domains III and IV. The LSH domain is encoded by conserved domains II and V and in combination with the two loops it forms the DNA binding surface of p53. The direct contact with the DNA is through residues R248 and R273, while other residues are necessary for the integrity of the structure. A zinc atom interacts with four metal-binding residues and stabilizes the structure of the loops. Such a role of zinc is demonstrated by the fact that metal chelating agents inhibit the sequence-specific DNA binding of p53 and change its conformation as detected by antibodies (Hainaut and Milner 1993a,b).

The important role of the structural integrity for the DNA binding function of the central core explains the evolutionary pressure for the conservation of critical amino acids in domains II–IV. Indeed, it is within these domains that most of the residues mutated in human cancer are found, and these mutations affect either the direct contact with DNA or the conformation of the core structure (Cho et al. 1994; Fig. 1).

The transcriptional activation domain is located at the N terminus of the protein (Raycroft et al. 1990; Unger et al. 1992), within the first 42 amino acids (Fields and Jang 1990; Raycroft et al. 1990; Unger et al. 1992; Fig. 1). The transactivation activity of p53 may also depend on neighboring residues (Liu et al. 1993; Chang et al. 1995). The conserved domain I is located between amino acids 13–23, and residues F19, L22 and W23 were shown to be required for the transcriptional activation function of p53 (Lin et al. 1994).

Several cellular and viral proteins can bind to the N-terminal domain (Fig. 1). Some, like the cellular protein mdm-2 and the adenovirus E1B-55kDa, will inhibit the transcriptional activity of p53 (Braithwaite et al. 1991; Momand et al. 1992; Oliner et al. 1993; Blaydes et al. 1997). Mdm-2 was reported to promote the rapid degradation of p53, and this may be the mechanism through which it can regulate the level of p53 and its activity (Haupt et al. 1997; Kubbutat et al. 1997). Binding of other proteins, such as the two subunits of the TFIID complex, $dTAF_{II}40/hTAF_{II}31$ and $TAF_{II}60/hTAF_{II}70$, may be important for transmitting activation signals between p53 and the initiation complex (Lu and Levine 1995; Thut et al. 1995; Levine 1997). TBP also binds to the N-terminal region of p53, and this binding may be involved in both activation and repression of transcription (Ginsberg et al. 1991; Subler et al. 1992; Chen et al. 1993; Mack et al. 1993; Ragimov et al. 1993). Additional proteins which were reported to bind to the N-terminal domain of p53 include the single-stranded DNA-binding protein RP-A and the p62 subunit of the transcription/repair factor TFIIH (Dutta et al. 1993; He et al. 1993; Li and Botchan 1993; Xiao et al. 1994; X. W. Wang et al. 1995; Leveillard et al. 1996), and the transcription coactivators CBP/p300 which bind through the C-terminal region of CBP and modulate p53 growth arrest and apoptosis activities (Gu et al. 1997; Lill et al. 1997).

The p53 in its native form exists as tetramers, and the oligomerization domain of p53 is located to its C-terminus, more specifically to residues 323–355 (Stürzbecher et al. 1992; Pavletich et al. 1993; Fig. 1). This domain is composed of a turn (residues 324–326), a β strand (residues 326–334), another turn (residues 335–336), and an α helix (residues 337–355). Each monomer interacts with another monomer to form a dimer in a way that the β strands and the helices are antiparallel, and two such dimers are held together by a large hydrophobic surface of each helix pair, forming a four-helix bundle. The resulting tetramer is thus composed of a dimer of dimers (Clore et al. 1994, 1995; Lee et al. 1994; Jeffrey et al. 1995). The tetramerization domain is linked to the DNA-binding core domain by a flexible linker (residues 287–323) (Jeffrey et al. 1995). The C-terminus appears to stabilize the sequence-specific binding activity (Shaulian et al. 1993; Tarunina and Jenkins 1993), and was reported to be indispensable for sequence-specific transactivation in some cases (Halazonetis and Kandil 1993; Pietenpol et al. 1994) but not in others (Shaulian et al. 1993; Slingerland et al. 1993; Tarunina and Jenkins 1993). Such differences may be explained by the requirement for tetramerization and stable binding to certain p53 target genes but not to others.

The extreme C-terminus of p53 (residues 368–393) is highly basic and is connected to the tetramerization domain by a flexible linker (Pavletich et al.

1993). It is capable of binding nonspecifically to various forms of DNA, in particular damaged DNA (Wang et al. 1993; Lee et al. 1995). It may catalyze renaturation of complementary single stranded DNA or RNA (Balkalkin et al. 1994; Lee et al. 1995; Wu et al. 1995; Ko and Prives 1996 and references therein), an activity which can be linked to the possible role of p53 in DNA repair. In this context, binding of XPD and XPB, which are part of the TFIIH transcription-repair complex, and of the strand-specific repair factor CSB to p53 was located to the C-terminus of p53 (Xiao et al. 1994; X. W. Wang et al. 1995; Fig. 1). Finally, the basic region in the extreme C-terminus was shown to have negative-regulatory functions (Hupp et al. 1993; Wu et al. 1994; Bayle et al. 1995). Thus, specific binding of p53 to DNA is enhanced by deletion, phosphorylation or O-glycosylation of this domain or when it is bound by antibody or dnaK (Hupp and Lane 1994; Takenaka et al. 1995; Shaw et al. 1996), and a peptide containing the last 30 amino acids can stimulate specific DNA binding by the full-length p53 (Hupp et al. 1995). Activation of specific p53-DNA binding is also achieved by interaction of short single strands of DNA (20–39 nucleotides) with the C-terminal domain, while interaction of this domain with longer, double stranded DNA inhibits the sequence-specific binding (Jayaraman and Prives 1995). Recently, a synthetic peptide corresponding to residues 361–382 of p53 was shown to activate specific DNA binding of wt p53 *in vitro* and to restore the transcriptional transactivation function of some mutant p53 proteins in living cells (Selivanova et al. 1997). In addition, the murine C-terminal alternatively spliced form of p53 was shown to induce attenuated apoptosis in myeloid cells, suggesting that the C terminus controls the apoptotic activity of the p53 molecule (Almog et al. 1997). Thus, the C-terminus of p53 plays an important role in regulating the activity of the whole protein.

The activity of p53 may also be modified by its interaction with many cellular and viral proteins (Table 1; reviewed by Levine 1997; Ko and Prives 1996). For example, binding of SV40 T antigen to the core domain or of HBV Hbx, adenovirus E4orf6 and human cytomegalovirus IE2 proteins to the C-terminal region of p53 results in inhibition of transactivation by p53 (Jiang et al. 1993; Wang et al. 1994; Dobner et al. 1996; Tsai et al. 1996), while binding of c-Abl to p53 was reported to enhance p53-dependent transcription (Goga et al. 1995). Recently, the redox/repair protein Ref-1 was identified as a potent activator of p53 (Jayaraman et al. 1997). Other cellular proteins which have been reported to interact with p53, such as 53BP1 and 53BP2 (Iwabuchi et al. 1994; Thukral et al. 1994), may also have regulatory activity (Fig. 1).

Finally, p53 is a phosphoprotein which is modified and regulated by multisite phosphorylation (reviewed by Meek et al. 1997; Milczarek et al. 1997). The major phosphorylation sites are located within the N- and C-terminal regions of the protein, but additional sites in the central core region were also reported (Milne et al. 1996; Adler et al. 1997; Fig. 1). Several protein kinases were shown to be capable of phosphorylating p53, including DNA-PK, a CK1-like enzyme, MAP kinase, JNK1 and 2 at the N-terminal sites (reviewed by

Table 1. Cellular and viral proteins which have been reported to interact with p53

Cellular proteins	Viral proteins
TFIID	SV40 TAg
TBP	Adenovirus E1B-55 kDa
hTAFII31	Adenovirus E4orf6
hTAFII70	Human papillomavirus E6
TFIIH	Hepatitis B virus HBx
XPD/ERCC2	Epstein–Barr virus BZLF1
XPB/ERCC3	Epstein–Barr virus EBNA-5
RP-A	Human cytomegalovirus IE2
CSB	
CSP/p300	
Sp1	
WT1	
CBF	
c-Abl	
S100b	
53BP1	
53BP2	

See Fig. 1 for localization of some of the interactions within the p53 domains and text for outcome of interactions. References are within the text and reviewed by Ko and Prives (1996).

Meek 1997). The biological effects of phosphorylation by these kinases is still not clear. Mutation of the serine 15 of human p53 – a site for DNA-PK phosphorylation – was reported to reduce the ability of p53 to inhibit cell cycle progression (Fiscella et al. 1993). On the other hand, p53 accumulation following irradiation was shown to be normal in fibroblasts from DNA-PK-deficient mice, as well as induction of p21 and cell cycle arrest, suggesting that DNA-PK may not be required for p53-dependent cell cycle arrest after DNA damage (Rathmell et al. 1997). Thus, the role of phosphorylation by DNA-PK may depend on environmental conditions and/or cell type. Another potential kinase which is similar to DNA-PK is the product of the ataxia-telangiectasia gene (ATM), which was reported to be upstream of p53 in the DNA damage-response pathway (Kastan et al. 1992; Xu and Baltimore 1996). Recent studies identified the c-Abl tyrosine kinase as a target of phosphorylation by ATM which contributes to p53-dependent responses to DNA damage (Yuan et al. 1996a,b; Baskaran et al. 1997). The C-terminal phosphorylation sites of p53 were reported to be phosphorylated by PKC, CKII and Cdks, including cyclin B/Cdc2 and cyclin A/Cdk2 (Wang and Prives 1995). Phosphorylation of the C-terminus appears to play a role in regulation of sequence-specific DNA binding and possibly in conferring binding preference (Wang and Prives 1995; Hall et al. 1996; Martinez et al. 1997), but the mechanisms of such regulations remain to be elucidated. Inhibition of protein phosphatase activity by okadaic acid was shown to enhance the binding of p53 to consensus DNA target sequence and to cause an increase in p53 transcriptional activity (Yan et al. 1997). Yet okadaic acid induced p53-dependent apoptosis in the absence of p53 transactivation (Yan et al. 1997). Thus, phosphorylation and dephosphorylation of p53 may synergize with other factors to regulate the biological activity of the protein.

3
Suppression of Cell Growth: p53 Induces a Cell Cycle Arrest and/or Apoptosis

The main function of p53 as a tumor suppressor appears to be in mediating the cellular response to DNA damage, and helping to maintain genomic stability (Kastan et al. 1991; Kuerbitz et al. 1992; Lane 1992). p53 is normally a short-lived protein which is thought to be degraded by ubiquitin-mediated proteolysis (Ciechanover et al. 1994), although calpain was recently reported to be a potential regulator of p53 stability (Gonen et al. 1997; Kubbutat and Vousden 1997). A rapid increase in p53 protein levels is induced by genotoxic stress, such as ultraviolet light, γ-irradiation and genotoxic chemicals (Kastan et al. 1992; Lu and Lane 1993; Zhan et al. 1993). Such an increase occurs mainly through stabilization of the protein (Maltzman and Czyzyk 1984; Kastan et al. 1992), although increased p53 mRNA level has also been reported (Sun et al. 1995). Proteasome inhibitors can also induce accumulation of the protein (Lopes et al. 1997). The main outcome of p53 activation appears to be growth-suppression. However, other functions of p53 have been reported, namely a role in differentiation (Aloni-Grinstein et al. 1995), senescence (Atadja et al. 1995), inhibition of angiogenesis (Dameron et al. 1994) and in development (Sah et al. 1995).

The ability of p53 to suppress cell growth can be explained by an induction of growth arrest and/or of apoptosis. By reintroducing and overexpressing wt p53 in cells which have lost it, the protein was first reported to induce a growth arrest in the G1 phase of the cell cycle (Diller et al. 1990; Mercer et al. 1990; Michalovitz et al. 1990; Martinez et al. 1991). Similarly, a G1 arrest after γ-irradiation was acquired following transfection of wt p53 into cells lacking it (Kuerbitz et al. 1992). On the other hand, cells from patients with the cancer-prone disease ataxia telangiectasia were shown to be deficient in irradiation-activated G1 checkpoint as well as in the induction of p53 protein following irradiation (Kastan et al. 1992). Thus, wt p53 can impose a G1 growth arrest under a variety of circumstances. In addition, several observations have suggested that p53 participates in a G2 cell cycle arrest and spindle checkpoint (Vikhanskaya et al. 1994; Cross et al. 1995).

The involvement of wt p53 in the control of apoptosis has been reported for many systems, both *in vitro* and *in vivo* (Yonish-Rouach et al. 1991; Shaw et al.

1992; Clarke et al. 1993; Lotem and Sachs 1993; Lowe et al. 1993b; Ryan et al. 1993). The first experiments that showed the ability of wt p53 to induce apoptosis were performed employing a clone of the mouse myeloid cell line M1 which is devoid of p53 expression. Cells of this clone were stably transfected with a temperature-sensitive (ts) mutant of p53, which acquires the conformation of the wt protein at the permissive temperature (32 °C). Upon downshift to the permissive temperature the transfectants underwent a rapid loss of viability which had the characteristics of apoptosis (Yonish-Rouach et al. 1991). Interestingly, although the same ts mutant imposed a reversible growth arrest in fibroblasts at the permissive temperature (Michalovitz et al. 1990), in the M1 cells p53-induced apoptosis was not preceded by growth arrest (Yonish-Rouach et al. 1993). However, the commitment to death was cell-cycle dependent and occurred in G1 (Yonish-Rouach et al. 1993).

Somewhat different results were obtained with mouse erythroleukemia cells DP16-1 lacking endogenous expression of p53 and which were transfected by the ts mutant (Ryan et al. 1993). In these cells, activation of wt p53 at the permissive temperature resulted in a G1 arrest, terminal differentiation and apoptosis (Johnson et al. 1993). However, in the same cells p53-mediated apoptosis was shown to be uncoupled from p53-mediated growth arrest, since the addition of cytokines inhibited the former but not the latter (Abrahamson et al. 1995; Lin and Benchimol 1995). A G1 arrest followed by apoptosis was also observed in the T-cell lymphoma line J3D expressing ts p53 upon temperature downshift (Y. S. Wang et al. 1995). Bcl-2 inhibited the p53-triggered apoptosis but not the G1 arrest, demonstrating that the two functions of p53 are independent.

p53 was also shown to mediate E1A-induced apoptosis (Debbas and White 1993; Lowe and Ruley 1993). Rodent cells transformed with E1A plus the ts p53 underwent apoptosis at the permissive temperature, and there was no growth arrest prior to apoptosis (Chiou et al. 1994). Bcl-2 prevented the p53-mediated apoptosis while keeping the cells in a predominantly growth-arrested state (Chiou et al. 1994). Thus, induction of apoptosis and induction of growth arrest by p53 were shown to be two separable functions.

Additional data supporting the dissociation between cell cycle arrest and apoptosis comes from Li-Fraumeni cells heterozygous for p53 gene mutations. A lymphoblastoid cell line harboring heterozygous mutation of p53 at codon 282 was capable of arresting at G1 following irradiation while it was relatively resistant to irradiation-induced apoptosis (Delia et al. 1997).

p53 is required not only for irradiation-induced G1 arrest, but also for irradiation-induced apoptosis in several *in vivo* and *in vitro* cell systems. Thus, thymocytes from genetically manipulated p53 null mice were resistant to irradiation-induced apoptosis (Clarke et al. 1993; Lotem and Sachs 1993; Lowe et al. 1993b). In addition, crypt cells from the small and large intestine of these mice were resistant to γ-irradiation *in vivo* while cells from normal mice underwent apoptosis (Clarke et al. 1994).

Exposure of cells of the interleukin-3 (IL-3)-dependent murine lymphoid cell line Baf-3 to irradiation resulted in rapid apoptosis if IL-3 has been withdrawn, while no apoptosis after irradiation was observed in the presence of IL-3 (Collins et al. 1992). Instead, in the presence of IL-3, Baf-3 cells arrested transiently both in the G1 and G2/M phases of the cell cycle after irradiation, and the G1 arrest was dependent on p53 function (Canman et al. 1995). Thus, a survival factor could modify the outcome of p53 activation following irradiation from apoptosis to growth arrest.

Similarly, in the M1 cells activation of wt p53 by temperature downshift in the presence of interleukin-6 (IL-6) did not result in cell death but rather induced a cell cycle exit into a G0-like quiescent state (Levy et al. 1993). A G1 growth arrest was also observed in the ts p53-expressing M1 cells prior to apoptosis when ectopic expression of Bcl-2 delayed cell death following wt p53 activation at the permissive temperature (Guillouf et al. 1995).

Induction of p53-mediated apoptosis following DNA damage was demonstrated in Epstein–Barr virus-immortalized human B lymphoid cell lines (Allday et al. 1995). Treatment with cisplatin rapidly induced apoptosis in these cells, which was initiated in the G1 phase and was preceded by an accumulation of cells in G1. A progression of the cells through G1/S appeared to be required, and a G1 block by starvation prevented cell death to a large extent.

Transiently transfected p53 was shown to induce apoptosis in the human transformed cell lines HeLa and Saos-2 (Yonish-Rouach et al. 1994), and cell death was not preceded by growth arrest (Haupt et al. 1995a; Yonish-Rouach et al. 1995). Cotransfection of Rb could rescue HeLa cells from p53-induced apoptosis, although cells were not arrested at the G1 phase (Haupt et al. 1995a). A protective effect of ectopically expressed p21/Waf1 against p53-induced apoptosis was observed in human melanoma cells infected with wt p53 and in p21-deficient mouse embryo fibroblasts (MEFs) overexpressing p53 (Gorospe et al. 1997). Thus, growth arrest signals may prevent p53-induced apoptosis.

While the outcome of p53 activation can be modified from apoptosis to growth arrest, it can also be modified from growth arrest to apoptosis. In a mouse embryo fibroblast line expressing the ts p53, activation of wt p53 following temperature downshift resulted in growth arrest in the G1 phase, but coexpression of wt p53 with transfected E2F-1 resulted in rapid apoptosis (Wu and Levine 1994).

4
Mechanism of p53-Induced Growth Arrest

Induction of a growth arrest by p53 was shown to depend on its ability to act as a sequence-specific transcriptional activator (Crook et al. 1994; Pietenpol et al. 1994). A number of genes have been identified which can be induced by p53, and an important target gene in the growth arrest pathway is WAF1/CIP1 (El-

Deiry et al. 1993; Harper et al. 1993). The protein product of this gene, p21, binds to cyclin-dependent kinases and inhibits their action, thereby blocking cell proliferation (Xiong et al. 1993). In addition, p21 also binds to and inhibits the proliferating-cell nuclear antigen (PCNA), a regulatory subunit of DNA polymerase δ (Waga et al. 1994). Thus, radiation-induced G1 arrest was shown to be selectively mediated by the p53-WAF1/CIP1 pathway in human thyroid cells (Namba et al. 1995), and a correlation was observed between G1 arrest and the stability of the p53 and p21 proteins following γ-irradiation of human lymphoma cells (Bae et al. 1995). Moreover, mice lacking p21 were shown to be defective in the G1 checkpoint control (Brugarolas et al. 1995; Deng et al. 1995). However, the G1 checkpoint was only partially impaired, indicating that p21 does not play an exclusive role in this pathway. An alternative mechanism for p53-induced growth arrest was confirmed in *in vitro* experiments: a human ts p53 mutant could arrest rat embryo fibroblasts in G1 at the permissive temperature without induced expression of WAF1/CIP1 (Hirano et al. 1995); and the extent of growth arrest in Saos-2 cells expressing inducible p53 was greater than in cells expressing inducible p21 (Chen et al. 1996).

Other p53-target genes may also play a role in the induction of growth arrest. Gadd45, which was originally identified through its induction by DNA damage (Forance et al. 1989), was shown to be directly transactivated by p53 (Kastan et al. 1992) and to suppress growth arrest in human tumor cell lines (Zhan et al. 1994b). Similarly to p21, Gadd45 can bind PCNA (Smith et al. 1995) and may also directly interact with p21 (Kearsey et al. 1995), thereby competing with it for interaction with PCNA (Chen et al. 1995). Induction of Gadd45 by p53 may therefore play a role in the regulation of the cell cycle.

Additional potential players in the control of cell cycle which are transactivated by p53 are cyclin G (Okamoto and Beach 1994; Zauberman et al. 1995), a possible subunit of a protein-dependent kinase, and mdm-2, which binds directly to the N-terminus of p53 and can serve as a negative feedback regulator by blocking transactivation and permitting reentry to the cell cycle (Barak et al. 1993; Chen et al. 1994).

Mechanisms other than transactivation of genes may also contribute to this function of p53, such as repression of specific target genes. Thus, p53-induced G2 arrest may be linked to its ability to downregulate the expression of DNA topoisomerase II alpha, an enzyme involved in DNA replication, transcription, recombination and mitosis (Wang et al. 1997).

5
p53-Induced Apoptotic Pathways

While transcriptional activation appears to be essential for the induction of growth arrest by p53, whether it is necessary at all for the induction of apoptosis is still a controversial issue. It has been reported that p53 may induce apoptosis in the absence of *de novo* RNA and protein synthesis (Caelles et al.

1994; Wagner et al. 1994), suggesting that transactivation of target genes is dispensable in this pathway. Moreover, a truncated murine p53 containing only the first 214 residues (p53dl214) or its human equivalent Trunc223, both of which are transactivation-deficient, were able to induce apoptosis in HeLa cells (Haupt et al. 1995b; Yonish-Rouach et al. 1996), while the human p53 mutant p53175P which retains the transcriptional activation function for certain target genes could not induce apoptosis in Saos-2 cells and in transformed REFs (Rowan et al. 1996). Direct microinjection of a p53 carboxy-terminal-derived peptide (amino acid residues 319–393) resulted in apoptosis of primary normal human fibroblasts which was suggested to be mediated through the binding of the p53 peptide to XPB and XPD (Wang et al. 1996) and to be independent of transcriptional activation. Finally, a function in apoptosis other than transactivation was reported to be inherent in the N-terminal domain of p53 since replacing this domain by the VP16 transactivation domain did not restore the ability of the protein to induce apoptosis, although the hybrid was fully competent for transactivation of at least some genes (Theis et al. 1997). On the other hand, several reports have suggested that transcriptional activation may play a role in p53-mediated apoptosis. Thus, inhibition of *de novo* protein synthesis partially blocked cell death induced in the ts p53-M1 cells at the permissive temperature (Yonish-Rouach et al. 1991). In baby rat kidney (BRK) cell lines transformed by E1A and the ts p53, temperature downshift induced rapid apoptosis. In BRK cell lines expressing E1A and a transcriptionally defective ts p53, mutated in codons 22 and 23 (Lin et al. 1994), apoptosis was practically abolished, demonstrating that p53-mediated transcription plays an essential role in E1A-induced apoptosis (Sabbatini et al. 1995). In HeLa cells, inhibition of protein synthesis by cycloheximide or of transcription by actinomycin D significantly reduced the ability of wt p53 to induce apoptosis (Yonish-Rouach et al. 1995). Finally, microinjection of different p53 mutants into p53 null primary mouse embryo fibroblasts (MEFs) has demonstrated that transcriptional activation was critical for the induction of apoptosis (Attardi et al. 1996). It therefore appears that p53 may mediate apoptosis through transactivation-dependent and transactivation-independent pathways.

The Bax gene, a member of the Bcl-2 family which accelerates apoptosis by forming heterodimers with Bcl-2 (Oltvai et al. 1993), was suggested as a possible target in this apoptotic pathway, since it was upregulated by p53 in some cells (Selvakumaran et al. 1994; Zhan et al. 1994a; Miyashita and Reed 1995). Ectopic Bax expression induced apoptosis in a cell system where p53 was in the mutant conformation, while the E1B 19K protein could neutralize Bax function by binding to it and could block p53-mediated apoptosis (Han et al. 1996). In addition, Bax-deficiency was shown to promote drug resistance and oncogenic transformation by attenuating p53-dependent apoptosis in primary fibroblasts expressing E1A (McCurrach et al. 1997). Finally, in a transgenic mouse brain tumor, p53-dependent expression of Bax was shown to be induced in slow-growing apoptotic tumors while tumor growth was accelerated and apoptosis dropped by 50% in Bax-deficient mice, indicating that Bax was required for a full p53-mediated response (Yin et al. 1997). However, induction of apoptosis by p53 was shown to occur in several cell systems without upregulation of Bax (Allday et al. 1995; Canman et al. 1995; Rowan et al. 1996), and overexpression of Bax failed to restore apoptosis after DNA damage in the absence of p53 (Brady et al. 1996). In addition, analysis of cells from Bax null mice indicated that thymocyte death induced by γ-irradiation, which is p53-dependent, was normal in these mice (Knudson et al. 1995). Thus, the involvement of Bax as a mediator of p53-induced apoptosis may depend on the cell type and the apoptotic stimulus, and in any case other p53-controlled factors would act in synergy with Bax to promote a full apoptotic response. It is also noteworthy that Bcl-2 was shown to be downregulated by p53 in some cells (Hadlar et al. 1994; Miyashita et al. 1994), and the equilibrium between Bcl-2 and Bax may be orchestrated by p53 in cells undergoing apoptosis. Indeed, both downregulation of Bcl-2 and upregulation of Bax were observed 2–5h post-irradiation in childhood ALL cell lines expressing wt p53 (Findley et al. 1997). Bcl-2 was also reported to compete with p53 for 53BP2 binding, suggesting a role for both Bcl-2 and 53BP2 in p53-mediated apoptosis (Naumovski and Cleary 1996). Another member of the Bcl-2 family, Bcl-XL, was shown to protect cells from p53-mediated apoptosis (Schott et al. 1995), although the levels of Bcl-XL are apparently independent of p53 expression (Findley et al. 1997).

Another potential mediator of the apoptotic pathway induced by p53 is Fas/APO-1 (CD95), a member of the tumor necrosis factor receptor superfamily. CD95 is able to transduce a signal for apoptosis upon engagement by its ligand or specific antibodies (reviewed by Nagata and Golstein 1995), and its expression was shown to be induced by p53 (Owen-Schaub et al. 1995). In this context, irradiation and chemotherapeutic agents have been reported to enhance CD95 expression and/or alter the sensitivity to anti-Fas mediated killing (Morimoto et al. 1993; Owen-Schaub et al. 1995), and drug-induced apoptosis in hepatoma cells was shown to be mediated by the CD95 receptor/ligand system and to involve p53 activation (Muller et al. 1997).

A third p53 target which might play a role in apoptosis is IGFBP-3 (Buckbinder et al. 1995), which is an antagonist of insulin-like growth factor-1 (IGF-1) (Baserga 1994). Downregulation of IGF-1 activity or of IGR-1 receptor levels has been shown to correlate with an apoptotic response in some cases (Baserga 1994). Endogenous IGF-1 receptor was recently shown to have a specific antiapoptotic signaling capacity and to protect fibroblasts from UV-induced apoptosis (Kulik et al. 1997). In this context, wt p53 was reported to repress transcription from the IGF-1R promoter (Werner et al. 1996), and to decrease the level of IGF-1R in 32D cells undergoing apoptosis upon IL-3 withdrawal (Prisco et al. 1997). Thus, the IGF-1R signaling pathway may participate in p53-mediated apoptosis.

Other p53 target genes which may be involved in the apoptotic pathway include EI24, which is induced by p53 in etoposide-treated murine NIH3T3 cells as well as in primary thymocytes after irradiation (Lehar et al. 1996); PAG608 which encodes a nuclear zinc finger protein and which is induced by DNA damage in a p53-dependent manner (Israeli et al. 1997); and Wip1, a novel human protein phosphatase which accumulates following irradiation in the presence of wt p53 (Fiscella et al. 1997).

The downstream biochemical events in p53-induced apoptosis may be common to other apoptotic pathways. Thus, a family of cysteine proteases (designated caspases) homologous to the nematode worm *Caenorhabditis elegans* death-promoting gene Ced-3 was shown to play an important role in apoptosis (reviewed by White 1996). Caspases exist in normal cells mainly as inactive proforms which require processing at critical aspartate residues for activation, and several cellular proteins have been identified as potential substrates (reviewed by Martin and Green 1995; Alnemri 1997). Caspase activity was reported to be essential for p53-mediated transcriptionally dependent apoptosis in BRK cells transformed by E1A and the ts p53 (Sabbatini et al. 1997), and lamins were identified as substrates for cleavage by the activated caspase (Rao et al. 1996). Activation of caspase-3 and caspase-7 was demonstrated in M1 cells expressing the ts p53 upon temperature downshift (Chandler et al. 1997). A possible substrate for caspase-3 is a novel protein designated DNA fragmentation factor (DFF), which was reported to be activated in apoptotic cells after being cleaved by caspase-3 and to produce DNA fragmentation characteristic of apoptosis (Liu et al. 1997). Additional substrates for caspases during p53-mediated apoptosis include actin (Guenal et al. 1997) and mdm2 (Erhardt et al. 1997).

6
Making Decisions: Growth Arrest or Apoptosis?

Whether a cell would undergo growth arrest or apoptosis following p53 activation appears to depend on a variety of factors, such as environmental conditions and the cell type. Irradiation of normal human fibroblasts resulted in a prolonged G1 arrest (Di Leonardo et al. 1994), while a similar dose of irradiation induced apoptosis in thymocytes (Clarke et al. 1993; Lowe et al. 1993b). On the other hand, fibroblasts expressing the adenovirus E1A gene were sensitized to irradiation-induced apoptosis (Lowe et al. 1993a), and the presence of survival factors could shift the response from apoptosis to growth arrest as demonstrated in the various examples above. The loss of the tumor suppressor pRb function may contribute to p53-induced apoptosis, since the activity of pRb and/or other pRb-related proteins was shown to be necessary for the induction of a G1 arrest by p53 following DNA damage (Demers et al. 1994; Hickman et al. 1994; Slebos et al. 1994; White et al. 1994), and pRb may have a protective effect on p53-induced apoptosis (Qin et al. 1994; Haupt et al. 1995a). In cells having a functional pRb, induction of p21 would lead to inactivation of cyclin-dependent kinases and therefore to inhibition of pRb phosphorylation. The hypophosphorylated pRb retains transcription factors of the E2F family, which are necessary for the G1/S transition, thus imposing a p53-induced G1 arrest. In the absence of functional pRb, p21 will still be induced by p53 activation, but cells will be unable to growth arrest and may therefore be "forced" to die by entering into S phase. Interestingly, pRb cleavage following caspases activation in several apoptotic pathways was reported (An and Dou 1996; Tan et al. 1997), suggesting that apoptosis may be incompatible with functional Rb. In this context it was reported that p53 induction by antineoplastic drugs led to apoptosis in Rb−/− MEFs, while Rb+/− and Rb+/+ MEFs underwent cell cycle arrest without apoptosis (Almasan et al. 1995). Along with this concept, deregulated expression of E2F was shown to induce p53-mediated apoptosis (Qin et al. 1994; Shan and Lee 1994; Wu and Levine 1994; Almasan et al. 1995; Logan et al. 1995). E2F1-DP1 complex was reported to bind to and to induce p53, thereby overriding survival factors to induce apoptosis (Hiebert et al. 1995; O'Connor et al. 1995). Without both Rb and p53, E2F activation would stimulate cell proliferation and permit tumor formation, as was demonstrated by the development of retinal tumors in HPV E7 transgenic mice (Howes et al. 1994; Pan and Griep 1994). Binding of WT1 to p53, on the other hand, was reported to stabilize it, increase transactivation and permit growth arrest, yet to inhibit transrepression and apoptosis (Maheswaran et al. 1995).

However, although in some cases the cell may "try" to growth arrest before undergoing apoptosis if it cannot, this may not always be the case. Cells from the p21 null mice were shown to undergo a normal p53-induced apoptosis, suggesting that this function – unlike the p53-induced G1 arrest which was impaired – does not require activation of this kinase inhibitor (Brugarolas et al. 1995; Deng et al. 1995). The observation that the p21 null mice did not show an enhanced incidence of tumor development suggests that the role of p53 in the induction of apoptosis is more important for its tumor-suppressing activity than the G1 arrest. A support to this model comes from a recent study in which the tumor-derived mutant of p53 175P was shown to retain the ability to activate p21 and to induce growth arrest, while it was unable to induce apoptosis (Rowan et al. 1996). This mutant was also unable to suppress transformation of rat embryo fibroblasts by E7 and ras, demonstrating the correlation between induction of apoptosis and the tumor suppressing function of p53 (Rowan et al. 1996). An extension of the study by the same group demonstrated that the p53 175P mutant was unable to transactivate the Bax and the IGF-BP3 gene promoters, while the p53 181L mutant, which is also unable to induce apoptosis yet capable of transactivating the Waf1 promoter, was defective in activation of a promoter containing IGF-BP3 box B sequences (Ludwig et al. 1996). Similar results were observed using the human ts mutant p53Ala143, which was capable of transactivating several p53-responsive promoters at the permissive temperature but unable to transactivate the Bax and

IGF-BP3 gene promoters and failed to induce apoptosis (Friedlander et al. 1996). Thus, p53 may differentially transactivate distinct families of target genes in the context of specific signals, and the outcome of such transactivation will depend on the responsive genes.

The effect of external signals on the outcome of p53 activation was demonstrated in a recent report, in which it was shown that the cellular level of p53 could dictate the response of the cells. Lower levels of p53 resulted in growth arrest while higher levels resulted in apoptosis, yet DNA damage could favor an apoptotic response without altering the protein level of p53 in the cells (Chen et al. 1996). It is possible that higher levels of p53 are necessary for an efficient binding and transactivation of apoptosis-specific target genes and/or for association with other proteins in a way which would lead to apoptosis. One could imagine a scenario in which a tight association of p53 with, for example, inhibitors of proteases, would require an elevated expression of the protein to set on the apoptotic pathway.

The complexity of the decision between growth arrest and apoptosis is demonstrated in another report, in which low levels of wild-type p53 were shown to have antiapoptotic activity in immortalized fibroblasts undergoing apoptosis induced by serum withdrawal (Lassus et al. 1996). This activity may be related to a possible role of p53 in embryonic development (Lassus et al. 1996).

Genetic studies in colorectal cancer cell lines indicated that the apoptotic response to p53 in these cells was modulated by at least two factors: p21-mediated growth arrest that could protect cells from apoptosis in lines which respond by growth arrest, and *trans*-acting factor(s) that could overcome this protection, resulting in apoptotic cell death in the death-sensitive cells (Polyak et al. 1996). These results are consistent with the possibility of transactivation of a death gene by p53 in some cell types undergoing p53-mediated apoptosis.

The reports that p53 may induce apoptosis in the absence of transcriptional activation in some cells (Caelles et al. 1994; Wagner et al. 1994; Haupt et al. 1995b), while activation of specific target genes may play a role in the induction of apoptosis in other cells (Yonish-Rouach et al. 1991, 1995; Sabbatini et al. 1995; Attardi et al. 1996), or even in the same cells under some other conditions (Haupt et al. 1995b; Yonish-Rouach et al. 1995; Rowan et al. 1996), suggest that several pathways may be involved in p53-mediated apoptosis (Fig. 2). Thus, p53 may activate a G1 checkpoint which will be retained in case of a functional pRb and p21, or will result in apoptosis in case of signals for progression into the cell cycle or due to dominant death factor(s). It may also directly activate apoptosis through the transactivation of specific genes, downregulation of some other genes or interaction with proteins. Such signals may be controlled by cytokines or by members of the Bcl-2 family. The choice between a G1 arrest and one of the apoptotic pathways, and the final outcome, will depend on a complex network of regulatory signals.

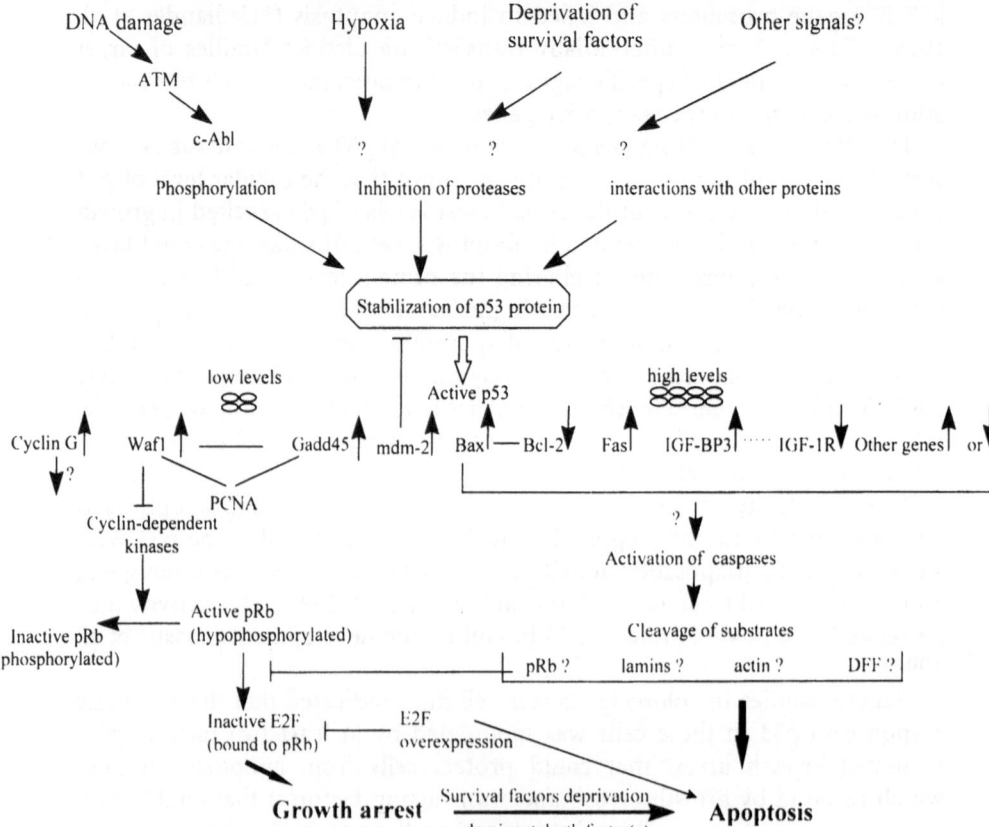

Fig. 2. Possible models for p53-mediated G1 arrest and/or apoptosis. The p53 protein has a short half-life and is stabilized following external stimuli such as DNA damage. The active protein may act as a transcription factor to exert its biological functions. Transcriptional activation of WAF1 plays a role in activation of G1 arrest via inhibition of cyclin-dependent kinases by p21[WAF1] protein, activation of pRb (and related proteins) and inhibition of E2F. Transcriptional activation of Gadd45 may also contribute to G1 arrest through its interaction with proliferating cell nuclear antigen (PCNA) and/or Waf1, and a potential role may be attributed to cyclin G. Several pathways may be involved in induction of apoptosis, some of which can be related to G1 arrest. pRb inactivation by HPV E7, or E2F overexpression, may result in inability to implement p53-mediated G1 arrest and result in apoptosis. Other pathways may lead directly to apoptosis regardless of cell cycle function, presumably when p53 is expressed to high levels, and may involve up- or downregulation of genes or interaction of p53 with other proteins. Potential target genes are listed and referred to in the text. The apoptotic pathway involves activation of caspases and cleavage of specific substrates

References

Abrahamson JL, Lee JM, Bernstein A (1995) Regulation of p53-mediated apoptosis and cell cycle arrest by Steel factor. Mol Cell Biol 15:6953–6960

Adler V, Pincus MR, Minamoto T, Fuchs SY, Bluth MJ, Brandt-Rauf PW, Friedman FK, Robinson RC, Chen JM, Wang XW, Harris CC, Ronai Z (1997) Conformation-dependent phosphorylation of p53. Proc Natl Acad Sci USA 94:1686–1691

Allday MJ, Inman GJ, Crawford DH, Farrell PJ (1995) DNA damage in human B cells can induce apoptosis, proceeding from G1/S when p53 is transactivation competent and G2/M when it is transactivation defective. EMBO J 14:4994–5005

Almasan A, Yin Y, Kelly RE, Lee EY, Bradley A, Li W, Bertino JR, Wahl GM (1995) Deficiency of retinoblastoma protein leads to inappropriate S-phase entry, activation of E2F-responsive genes, and apoptosis. Proc Natl Acad Sci USA 92:5436–5440

Almog N, Li R, Peled A, Schwartz D, Wolkowicz R, Goldfinger N, Pei H, Rotter V (1997) The murine C'-terminally alternatively spliced form of p53 induces attenuated apoptosis in myeloid cells. Mol Cell Biol 17:713–722

Alnemri ES (1997) Mammalian cell death proteases: a family of highly conserved aspartate specific cysteine proteases. J Cell Biochem 64:33–42

Aloni-Grinstein R, Schwartz D, Rotter V (1995) Accumulation of wild-type p53 protein upon γ-irradiation induces a G2 arrest-dependent immunoglobulin k light chain gene expression. EMBO J 14:1393–1401

An B, Dou QP (1996) Cleavage of retinoblastoma protein during apoptosis: an interleukin 1 beta-converting enzyme-like protease as candidate. Cancer Res 56:438–442

Atadja P, Wong IH, Garkavtsev Veillette C, Riabowol K (1995) Increased activity of p53 in senescing fibroblasts. Proc Natl Acad Sci USA 92:8348–8352

Attardi LD, Lowe SW, Brugarolas J, Jacks T (1996) Transcriptional activation by p53, but not induction of the p21 gene, is essential for oncogene-mediated apoptosis. EMBO J 15:3702–3712

Bae I, Fan S, Bhatia K, Kohn KW, Fornace AJ Jr, O'Connor PM (1995) Relationships between G1 arrest and stability of the p53 and p21Cip1/Waf1 proteins following gamma-irradiation of human lymphoma cells. Cancer Res 55:2387–2393

Baker SJ, Fearon ER, Nigro JM, Hamilton SR, Preisinger AC, Jessup JM, vanTuinen P, Ledbetter DH, Barker DF, Nakamura Y, et al. (1989) Chromosome 17 deletions and p53 gene mutations in colorectal carcinomas. Science 244:217–221

Balkalkin G, Yakovleva T, Selivanova G, Magnusson KP, Szekely L, Kiseleva E, Klein G, Terenius L, Wiman KG (1994) p53 binds single-stranded DNA ends and catalyzes DNA renaturation and strand transfer. Proc Natl Acad Sci USA 91:413–417

Barak Y, Juven T, Haffner R, Oren M (1993) Mdm-2 expression is induced by wild-type p53 activity. EMBO J 12:461–468

Bargonetti J, Manfredi JJ, Chen X, Marshak DR, Prives CA (1993) Proteolytic fragment from the central region of p53 has marked sequence-specific DNA-binding activity when generated from wild-type but not from oncogenic mutant p53 protein. Genes Dev 7:2565–2574

Baserga R (1994) Oncogenes and the strategy of growth factors. Cell 79:927–939

Baskaran R, Wood LD, Whitaker LL, Canman CE, Morgan SE, Xu Y, Barlow C, Baltimore D, Wynshaw-Boris A, Kastan MB, Wang JY (1997) Ataxia telangiectasia mutant protein activates c-Abl tyrosine kinase in response to ionizing radiation. Nature 387:516–519

Bayle JH, Elenbaas B, Levine AJ (1995) The carboxyl-terminal domain of the p53 protein regulates sequence-specific DNA binding through its nonspecific nucleic acid-binding activity. Proc Natl Acad Sci USA 92:5729–5733

Blaydes JP, Gire V, Rowson JM, Wynford-Thomas D (1997) Tolerance of high levels of wild-type p53 in transformed epithelial cells dependent on auto-regulation by mdm-2. Oncogene 14:1859–1868

Bourdon JC, Deguin-Chambon V, Lelong JC, Dessen P, May P, Debuire B, May E (1997) Further characterisation of the p53 responsive element – identification of new candidate genes for trans-activation by p53. Oncogene 14:85–94

Brady HJ, Salomons GS, Bobeldijk RC, Berns AJ (1996) T cells from baxalpha transgenic mice show accelerated apoptosis in response to stimuli but do not show restored DNA damage-induced cell death in the absence of p53 gene product. EMBO J 15:1221–1230

Braithwaite AW, Blair GE, Nelson CC, McGoven J, Bellett AJD (1991) Adenovirus E1b-58kD antigen binds to p53 during infection of rodent cells: evidence for an N-terminal binding site on p53. Oncogene 6:781–787

Brugarolas J, Chandrasekaran, C, Gordon JI, Beach D, Jacks T, Hannon JG (1995) Radiation-induced cell cycle arrest compromised by p21 deficiency. Nature 377:552–557

Buckbinder L, Talbott, R, Velasco-Miguel S, Takenaka I, Faha B, Seizinger R, Kley N (1995) Induction of the growth inhibitor IGF-binding protein 3 by p53. Nature 377:646–649

Caelles C, Helmberg A, Karin M (1994) p53-dependent apoptosis in the absence of transcriptional activation of p53-target genes. Nature 370:220–223

Canman CE, Gilmer TM, Coutts SB, Kastan MB (1995) Growth factor modulation of p53-mediated growth arrest versus apoptosis. Genes Dev 9:600–611

Chandler JM, Alnemri ES, Cohen GM, MacFarlane M (1997) Activation of CPP32 and Mch3 alpha in wild-type p53-induced apoptosis. Biochem J 322:19–23

Chang J, Kim DH, Lee SW, Choi KY, Sung YC (1995) Transactivation ability of p53 transcriptional activation domain is directly related to the binding affinity to TATA-binding protein. Biol Chem 270:25014–25019

Chen CY, Oliner JD, Zhan Q, Forance AJ, Vogelstein B, Kastan MB (1994) Interactions between p53 and MDM2 in a mammalian cell cycle checkpoint pathway. Proc Natl Acad Sci USA 91:2684–2688

Chen IT, Smith ML, O'Connor PM, Fornace AJ (1995) Direct interaction of Gadd45 with PCNA and evidence for competitive interaction of Gadd45 and p21Wafl/Cip1 with PCNA. Oncogene 11:1931–1937

Chen X, Farmer G, Zhu H, Prywes R, Prives C (1993) Cooperative DNA binding of p53 with TFIID (TBP): a possible mechanism for transcriptional activation. Genes Dev 7:1837–1849

Chen X, Ko LJ, Jayaraman L, Prives C (1996) p53 levels, functional domains, and DNA damage determine the extent of the apoptotic response of tumor cells. Genes Dev 10:2438–2451

Chiou S-K, Rao L, White E (1994) Bcl-2 blocks p53-dependent apoptosis. Mol Cell Biol 14:2556–2563

Cho Y, Gorina S, Jeffrey PD, Pavletich NP (1994) Crystal structure of a p53 tumor suppressor-DNA complex: understanding tumorigenic mutations. Science 265:346–355

Ciechanover A, Shkedy D, Oren M, Bercovich B (1994) Degradation of the tumor suppressor protein p53 by the ubiquitin-mediated proteolytic system requires a novel species of ubiquitin-carrier protein, E2. J Biol Chem 269:9582–9589

Clarke AR, Purdie CA, Harrison DJ, Morris RG, Bird CC, Hooper ML, Wyllie AH (1993) Thymocyte apoptosis induced by p53-dependent and independent pathways. Nature 362:849–852

Clarke AR, Gledhill S, Hooper ML, Bird CC, Wyllie AH (1994) p53 dependence of early apoptotic and proliferative responses within the mouse intestinal epithelium following gamma-irradiation. Oncogene 9:1767–1773

Clore GM, Omichinski JG, Sakaguchi K, Zambrano N, Sakamoto H, Appella E, Gronenborn AM (1994) High-resolution structure of the oligomerization domain of p53 by multidimensional NMR. Science 265:386–391

Clore GM, Omichinski JG, Sakaguchi K, Zambrano N, Sakamoto H, Appella E, Gronenborn AM (1995) Interhelical angles in the solution structure of the oligomerization domain of p53: correction. Science 267:1515–1516

Collins MKL, Marvel J, Malde P, Lopez-Rivas A (1992) Interleukin 3 protects murine bone marrow cells from apoptosis induced by DNA damaging agents. J Exp Med 171:1043–1051

Crook T, Marston NJ, Sara EA, Vousden KH (1994) Transcriptional activation by p53 correlates with suppression of growth but not transformation. Cell 79:817–827

Cross SM, Sanchez CA, Morgan CA, Schimke MK, Ramel S, Idzerda RL, Raskind WH, Reid BJ (1995) A p53-dependent mouse spindle checkpoint. Science 267:1353–1356

Culotta E, Koshland DE Jr (1993) p53 sweeps through cancer research. Science 262:1958–1961

Dameron KM, Volpert OV, Tainsky MA, Bouck N (1994) Control of angiogenesis in fibroblasts by p53 reglation of thrombospondin-1. Science 265:1582–1584

Debbas M, White E (1993) Wild-type p53 mediates apoptosis by E1A, which is inhibited by E1B. Genes Dev 7:546–554

DeLeo AB, Jay G, Appella E, Dubois GC, Law LW, Old LJ (1979) Detection of a transformation-related antigen in chemically induced sarcomas and other transformed cells of the mouse. Proc Natl Acad Sci USA 76:2420–2424

Delia D, Goi K, Mizutani S, Yamada T, Aiello A, Fontanella E, Lamorte G, Iwata S, Ishioka C, Krajewski S, Reed JC, Pierotti MA (1997) Dissociation between cell cycle arrest and apoptosis can occur in Li-Fraumeni cells heterozygous for p53 gene mutations. Oncogene 14:2137–2147

Demers GW, Foster SA, Halbert CL, Galloway DA (1994) Growth arrest by induction of p53 in DNA damaged keratinocytes is bypassed by human papillomavirus 16 E7. Proc Natl Acad Sci USA 91:4382–4386

Deng C, Zhang P, Harrper JW, Elledge SJ, Leder P (1995) Mice lacking p21$^{PCII/WAFI}$ undergo normal development, but are defective in G1 checkpoint control. Cell 82:675–684

Di Leonardo A, Linke SP, Clarkin K, Wahl GM (1994) DNA damage triggers a prolonged p53-dependent G1 arrest and long-term induction of Cip1 in normal human fibroblasts. Genes Dev 8:2540–2551

Diller L, Kassel J, Nelson CE, Gryka MA, Litwak G, Gebhardt M, Bressac B, Ozturk M, Baker SJ, Vogelstein B, Friend SH (1990) p53 functions as a cell cycle control protein in osteosarcomas. Mol Cell Biol 10:5772–5781

Dobner T, Horikoshi N, Rubenwolf S, Shenk T (1996) Blockage by adenovirus E4orf6 of transcriptional activation by the p53 tumor suppressor. Science 272:1470–1473

Donehower LA, Bradley A (1993) The tumor suppressor p53. Biochim Biophys Acta 1155:181–205

Donehower LA, Harvey M, Slagle BL, McArthur MJ, Montgomery CA Jr, Butel JS, Bradley A (1992) Mice deficient for p53 are developmentally normal but susceptible to spontaneous tumours. Nature 356:215–221

Dutta A, Ruppert JM, Aster JC, Winchester E (1993) Inhibition of DNA replication factor RPA by p53. Nature 365:79–82

El-Deiry WS, Kern SE, Pietenpol JA, Kinzler KW, Vogelstein B (1992) Definition of a consensus binding site for p53. Nat Genet 1:45–49

El-Deiry WS, Tokino T, Velculenscu VE, Levy DB, Parsons R, Trent JM, Lin D, Mercer EW, Kinzler KW, Vogelstein B (1993) WAF1, a potential mediator of p53 tumor suppression. Cell 71:817–825

Eliyahu D, Raz A, Gruss P, Givol D, Oren M (1984) Participation of p53 cellular tumour antigen in transformation of normal embryonic cells. Nature 312:646–649

Eliyahu D, Goldfinger N, Pinhasi-Kimhi O, Shaulsky G, Skurnik Y, Arai N, Rotter V, Oren M (1988) Meth A fibrosarcoma cells express two transforming mutant p53 species. Oncogene 3:313–321

Erhardt P, Tomaselli KJ, Cooper GM (1997) Identification of the MDM2 oncoprotein as a substrate for CPP32-like apoptotic proteases. J Biol Chem 272:15049–15052

Farmer G, Bargonetti J, Zhu H, Friedman P, Prywes R, Prives C (1992) Wild-type p53 activates transcription in vitro. Nature 358:83–86

Fields S, Jang SK (1990) Presence of a potent transcription activating sequence in the p53 protein. Science 249:1046–1049

Findley HW, Gu L, Yeager AM, Zhou M (1997) Expression and regulation of Bcl-2, Bcl-xl, and Bax correlate with p53 status and sensitivity to apoptosis in childhood acute lymphoblastic leukemia. Blood 89:2986–2993

Finlay CA, Hinds PW, Tan TH, Eliyahu D, Oren M, Levine AJ (1988) Activating mutations for transformation by p53 produce a gene product that forms an hsc70-p53 complex with an altered half-life. Mol Cell Biol 8:531–539

Fiscella M, Ullrich SJ, Zambrano N, Shields MT, Lin D, Lees-Miller SP, Anderson CW, Mercer WE, Appella E (1993) Mutation of the serine 15 phosphorylation site of human p53 reduces the ability of p53 to inhibit cell cycle progression. Oncogene 8:1519–1528

Fiscella M, Zhang H, Fan S, Sakaguchi K, Shen S, Mercer WE, Vande Woude GF, O'Connor PM, Appella E (1997) Wip1, a novel human protein phosphatase that is induced in response to ionizing radiation in a p53-dependent manner. Proc Natl Acad Sci USA 94:6048–6053

Forance AJ, Nebert DW, Hollander MC, Luethy JD, Papathanasiou M, Fargnoli J, Holbrook NJ (1989) Mammalian genes coordinately regulated by growth arrest signals and DNA-damaging agents. Mol Cell Biol 9:4196–4203

Friedlander P, Haupt Y, Prives C, Oren M (1996) A mutant p53 that discriminates between p53-responsive genes cannot induce apoptosis. Mol Cell Biol 16:4961–4971

Funk WD, Pak DT, Karas RH, Wright WE, Shay JW (1992) A transcriptionally active DNA-binding site for human p53 protein complexes. Mol Cell Biol 12:2866–2871

Ginsberg D, Mechta F, Yaniv M, Oren M (1991) Wild-type p53 can down-modulate the activity of various promoters. Proc Natl Acad Sci USA 88:9979–9983

Goga A, Liu X, Hambuch TM, Senechal K, Major E, Berk AJ, Witte ON, Sawyers CL (1995) p53 dependent growth suppression by the c-Abl nuclear tyrosine kinase. Oncogene 11:791–799

Gonen H, Shkedy D, Barnoy S, Kosower NS, Ciechanover A (1997) On the involvement of calpains in the degradation of the tumor suppressor protein p53. FEBS Lett 406:17–22

Gorospe M, Cirielli C, Wang X, Seth P, Capogrossi MC, Holbrook NJ (1997) p21(Waf1/Cip1) protects against p53-mediated apoptosis of human melanoma cells. Oncogene 14:929–935

Gu W, Shi XL, Roeder RG (1997) Synergistic activation of transcription by CBP and p53. Nature 387:819–823

Guenal I, Risler Y, Mignotte B (1997) Down-regulation of actin genes precedes microfilament network disruption and actin cleavage during p53-mediated apoptosis. J Cell Sci 110:489–495

Guillouf C, Grana X, Selvakumaran M, De Luca XA, Giordano A, Hoffman B, Liebermann DA (1995) Dissection of the genetic programs of p53-mediated G1 growth arrest and apoptosis: blocking p53-induced apoptosis unmasks G1 arrest. Blood 85:2691–2698

Hadlar S, Negrini M, Monne M, Sabbioni S, Croce CM (1994) Down regulation of bcl-2 by p53 in breast cancer cells. Cancer Res 54:2095–2097

Hainaut P, Milner J (1993a) A structural role for metal ions in the "wild-type" conformation of the tumor suppressor protein p53. Cancer Res 53:1739–1742

Hainaut P, Milner J (1993b) Redox modulation of p53 conformation and sequence-specific DNA binding in vitro. Cancer Res 53:4469–4473

Halazonetis TD, Kandil AN (1993) Wild-type p53 adopts a "mutant"-like conformation when bound to DNA. EMBO J 12:1021–1028

Hall SR, Campbell LE, Meek DW (1996) Phosphorylation of p53 at the casein kinase II site selectively regulates p53-dependent transcriptional repression but not transactivation. Nucleic Acids Res 24:1119–1126

Han J, Sabbatini P, Perez D, Rao L, Modha D, White E (1996) The E1B 19K protein blocks apoptosis by interacting with and inhibiting the p53-inducible and death-promoting Bax protein. Genes Dev 10:461–477

Hansen R, Reddel R, Braithwaite A (1995) The transforming oncoproteins determine the mechanism by which p53 suppresses cell transformation: pRb-mediated growth arrest or apoptosis. Oncogene 11:2535–2545

Harper JW, Adami GR, Wei N, Keyomarsi K, Elledge SJ (1993) The p21 Cdk-interacting protein Cip 1 is a potent inhibitor of G1 cyclin-dependent kinases. Cell 75:805–816

Harvey M, Vogel H, Moris D, Bradeley A, Bernstein A, Donehower LA (1995) A mutant p53 transgene accelerates tumour development in heterozygous but not nullizygous p53-deficient mice. Nat Genet 9:305-311

Haupt Y, Rowan S, Oren M (1995a) p53-mediated apoptosis in HeLa cells can be overcome by excess pRb. Oncogene 10:1563-1571

Haupt Y, Rowan S, Shaulian E, Vousden KH, Oren M (1995b) Induction of apoptosis in HeLa cells by *trans*-activation-deficient p53. Genes Dev 9:2170-2183

Haupt Y, Maya R, Kazaz A, Oren M (1997) Mdm2 promotes the rapid degradation of p53. Nature 387:296-299

He Z, Brinton BT, Greenblatt J, Hassell JA, Ingles CJ (1993) The transactivator proteins VP16 and GAL4 bind replication factor A. Cell 73:1223-1232

Hickman ES, Picksley SM, Vousden KH (1994) Cells expressing HPV16 E7 continue cell cycle progression following DNA damage induced p53 activation. Oncogene 9:2177-2181

Hiebert SW, Packham G, Strom DK, Haffner R, Oren M, Zambetti G, Cleveland JL (1995) E2F-1:DP-1 induces p53 and overrides survival factors to trigger apoptosis. Mol Cell Biol 15:6864-6874

Hinds P, Finlay C, Levine AJ (1989) Mutation is required to activate the p53 gene for cooperation with the ras oncogene and transformation. J Virol 63:739-746

Hirano Y, Yamato K, Tsuchida N (1995) A temperature sensitive mutant of the human p53, Val138, arrests rat cell growth without induced expression of cip1/waf1/sdi1 after temperature shift-down. Oncogene 10:1879-1885

Hollstein M, Sidranski D, Vogelstein B, Harris CC (1991) p53 mutations in human cancers. Science 253:49-53

Hollstein M, Rice K, Greenblatt MS, Soussi T, Fuchs R, Sorlie T, Hovig E, Smith-Sorensen B, Montesano R, Harris CC (1994) Database of p53 gene somatic mutations in human tumors and cell lines. Nucleic Acids Res 22:3551-3555

Howes KA, Ransom LN, Papermaster DS, Lasudry JGH, Albert DM, Windle JJ (1994) Analysis or retinoblastoma: alternative fates of photoreceptors expressing the HPV-16 E7 gene in the presence or absence of p53. Genes Dev 8:1300-1310

Hupp TR, Lane DP (1994) Allosteric activation of latent p53 tetramers. Curr Biol 4:865-875

Hupp TR, Meek DW, Midgley CA, Lane DP (1993) Activation of the cryptic DNA binding function of mutant forms of p53. Nucleic Acids Res 21:3167-3174

Hupp TR, Sparks A, Lane DP (1995) Small peptides activated the latent sequence-specific DNA binding function of p53. Cell 83:237-245

Israeli D, Tessler E, Haupt Y, Elkeles A, Wilder S, Amson R, Telerman A, Oren M (1997) A novel p53-inducible gene, PAG608, encodes a nuclear zinc finger protein whose overexpression promotes apoptosis. EMBO J 16:4384-4392

Iwabuchi K, Bartel PL, Li B, Marraccino R, Fields S (1994) Two cellular proteins that bind to wild-type but not mutant p53. Proc Natl Acad Sci USA 91:6098-6102

Jacks TL, Remington L, Williams BO, Schmitt EM, Halachmi S, Bronson RT, Weinberg RA (1994) Tumor spectrum analysis in p53-mutant mice. Curr Biol 4:1-7

Jayaraman J, Prives C (1995) Activation of p53 sequence-specific DNA binding by short single strands of DNA requires the p53 C-terminus. Cell 81:1021-1029

Jayaraman L, Murthy KG, Zhu C, Curran T, Xanthoudakis S, Prives C (1997) Identification of redox/repair protein Ref-1 as a potent activator of p53. Genes Dev 11:558-570

Jeffrey PD, Gorina S, Pavletich NP (1995) Crystal structure of the tetramerization domain of the p53 tumor suppressor at 1.7 ångströms. Science 267:1498-1502

Jiang D, Srinivasan A, Lozano G, Robbins PD (1993) SV40 T antigen abrogates p53-mediated transcriptional activity. Oncogene 8:2805-2812

Johnson P, Chung S, Benchimol S (1993) Growth suppression of Friend virus-transformed erythroleukemia cells by p53 protein is accompanied by hemoglobin production and is sensitive to erythropoietin. Mol Cell Biol 13:1456-1463

Kastan MB, Onyekwere O, Sidransky D, Vogelstein B, Craig RW (1991) Participation of p53 protein in the cellular response to DNA damage. Cancer Res 51:6304–6311

Kastan MB, Zhan Q, El-Deiry WS, Carrier F, Jacks T, Walsh WV, Plunkett BS, Vogelstein B, Fornace AJ (1992) A mammalian cell cycle checkpoint pathway utilizing p53 and GADD45 is defective in ataxia-telangiectasia. Cell 71:587–597

Kearsey JM, Coates PJ, Prescott AR, Warbrick E, Hall PA (1995) Gadd45 is a nuclear cell cycle regulated protein which interacts with p21Cip1. Oncogene 11:1675–1683

Kern SE, Kinzler KW, Bruskin A, Jarosz D, Friedman P, Prives C, Vogelstein B (1991) Identification of p53 as a sequence-specific DNA binding protein. Science 252:1708–1711

Knudson C, Tunk K, Tourtellotte W, Brown G, Korsmeyer S (1995) Bax-deficient mice with lymphoid hyperplasia and male germ cell death. Science 270:96–99

Ko LJ, Prives C (1996) p53: puzzle and paradigm. Genes Dev 10:1054–1072

Komarova EA, Chernov MV, Franks R, Wang K, Armin G, Zelnick CR, Chin DM, Bacus SS, Stark GR, Gudkov AV (1997) Transgenic mice with p53-responsive lacZ: p53 activity varies dramatically during normal development and determines radiation and drug sensitivity in vivo. EMBO J 16:1391–1400

Kress M, May E, Cassingena R, May P (1979) Simian virus 40-transformed cells express new species of proteins precipitable by anti-simian virus 40 tumor serum. J Virol 31:472–483

Kubbutat MH, Vousden KH (1997) Proteolytic cleavage of human p53 by calpain: a potential regulator of protein stability. Mol Cell Biol 17:460–468

Kubbutat MH, Jones SN, Vousden KH (1997) Regulation of p53 stability by Mdm2. Nature 387:299–303

Kuerbitz SJ, Plunkett BS, Walsh WV, Kastan MB (1992) Wild-type p53 is a cell cycle checkpoint determinant following irradiation. Proc Natl Acad Sci USA 89:7491–7495

Kulik G, Klippel A, Weber MJ (1997) Antiapoptotic signalling by the insulin-like growth factor I receptor, phosphatidylinositol 3-kinase, and Akt. Mol Cell Biol 17:1595–1606

Lane DP (1992) p53, guardian of the genome. Nature 358:15–16

Lane DP, Crawford LV (1979) T antigen is bound to a host protein in SV40-transformed cells. Nature 278:261–263

Lassus P, Ferlin M, Piette J, Hibner U (1996) Anti-apoptotic activity of low levels of wild-type p53. EMBO J 15:4566–4573

Lee S, Elenbaas B, Levine A, Griffith J (1995) p53 and its 14 kDa C-terminal domain recognize primary DNA damage in the form of insertion/deletion mismatches. Cell 81:1013–1020

Lee W, Harvey TS, Yin Y, Yau P, Litchfield D, Arrowsmith CH (1994) Solution structure of the tetrameric minimum transforming domain of p53. Nat Struct Biol 1:877–890

Lehar SM, Nacht M, Jacks T, Vater CA, Chittenden T, Guild BC (1996) Identification and cloning of EI24, a gene induced by p53 in etoposide-treated cells. Oncogene 12:1181–1187

Leveillard T, Andera L, Bissonnette N, Schaeffer L, Bracco L, Egly JM, Wasylyk B (1996) Functional interactions between p53 and the TFIIH complex are affected by tumour-associated mutations. EMBO J 15:1615–1624

Levine AJ (1997) p53, the cellular gatekeeper for growth and division. Cell 88:323–331

Levy N, Yonish-Rouach E, Oren M, Kimchi A (1993) Complementation by wild-type p53 of interleukin-6 effects on M1 cells: induction of cell cycle exit and cooperativity with c-myc suppression. Mol Cell Biol 13:7942–7952

Lill NL, Grossman SR, Ginsberg D, DeCaprio J, Livingston DM (1997) Binding and modulation of p53 by p300/CBP coactivators. Nature 387:823–827

Lin J, Chen J, Elenbaas B, Levine AJ (1994) Several hydrophobic amino acids in the p53-amino-terminal domain are required for transcriptional activation, binding to mdm-2 and the adenovirus 5 E1B 55-kD protein. Genes Dev 8:1235–1246

Lin Y, Benchimol S (1995) Cytokines inhibit p53-mediated apoptosis but not p53-mediated G1 arrest. Mol Cell Biol 15:6045–6054

Linzer DI, Levine AJ (1979) Characterization of a 54k dalton cellular SV40 tumor antigen present in SV40-transformed cells and uninfected embryonal carcinoma cells. Cell 17:43–52

Liu X, Miller CW, Koeffler PH, Berk AJ (1993) The p53 activation domain binds the TATA box-binding polypeptide in Holo-TFIID, and a neighboring p53 domain inhibits transcription. Mol Cell Biol 13:3291-3300

Liu X, Zou H, Slaughter C, Wang X (1997) DFF, a heterodimeric protein that functions downstream of caspase-3 to trigger DNA fragmentation during apoptosis. Cell 89:175-184

Logan TJ, Evans DL, Mercer WE, Bjornsti MA, Hall DJ (1995) Expression of a deletion mutant of the E2F1 transcription factor in fibroblasts lengthens S phase and increases sensitivity to S phase-specific toxins. Cancer Res 55:2883-2891

Lopes UG, Erhardt P, Yao R, Cooper GM (1997) p53-dependent induction of apoptosis by proteasome inhibitors. J Biol Chem 272:12893-12896

Lotem J, Sachs L (1993) Hematomoietic cells from mice deficient in wild-type p53 are more resistant to induction of apoptosis by some agents. Blood 83:1092-1096

Lowe SW, Ruley HE (1993) Stabilization of the p53 tumor suppressor is induced by the adenovirus 5 E1A and accompanies apoptosis. Genes Dev 7:535-545

Lowe SW, Ruley HE, Jacks T, Huosman DE (1993a) p53-dependent apoptosis modulates the cytotoxicity of anticancer agents. Cell 74:957-967

Lowe SW, Schmitt EM, Smith SW, Osborne BA, Jacks T (1993b) p53 is required for radiation-induced apoptosis in mouse thymocytes. Nature 362:847-849

Lu H, Levine AJ (1995) Human TAFII31 protein is a transcriptional coactivator of the p53 protein. Proc Natl Acad Sci USA 92:5154-5158

Lu X, Lane DP (1993) Differential induction of transcriptionally active p53 following UV or ionizing radiation: defects in chromosome instability syndromes? Cell 75:765-778

Ludwig R, Bates S, Vousden KH (1996) Differential activation of target cellular promoters by p53 mutants with impaired apoptotic function. Mol Cell Biol 16:4952-4960

Mack DH, Vartikar J, Pipas JM, Laimins LA (1993) Specific repression of TATA-mediated but not initiator-mediated transcription by wild type p53. Nature 363:281-283

Maheswaran S, Englert C, Bennett P, Heinrich G, Haber DA (1995) The WT1 gene product stabilizes p53 and inhibits p53-mediated apoptosis. Genes Dev 9:2143-2156

Maltzman W, Czyzyk L (1984) UV irradiation stimulates levels of p53 cellular tumor antigen in nontransformed mouse cells. Mol Cell Biol 4:1689-1694

Martin SJ, Green DR (1995) Protease activation during apoptosis: death by a thousand cuts? Cell 82:349-352

Martinez J, Georgoff I, Martinez J, Levine AJ (1991) Cellular localization and cell cycle regulation by a temperature sensitive p53 protein. Genes Dev 5:151-159

Martinez JD, Craven MT, Joseloff E, Milczarek G, Bowden GT (1997) Regulation of DNA binding and transactivation in p53 by nuclear localization and phosphorylation. Oncogene 14:2511-2520

McCurrach ME, Connor TM, Knudson CM, Korsmeyer SJ, Lowe SW (1997) Bax-deficiency promotes drug resistance and oncogenic transformation by attenuating p53-dependent apoptosis. Proc Natl Acad Sci USA 94:2345-2349

Meek DW (1997) Post-translational modification of p53 and the integration of stress signals. Pathol Biol 45:804-814

Meek DW, Campbell LE, Jardine LJ, Knippschild U, McKendrick L, Milne DM (1997) Multi-site phosphorylation of p53 by protein kinases inducible by p53 and DNA damage. Biochem Soc Trans 25:416-419

Mercer WE, Shields MT, Amin M, Sauve MGJ, Apella E, Romano JW, Ullrich SJ (1990) Negative growth regulation in a glioblastoma tumor cell line that conditionally expresses human wild-type p53. Proc Natl Acad Sci USA 87:6166-6170

Michalovitz D, Halevy O, Oren M (1990) Conditional inhibition of transformation and of cell proliferation by a temperature-sensitive mutant of p53. Cell 62:671-680

Milczarek GJ, Martinez J, Bowden GT (1997) p53 phosphorylation: biochemical and functional consequences. Life Sci 60:1-11

Milne DM, McKendrick L, Jardine LJ, Deacon E, Lord JM, Meek DW (1996) Murine p53 is phosphorylated within the PAb421 epitope by protein kinase C in vivo, even after stimulation with the phorbol ester o-tetradecanoylphorbol 13-acetate. Oncogene 13: 205–211

Miyashita T, Reed JC (1995) Tumor suppressor p53 is a direct transcriptional activator of the human *bax* gene. Cell 80:293–299

Miyashita T, Krajewski S, Krajewska M, Wang HG, Lin HK, Hoffman B, Lieberman D, Reed JC (1994) Tumor suppressor p53 is a regulator of *bcl-2* and *bax* gene expression in vitro and in vivo. Oncogene 9:1799–1805

Momand J, Zambetti GP, Olson DC, George D, Levine AJ (1992) The *mdm-2* oncogene product forms a complex with the p53 protein and inhibits p53-mediated transactivation. Cell 69:1237–1245

Morimoto H, Yonehara S, Bonavida B (1993) Overcoming tumor necrosis factor and drug resistance of human tumor cell lines by combination treatment with anti-Fas antibody and drugs and toxins. Cancer Res 53:2591–2596

Muller M, Strand S, Hug H, Heinemann EM, Walczak H, Hofmann WJ, Stremmel W, Krammer PH, Galle PR (1997) Drug-induced apoptosis in hepatoma cells is mediated by the CD95 (APO-1/Fas) receptor/ligand system and involves activation of wild-type p53. J Clin Invest 99:403–413

Nagata S, Golstein P (1995) The Fas death factor. Science 267:1449–1456

Namba H, Hara T, Tukazaki T, Migita K, Ishikawa N, Ito K, Nagataki S, Yamashita S (1995) Radiation-induced G1 arrest is selectively mediated by the p53-WAF1/Cip1 pathway in human thyroid cells. Cancer Res 55:2075–2080

Naumovski L, Cleary ML (1996) The p53-binding protein 53BP2 also interacts with Bcl2 and impedes cell cycle progression at G2/M. Mol Cell Biol 16:3884–3892

O'Connor DJ, Lam EW, Griffin S, Zhong S, Leighton LC, Burbidge SA, Lu X (1995) Physical and functional interactions between p53 and cell cycle co-operating transcription factors, E2F1 and DP1. EMBO J 14:6184–6192

Okamoto K, Beach D (1994) Cyclin G is a transcriptional target of the p53 tumor suppressor protein. EMBO J 13:4816–4822

Oliner JD, Pietenpol JA, Thiagalingam S, Gyuris J, Kinzler KW, Vogelstein B (1993) Oncoprotein MDM2 conceals the activation domain of tumor suppressor p53. Nature 362:857–860

Oltvai Z, Milliman C, Korsmeyer SJ (1993) Bcl-2 heterodimerizes in vivo with a conserved homolog, Bax, that accelerates programmed cell death. Cell 74:609–619

Owen-Schaub LB, Zhang W, Cusack JC, Angelo LS, Santee SM, Fujiwara T, Roth JA, Deiseroth AB, Zhang W-W, Kruzel E, Radinsky R (1995) Wild-type human p53 and a temperature-sensitive mutant induce Fas/APO-1 expression. Mol Cell Biol 15:3032–3040

Pan H, Griep AE (1994) Altered cell cycle regulation in the lens of HPV-16 E6 or E7 transgenic mice: implications for tumor suppressor gene function in development. Genes Dev 8:1285–1299

Parada LF, Land H, Weinberg RA, Wolf D, Rotter V (1984) Cooperation between gene encoding p53 tumour antigen and ras in cellular transformation. Nature 312:649–651

Pavletich NP, Chambers KA, Pabo CO (1993) The DNA-binding domain of p53 contains the four conserved regions and the major mutation hot spots. Genes Dev 7:2556–2564

Pietenpol JA, Tokino T, Thiagalingam S, El-Deiry WS, Kinzler KW, Vogelstein B (1994) Sequence-specific transcriptional activation is essential for growth suppression by p53. Proc Natl Acad Sci USA 91:1998–2002

Polyak K, Waldman T, He T-C, Kinzler KW, Vogelstein B (1996) Genetic determinants of the p53-induced apoptosis and growth arrest. Genes Dev 10:1945–1952

Prisco M, Hongo A, Rizzo MG, Sacchi A, Baserga R (1997) The insulin-like growth factor I receptor as a physiologically relevant target of p53 in apoptosis caused by interleukin-3 withdrawal. Mol Cell Biol 17:1084–1092

Qin XQ, Livingston DM, Kaelin WG, Adams PD (1994) Deregulated transcription factor E2F-1 expression leads to S-phase entry and p53-mediated apoptosis. Proc Natl Acad Sci USA 91:10918–10922

Ragimov N, Krauskopf A, Navot N, Rotter V, Oren M, Aloni Y (1993) Wild-type but not mutant p53 can repress transcription initiation in vitro by interfering with the binding of basal transcription factors to the TATA motif. Oncogene 8:1183–1193

Rao L, Perez D, White E (Dec 1996) Lamin proteolysis facilitates nuclear events during apoptosis. J Cell Biol 135:1441–1455

Rathmell WK, Kaufmann WK, Hurt JC, Byrd LL, Chu G (1997) DNA-dependent protein kinase is not required for accumulation of p53 or cell cycle arrest after DNA damage. Cancer Res 57:68–74

Raycroft L, Wu H, Lozano G (1990) Transcriptional activation by wild-type but not transforming mutants of the p53 anti-oncogene. Science 249:1049–1051

Rowan S, Ludwig RL, Haupt Y, Bates S, Lu X, Oren M, Vousden KH (1996) Specific loss of apoptotic but not cell-cycle arrest function in a human tumor derived p53 mutant. EMBO J 15:827–838

Ryan JJ, Danish R, Gottlieb CA, Clarke MF (1993) Cell cycle analysis of p53-induced cell death in murine erythroleukemia cells. Mol Cell Biol 13:711–719

Sabbatini P, Lin J, Levine AJ, White E (1995) Essential role for p53-mediated transcription in E1A-induced apoptosis. Genes Dev 9:2184–2192

Sabbatini P, Han J, Chiou SK, Nicholson DW, White E (1997) Interleukin 1 beta converting enzyme-like proteases are essential for p53-mediated transcriptionally dependent apoptosis. Cell Growth Differ 8:643–653

Sah VP, Attardi LD, Mulligan GJ, Williams BO, Bronson RT, Jacks T (1995) A subset of p53-deficient embryos exhibit exencephaly. Nat Genet 10:175–180

Schott AF, Apel IJ, Nunez G, Clarke MF (1995) Bcl-XL protects cancer cells from p53-mediated apoptosis. Oncogene 11:1389–1394

Selivanova G, Iotsova V, Okan I, Fritsche M, Strom M, Groner B, Grafstrom RC, Wiman KG (1997) Restoration of the growth suppression function of mutant p53 by a synthetic peptide derived from the p53 C-terminal domain. Nat Med 3:632–638

Selvakumaran M, Lin HK, Miyashita T, Wang HG, Krajewski S, Reed JC, Hoffman B, Liebermann D (1994) Immediate early up-regulation of bax expression by p53 but not TGF-beta 1: a paradigm for distinct apoptotic pathways. Oncogene 9:1791–1798

Shan B, Lee WH (1994) Deregulated expression of E2F-1 induces S-phase entry and leads to apoptosis. Mol Cell Biol 14:8166–8173

Shaulian E, Zauberman A, Milner J, Davies EA, Oren M (1993) Tight DNA binding and oligomerization are dispensable for the ability of p53 to transactivate target genes and suppress transformation. EMBO J 12:2789–2797

Shaw P, Bovey R, Tardy S, Sahli R, Sordat B, Costa J (1992) Induction of apoptosis by wild-type p53 in human colon tumor-derived cell line. Proc Natl Acad Sci USA 89:4495–4499

Shaw P, Freeman J, Bovey R, Iggo R (1996) Regulation of specific DNA binding by p53: evidence for a role for O-glycosylation and charged residues at the carboxy-terminus. Oncogene 12:921–930

Slebos RJ Lee MH, Plunkett BS, Kessis TD, Williams BO, Jacks T, Hedrick L, Kastan MB, Cho KR (1994) p53-dependent G1 arrest involves pRB-related proteins and is disrupted by the human papillomavirus 16 E7 oncoprotein. Proc Natl Acad Sci USA 91:5320–5324

Slingerland JM, Jenkins JR, Benchimol S (1993) The transforming and suppressor functions of p53 alleles: effects of mutations that disrupt phosphorylation, oligomerization and nuclear translocation. EMBO J 12:1029–1037

Smith ML, Chen I-T, Zhan Q, Bae I, Chen C-Y, Glimer TM, Kastan MB, O'Conner PM, Forance AJ (1995) Interaction of the p53-regulated protein Gadd45 with proliferating cell nuclear antigen. Science 266:1376–1380

Soussi T, Caron de Fromentel C, May P (1990) Structural aspects of the p53 protein in relation to gene evolution. Oncogene 5:945–952

Stürzbecher H-W, Brain R, Addison C, Rudge K, Remm M, Grimaldi M, Keenan E, Jenkins JR (1992) A C-terminal α-helix plus basic region is the major structural determinant of p53 tetramerization. Oncogene 7:1513–1523

Subler MA, Martin DW, Deb S (1992) Inhibition of viral and cellular promoters by human wild-type p53. J Virol 66:4757–4762

Sun X, Shimizu H, Yamamoto K (1995) Identification of a novel p53 promoter element involved in genotoxic stress-inducible p53 gene expression. Mol Cell Biol 15:4489–4496

Takenaka I, Morin F, Seizinger BR, Kley N (1995) Regulation of sequence-specific DNA binding function of p53 by protein kinase C and protein phosphatases. J Biol Chem 270:5405–5411

Tan X, Martin SJ, Green DR, Wang JYJ (1997) Degradation of retinoblastoma protein in tumor necrosis factor- and CD95-induced cell death. J Biol Chem 272:9613–9616

Tarunina M, Jenkins JR (1993) Human p53 binds DNA as a protein homodimer but monomeric variants retain full transcription transactivation activity. Oncogene 8:3165–3173

Theis S, Atz J, Mueller-Lantzsch N, Roemer K (1997) A function in apoptosis other than transactivation inherent in the NH_2-terminal domain of p53. Int J Cancer 71:858–866

Thukral SK, Blain GC, Chang KK, Fields S (1994) Distinct residues of human p53 implicated in binding to DNA, simian virus 40 large T antigen, 53BP1, and 53BP2. Mol Cell Biol 14:8315–8321

Thut CJ, Chen J-L, Klemm R, Tjian R (1995) p53 transcriptional activation mediated by coactivators TAFII40 and TAFII60. Science 267:100–103

Tsai HL, Kou GH, Chen SC, Wu CW, Lin YS (1996) Human cytomegalovirus immediate-early protein IE2 tethers a transcriptional repression domain to p53. J Biol Chem 271:3534–3540

Unger T, Nau MM, Segal S, Minna JD (1992) p53: a *trans*-dominant regulator of transcription whose function is ablated by mutations occurring in human cancer. EMBO J 11:1383–1390

Vikhanskaya F, Erba E, D'Incalci M, Broggini M (1994) Introduction of wild-type p53 in a human ovarian cancer cell line not expressing endogenous p53. Nucleic Acids Res 22:1012–1017

Waga S, Hannon GJ, Beach D, Stillman B (1994) The p21 inhibitor of cyclin-dependent kinases controls DNA replication by interaction with PCNA. Nature 369:574–578

Wagner AJ, Kokontis JM, Hay N (1994) Myc-mediated apoptosis requires wild-type p53 in a manner independent of cell cycle arrest and the ability of p53 to induce $p21^{wafl/cipl}$. Genes Dev 8:2817–2830

Wang Q, Zambetti GP, Suttle DP (1997) Inhibition of DNA topoisomerase II alpha gene expression by the p53 tumor suppressor. Mol Cell Biol 17:389–397

Wang XW, Forrester K, Yeh H, Feitelson MA, Gu JR, Harris CC (1994) Hepatitis B virus X protein inhibits p53 sequence-specific DNA binding, transcriptional activity, and association with transcription factor ERCC3. Proc Natl Acad Sci USA 91:2230–2234

Wang XW, Yeh H, Schaeffer L, Roy R, Moncollin V, Egly JM, Wang Z, Friedberg EC, Evans MK, Taffe BG, Bohr VA, Hoeijmakers JHJ, Forrester K, Harris CC (1995) p53 modulation of TFIIH-associated nucleotide excision repair activity. Nat Genet 10:188–195

Wang XW, Vermeulen W, Coursen JD, Gibson M, Lupold SE, Forrester K, Xu G, Elmore L, Yeh H, Hoeijmakers JHJ, Harris CC (1996) The XPB and XPD DNA helicases are components of the p53-mediated apoptosis pathway. Genes Dev 10:1219–1232

Wang Y, Prives C (1995) Increased and altered DNA binding of human p53 by S and G2/M but not G1 cyclin-dependent kinases. Nature 376:88–91

Wang Y, Reed M, Wang P, Stenger JE, Mayr G, Anderson ME, Schweded JF, Tegtmeyer P (1993) p53 domains: identification and characterization of two autonomous DNA-binding regions. Genes Dev 7:2575–2586

Wang YS, Okan I, Szekely L, Klein G, Wiman KG (1995) *bcl-2* inhibits wild-type p53-triggered apoptosis but not G1 cell cycle arrest and transactivation of WAF1 and *bax*. Cell Growth Differ 6:1071–1075

Werner H, Karnieli E, Rauscher FJ III, LeRoith D (1996) Wild type and mutant p53 differentially regulate transcription of insulin-like growth factor I receptor gene. Proc Natl Acad Sci USA 93:8318–8323

White AE, Livanos EM, Tlsty TD (1994) Differential disruption of genomic integrity and cell cycle regulation in normal human fibroblasts by the HPV oncoproteins. Genes Dev 8:666–677

White E. (1996) Life, death, and the pursuit of apoptosis. Genes Dev 10:1–15

Williams BO, Remington L, Albert DM, Mukai S, Bronson RT, Jacks T (1994) Cooperative tumorigenic effects of germline mutations in Rb and p53. Nat Genet 7:480–484

Wu L, Bayle JH, Elenbaas B, Pavletich NP, Levine AJ (1995) Alternatively spliced forms in the carboxy-terminal domain of the p53 protein regulate its ability to promote annealing of complementary single strands of nucleic acids. Mol Cell Biol 15:497–504

Wu X, Levine AJ (1994) p53 and E2F-1 cooperate to mediate apoptosis. Proc Natl Acad Sci USA 91:3602–3606

Wu Y, Liu Y, Lee L, Miner Z, Kulesz-Martin M (1994) Wild-type alternatively spliced p53: binding to DNA and interaction with the major p53 protein in vitro and in cells. EMBO J 13:4823–4830

Xiao H, Pearson A, Coulombe B, Truant R, Zhang S, Regier JL, Triezenberg SJ, Reinberg D, Flores O, Ingles CJ et al. (1994) Binding of basal transcription factor TFIIH to the acidic activation domains of VP16 and p53. Mol Cell Biol 14:7013–7124

Xiong Y, Hannon GJ, Zhang H, Casso D, Kobayashi R, Beach D (1993) p21 is a universal inhibitor of cyclin kinases. Nature 366:701–704

Xu Y, Baltimore D (1996) Dual roles of ATM in the cellular response to radiation and in cell growth control. Genes Dev 10:2401–2410

Yan Y, Shay JW, Wright WE, Mumby MC (1997) Inhibition of protein phosphatase activity induces p53-dependent apoptosis in the absence of p53 transactivation. Biol Chem 272:15220–15226

Yin C, Knudson CM, Korsmeyer SJ, Van Dyke T (1997) Bax suppresses tumorigenesis and stimulates apoptosis in vivo. Nature 385:637–640

Yonish-Rouach E, Resnitzky D, Lotem J, Sachs L, Kimchi A, Oren M (1991) Wild-type p53 induces apoptosis of myeloid leukaemic cells that is inhibited by interleukin-6. Nature 352:345–347

Yonish-Rouach E, Grunwald D, Wilder S, Kimchi A, May E, Lawrence J-J, May P, Oren M (1993) p53-mediated cell death: relationship to cell cycle control. Mol Cell Biol 13:1415–1423

Yonish-Rouach E, Bordé J, Gotteland M, Mishal Z, Viron A, May E (1994) Induction of apoptosis by transiently transfected metabolically stable wt p53 in transformed cell lines. Cell Death Differ 1:39–47

Yonish-Rouach E, Deguin V, Zaitchouk T, Breugnot C, Mishal Z, Jenkins JR, May E (1995) Transcriptional activation plays a role in the induction of apoptosis by transiently transfected wild-type p53. Oncogene 11:2197–2205

Yonish-Rouach E, Deguin V, Choisy C, May E (1996) The role of p53 as a transcription factor in the induction of apoptosis. In: Seiler FR, Schulz G (eds) Programmed cell death. Behring Institute Research Communications. 97:60–71

Yuan ZM, Huang Y, Fan MM, Sawyers C, Kharbanda S, Kufe D (1996a) Genotoxic drugs induce interaction of the c-Abl tyrosine kinase and the tumor suppressor protein p53. J Biol Chem 271:26457–26460

Yuan ZM, Huang Y, Whang Y, Sawyers C, Weichselbaum R, Kharbanda S, Kufe D (1996b) Role for c-Abl tyrosine kinase in growth arrest response to DNA damage. Nature 382:272–274

Zauberman A, Lupo A, Oren M (1995) Identification of p53 target genes through immune selection of genomic DNA: the cyclin G gene contains two distinct p53 binding sites. Oncogene 10:2361–2366

Zhan Q, Carrier F, Forance AJ (1993) Induction of cellular p53 activity by DNA-damaging agents and growth arrest. Mol Cell Biol 13:4242–4250

Zhan Q, Fan S, Bael, Guillouf C, Liebermann DA, O'Connor PM, Forance AJ Jr (1994a) Induction of bax by genotoxic stress in human cells correlates with normal p53 status and apoptosis. Oncogene 9:3743–3751

Zhan Q, Lord KA Alamo I, Hollander MC, Carrier F, Ron D, Hohn K, Hoffman B, Liebermann DA, Forance AJ (1994b) The gadd and MyD genes define a novel set of mammalian genes encoding acidic proteins that synergistically suppress cell growth. Mol Cell Biol 14:2361–2371

The Bcl-2 Protein Family

Liam O'Connor and Andreas Strasser

1
Introduction

Apoptosis is an increasingly well described, but still poorly understood, mechanism for killing metazoan cells (Kerr et al. 1972; Wyllie et al. 1980). It is fundamental to processes as diverse as tissue remodelling during embryogenesis, maintenance of tissue homeostasis in the adult and to both innate and cognate immunity (Ellis et al. 1991; Cohen et al. 1992; Raff 1992; Ameisen et al. 1994; Vaux et al. 1994; Strasser 1995b; Golstein 1997; Jacobson et al. 1997; Nagata 1997). Cells undergoing apoptosis exhibit characteristic changes in their appearance, notably plasma membrane blebbing and chromatin condensation (Kerr et al. 1972; Wyllie et al. 1980). Molecular indicators of apoptosis include internucleosomal DNA cleavage (Wyllie 1980), proteolysis of vital cellular substrates (Casciola-Rosen et al. 1996), externalisation of phosphatidyl-serine in the plasma membrane and expression of receptors that facilitate engulfment by phagocytes (Savill et al. 1993). Apoptosis can be caused by a broad range of physiological stimuli or experimentally applied stress conditions, and it appears that cells from all tissue types and from all multi-cellular organisms studied thus far possess this evolutionarily conserved death programme (Ellis et al. 1991; Raff 1992; Hengartner and Horvitz 1994c; Steller 1995; Vaux and Strasser 1996). Apoptosis can be induced via multiple independent pathways with distinct signalling intermediates. These routes all converge upon activation of latent cysteine proteases (caspases), leading to the proteolysis of vital cellular substrates and ultimately collapse of the cell (Vaux and Strasser 1996; Jacobson et al. 1997). Proteins of the Bcl-2 family have been found to be key controllers of the cell death effector machinery.

This chapter presents an overview of the members of the Bcl-2 family of proteins, concentrating on their structural and functional differences and similarities. It is focused on those studies which elucidated the relationships between Bcl-2 family members and the role of these proteins in disease, and deals more briefly with the molecular mechanisms of action of Bcl-2 family

The Walter and Eliza Hall Institute of Medical Research, Post Office Royal Melbourne Hospital, Melbourne 3050, Australia

members. For additional information on the function of members of the Bcl-2 family, the reader is referred to previous reviews from our laboratory (Cory et al. 1994; Cory 1995; Strasser 1995b; Strasser et al. 1996, 1997) and by others (Korsmeyer 1992, 1995; Reed 1994, 1996, 1997; Reed et al. 1996; Yang and Korsmeyer 1996).

2
History

The *bcl*-2 gene was first discovered when several groups cloned the t(14;18) chromosomal translocation, a common feature of human follicular centre B cell lymphomas and presumably a cause of this cancer (Tsujimoto et al. 1984; Bakhshi et al. 1985; Cleary et al. 1986). This translocation was found to dysregulate expression of the *bcl*-2 gene by placing the coding region under the control of the immunoglobulin heavy chain enhancer (Eμ). This translocation usually results in aberrant overexpression of Bcl-2 (Tsujimoto and Croce 1986; Cleary and Sklar 1987; Graninger et al. 1987; Tsujimoto et al. 1987; Chen-Levy et al. 1989); translocations resulting in chimaeric proteins have not been observed to our knowledge.

Experiments in which Bcl-2 was overexpressed in growth factor-dependent cell lines gave the first clue to its function. Bcl-2 protected two interleukin-3 (IL-3)-dependent murine haemopoietic cell lines from apoptosis induced by cytokine withdrawal (Vaux et al. 1988). The surviving cells ceased proliferation and accumulated in the quiescent G_0 state of the cell cycle. These results demonstrated that cell proliferation and cell survival are subject to distinct genetic control and identified Bcl-2 as a mammalian inhibitor of apoptotic cell death. Since this observation, there have been numerous studies, in transformed cell lines and in cells of transgenic mice, investigating the consequences of overexpression or absence of Bcl-2. Homologues have been cloned and characterised in mammals, nematodes and viruses, leading to the description of a family of related proteins which is involved in the regulation of cell death.

3
Structure

A number of genes similar to *bcl*-2 has been discovered in mammals, some with extensive sequence homology and others more distantly related (Oltvai and Korsmeyer 1994; Cory 1995; White 1996; Yang and Korsmeyer 1996; Reed 1997). Some of these genes (*bcl*-x, *bcl*-w, *bak*) were cloned by virtue of their sequence similarity to conserved regions in the already known members of the *bcl*-2 gene family (Boise et al. 1993; Chittenden et al. 1995b; Farrow et al. 1995; Kiefer et al. 1995; Gibson et al. 1996), others (*bax, bad, bik/nbk, bid, harakiri-bim*) as a result of their encoded proteins' ability to bind to Bcl-2 (Oltvai et al.

1993; Boyd et al. 1995; Yang et al. 1995; Han et al. 1996b; Wang et al. 1996b; Inohara et al. 1997; O'Connor et al. 1998), and still others (*A1* and *mcl*-1) were initially discovered as genes induced by cytokine treatment (Lin et al. 1993; Kozopas et al. 1993). Ced-9 has been recognised as the *Caenorhabditis elegans* homologue of Bcl-2 (Hengartner et al. 1992; Vaux et al. 1992; Hengartner and Horvitz 1994b) and a number of Bcl-2 homologues have been discovered in viruses which infect mammalian cells. They include adenovirus protein E1B19kDa (White et al. 1992), Epstein–Barr virus protein BHRF1 (Henderson et al. 1993), African swine fever virus protein LMW5.1 (Neilan et al. 1993) and a number of herpes virus-encoded Bcl-2 homologues (Cheng et al. 1997; Nava et al. 1997; Sarid et al. 1997). So far no *bcl*-2-related gene has been found in *Drosophila melanogaster* or *Spodoptera frugiperda*, two organisms that are known to possess cysteine proteases (caspases) and thus at least one component of the conserved cell death effector machinery (Ahmad et al. 1997; Fraser and Evan 1997; Song et al. 1997). Detailed characterisation of the amino acid sequences of Bcl-2 family members, their biochemical action and their function has led to three key observations:

1. There are conserved regions in Bcl-2 family members, the four BH (for Bcl-2 Homology) domains, and the membrane-spanning domain (TM).
2. There are multiple types of physical interaction between proteins of the Bcl-2 family.
3. There are anti-apoptotic (Bcl-2, Bcl-x_L, Bcl-w, A1, Mcl-1, *C. elegans* Ced-9, all known viral homologues) and pro-apoptotic members [Bax, Bcl-x_S (a splice variant of the *bcl*-x gene), Bak, Bad, Bik/Nbk, Bid, Harakiri] Bim of the Bcl-2 family.

Figure 1 presents some of the structural features of Bcl-2 family members which are pertinent to the understanding of many of the following points.

3.1
Primary Structure

An important observation drawn from the discovery of new Bcl-2 family members was not only that they contained conserved regions, but also that not all of those regions were present in every family member. This meant that differences in function could be quickly assigned to conserved regions. Without this information, painstaking random mutagenesis would have been necessary for structure/function studies, since the sequences of Bcl-2 family members have none of the characteristic motifs which provide evidence for some enzymatic activity or indicate relatedness and thus similar function to other known proteins.

Four conserved domains have so far been identified in Bcl-2 family members (Borner et al. 1994; Yin et al. 1994). These regions have been denoted (from the N terminus) BH4, BH3, BH1 and BH2. The numbering reflects the

Protein		size (aa)

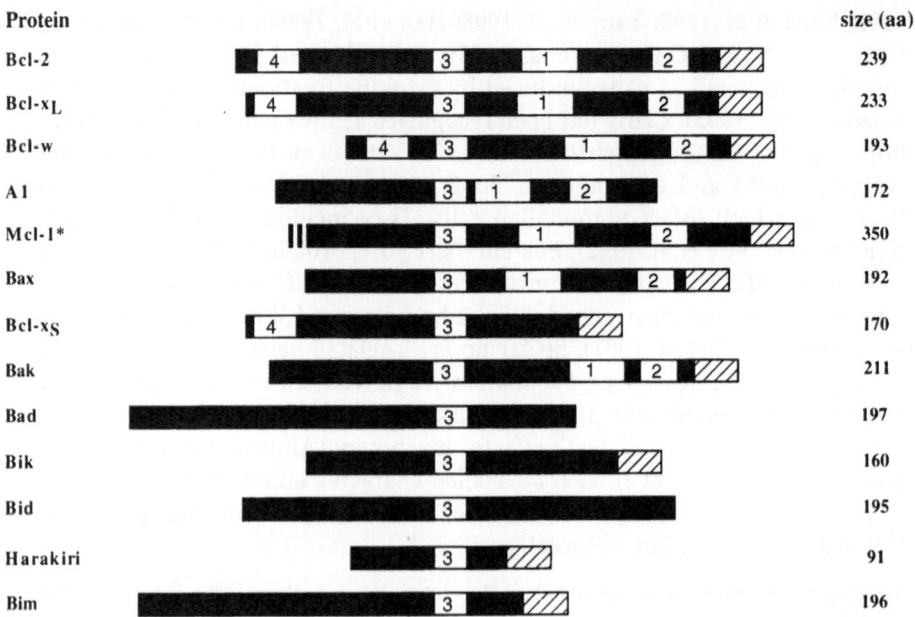

Fig. 1. The Bcl-2 family of proteins. *Numbered regions 1–4* indicate BH1–4; *hatched regions* indicate a transmembrane domain. *Asterisk* indicates that only the last 200 amino acids *aa* are shown

order of the discovery of these domains. The anti-apoptotic members of the Bcl-2 family characteristically share at least three, and usually four, of the BH regions, which is reflected in the significant overall similarity between these proteins. The fact that A1 and several of the viral Bcl-2 homologues function as anti-apoptotic proteins (Lin et al. 1996; Cheng et al. 1997; Nava et al. 1997; Sarid et al. 1997) without a recognisable BH3 region indicates that this region is not absolutely required for survival function.

Remarkably little overall sequence similarity is needed for similar function among the pro-apoptotic Bcl-2 family members. Since all the pro-apoptotic proteins so far discovered have a BH3 region, and not all of the anti-apoptotic proteins have this region, this may be a defining characteristic for the two subfamilies. For instance, Bik/Nbk, Bid Harakiri Bad and Bim have only the BH3 domain in common, yet all are pro-apoptotic when overexpressed and all bind to other Bcl-2 family members (Boyd et al. 1995; Yang et al. 1995; Han et al. 1996b; Wang et al. 1996b; Inohara et al. 1997; O'Connor et al. 1998). The limited common features of the pro-apoptotic members of the Bcl-2 family (all they share is a 16 amino acid BH3 domain) may mean that each has some specific activity, such as interaction with specific upstream signalling molecules, and that their common function is to bind and antagonise the anti-apoptotic effects of their ligands. This implies that the biochemical effects of the Bcl-2 family on apoptosis are mediated through the anti-apoptotic members. This has important implications when suggesting a model of how the Bcl-2 family may regulate apoptosis, and these are discussed below in Section 6.

Many Bcl-2 family members have a conserved C-terminal transmembrane region. This region localises proteins to the outer leaflet of the nuclear envelope, the outer mitochondrial membrane and the endoplasmic reticulum (Chen-Levy et al. 1989; Chen-Levy and Cleary 1990; de Jong et al. 1992, 1994; Krajewski et al. 1993; Monaghan et al. 1992; Lithgow et al. 1994). Subcellular localisation studies by electron microscopy and biochemical fractionation have established that the bulk of the Bcl-2 protein is on the cytosolic aspect of these membranes. Similar subcellular localisation has been reported for Bcl-x_L, A1, Mcl-1, Bax and Bak (Gonzalez-Garcia et al. 1994; Krajewski et al. 1994a,b, 1995, 1996). Bcl-2 family members naturally lacking a predicted transmembrane region or mutant forms which have it deleted can still localise to these membranes, albeit less efficiently, indicating that the sites of intrafamily interactions are at these membranes (Hockenbery et al. 1993; Borner et al. 1994; Zha et al. 1996b).

3.2
Tertiary Structure

The structure of Bcl-x_L has been solved by NMR-spectroscopy and X-ray crystallography (Muchmore et al. 1996) and while this has answered some questions, particularly regarding dimerisation, it has posed many others. Bcl-x_L protein consists of two hydrophobic central helices, surrounded by five amphipathic helices. The BH1, BH2 and BH3 regions are brought close to each other, and form a hydrophobic cleft. This cleft has subsequently been shown to be the binding site for a peptide from the BH3 region of the pro-apoptotic protein Bak (Sattler et al. 1997). Computer modelling of this interaction has suggested that the binding of the BH3-only proteins (Bik/Nbk, Bid, Harakiri Bim) to Bcl-x_L may not require as great a conformational shift as that of those pro-apoptotic Bcl-2 family members with more extensive similarity (Bax, Bad, Bak). This may mean that heterodimers of BH3-only proteins and anti-apoptotic Bcl-2 family members have binding characteristics distinct from heterodimers between anti-apoptotic Bcl-2 family proteins and those pro-apoptotic family members that have more than one BH domain.

A resemblance has been noted between the tertiary structure of Bcl-x_L and that of some bacterial toxins, notably diphtheria toxin (Muchmore et al. 1996). These toxins can form pores in the outer membranes of cells and translocate proteins to the interior (Parker and Pattus 1993). This has spawned the idea that Bcl-x_L and related proteins may also form membrane pores. In those studies published to date, however, Bcl-2 family members have only been shown to form membrane pores under non-physiological conditions of pH and tonicity (Minn et al. 1997; Schendel et al. 1997).

4
Intrafamily Protein Interactions

A second key observation drawn from the discovery of new Bcl-2 family members is reflected by the experimental technique for finding some of them: protein interaction traps (Oltvai et al. 1993; Boyd et al. 1995; Yang et al. 1995; Han et al. 1996b; Wang et al. 1996b; Inohara et al. 1997; O'Connor et al. 1998). Screening for proteins which physically interact with Bcl-2 and then finding that such interaction partners had sequence homology with Bcl-2 has led to a new avenue of inquiry, exploring intrafamily protein interactions. Table 1 shows the physical interactions found so far between Bcl-2 family members. Interactions between Bcl-2 family members are often described as dimerisations, and this might suggest to a reader that the complex in which Bcl-2 family members are found is limited to two proteins. There is actually no evidence that this is the case and complexes of Bcl-2 family members may well be made up of more than two members. There are several other proteins (not part of the Bcl-2 family) which have been reported to bind to members of the Bcl-2 family. They include r-Ras (Fernandez-Sarabia and Bischoff 1993), Raf (Wang et al. 1996a), Bag-1 (Takayama et al. 1995), Nip-1, Nip-2, Nip-3 (Boyd et al. 1994), 53BP2 (Naumovski and Cleary 1996) and calcineurin (Shibasaki et al. 1997), and any of them could equally well be part of a multi-protein complex containing several molecules of the Bcl-2 family.

Table 1. Intrafamily interactions of the Bcl-2 family of proteins

	Bcl-2	Bcl-x$_L$	A1	Mcl-1	Bax	Bcl-x$_S$	Bad	Bak	Bik	Bid	Hrk	Bim
Bcl-2	+[IP]											
Bcl-x$_L$	+[Y]	−[Y]										
A1	+[Y]	−[Y]	ND									
Mcl-1	−[Y]	−[Y]	−[Y]	−[Y]								
Bax	+[IP]	+[Y]	+[Y]	+[Y]	+[Y]							
Bcl-x$_S$	+[Y]	+[Y]	ND	−[Y]	−[Y]	ND						
Bad	+[IP]	+[IP]	ND	ND	ND	ND	ND					
Bak	−[Y]	+[Y]	ND	ND	−[Y]	−[Y]	ND	−[Y]				
Bik	+[IP]	+[IP]	ND	ND	ND	+[Y]	ND	ND	ND			
Bid	+[R]	+[R]	ND	ND	+[R]	ND	ND	ND	ND	−[R]		
Hrk	+[IP]	+[IP]	ND	ND	−[Y]	−[Y]	ND	−[Y]	ND	ND	ND	
Bim	+[IP]	+[IP]	+[IP]	ND	−[IP]	ND	ND	ND	ND	ND	ND	−[IP]

These data have been collated from a number of primary sources and reviews (including Oltvai et al. 1993; Boise et al. 1993; Oltvai and Korsmeyer 1994; Boyd et al. 1994, 1995; Chittenden et al. 1995a,b; Farrow et al. 1995; Hanada et al. 1995; Kiefer et al. 1995; Sedlak et al. 1995; Yang et al. 1995; Cheng et al. 1996; Wang et al. 1996b; Zha et al. 1996a; Inohara et al. 1997). O'Connor et al. 1998.

+, interaction demonstrated; −, no interaction; ND, no data (not tested); Y, tested in yeast two hybrid assay; IP, tested by co-immunoprecipitation in mammalian cells; R, tested by binding of recombinant proteins in vitro.

4.1
In Vitro Mutagenesis

Many of the results from in vitro mutagenesis of Bcl-2 family members were predictable from their primary sequences in that mutations in conserved regions of these proteins had drastic consequences. For example, in Bcl-2, point mutations G145E and W188A of conserved residues in BH1 and BH2 abrogated or severely impaired survival function (Yin et al. 1994). Anti-apoptotic activity was also lost as a result of deletion of the BH4 region (Borner et al. 1994). Alterations in the corresponding regions within Bcl-x$_L$ (Cheng et al. 1996) had similar consequences. Surprisingly, however, point mutations of highly conserved residues in BH4 had no effect on cell survival (Huang et al. 1998). Mutations in Bcl-2 outside the BH domains generally had no effect on anti-apoptotic activity (Borner et al. 1994; Yin et al. 1994; Hanada et al. 1995; Cheng et al. 1996; Simonian et al. 1996a,b; Huang et al. 1998). The exceptions known to date are that mutation of a non-conserved serine residue (S70) in Bcl-2 abolishes survival function (Ito et al. 1997) and that deletion of the non-conserved region between BH4 and BH3, which contains several potential phosphorylation sites, increases the cell survival activity of Bcl-2 (Chang et al. 1997). This suggests that phosphorylation and possibly other post-translational modifications can influence the function of the anti-apoptotic Bcl-2 family members.

When proteins lacking the transmembrane region were expressed in cells, they were not as potent in their pro- or anti-apoptotic functions as the native proteins (Hockenbery et al. 1993; Borner et al. 1994). It has been proposed that such mutant proteins only function when they are co-localised at a membrane with another Bcl-2 family member, and that the proportion of the mutant protein which is free in the cytosol is inactive (Hockenbery et al. 1993a).

Among the anti-apoptotic family members, mutations in the conserved BH regions or the transmembrane domain affect both their survival function and their ability to dimerise with pro-apoptotic members (Borner et al. 1994; Yin et al. 1994). There is not, however, a direct correspondence between ability to protect against cell death and ability to heterodimerise. A mutant form of Bcl-x$_L$ has been described which is incapable of binding to either Bax or Bak, but is unperturbed in its survival function (Cheng et al. 1996). Obversely, there is a mutant of Bcl-2 (ΔBH4) which has no survival function, but nevertheless binds to Bax (Borner et al. 1994). The existence of such mutant proteins indicates that those models of Bcl-2 family function which would have heterodimers of pro- and anti-apoptotic Bcl-2 family members as the active moiety must be overly simplistic. A more detailed discussion of models for the function of the Bcl-2 family can be found below in Section 4.3.

Homodimerisation and heterodimerisation of Bcl-2 family members are thought to involve different regions within these proteins, since there are mutations in the BH1 (G145A and G145E) and BH2 (W188A and Q190A) regions of Bcl-2 which abolish binding to Bax, but leave homodimerisation intact (Yin et al. 1994). However, recent quantitative biochemical analyses demonstrated that heterodimerisation and homodimeristaion of Bcl-2 family members can both be blocked by the same BH3-derived peptide (Diaz et al. 1997). This result therefore casts some doubt on the earlier results, which were obtained from less quantifiable analyses performed in the yeast-two-hybrid system or by co-immunoprecipitation of overexpressed proteins. To our knowledge, it has not yet been determined whether the ability to form homodimers is essential for the survival function of the anti-apoptotic Bcl-2 proteins. Clearly, homodimerisation of Bcl-2 is not sufficient for survival activity since the G145E loss of function mutation does not impair this interaction.

As a group, the pro-apoptotic Bcl-2 family members (particularly the BH3-only proteins) have not been so exhaustively tested in mutagenesis experiments as the anti-apoptotic proteins. Bax was the first pro-apoptotic Bcl-2 family member discovered (Oltvai et al. 1993), and so for historical reasons it is the best studied. There have been some surprises. The most striking of these is how little of the protein seems to be necessary for pro-apoptotic function. Pro-apoptotic capability seems to be conferred largely by the BH3 domain, since it is the only common feature of all members within this group and since a mutant of Bak, comprising only the BH3 domain and the transmembrane region, can kill cells (Chittenden et al. 1995a). That the other BH regions of Bax, Bad and Bak have been protected from evolutionary drift (Oltvai et al. 1993; Chittenden et al. 1995b; Farrow et al. 1995; Kiefer et al. 1995; Yang et al. 1995) indicates that they are also under selective pressure. For instance, the interaction between Bad and the 14.3.3 proteins, which are thought to act as a platform for several signalling molecules (Aitken 1995), has been shown to be dependent upon phosphorylation of serine residues outside the BH domains (Zha et al. 1996b). Mutagenesis showed that the subcellular localisation, pro-apoptotic activity and binding of Bad to Bcl-x_L were all affected by phosphorylation of these serine residues (Zha et al. 1996b). That regions other than the BH regions appear to us to be redundant for function may simply be a reflection of the preliminary state of our knowledge of the biochemistry of apoptosis.

4.2
Functional Relationship Between the Bcl-2 Family and Caspases

Recently, it has been demonstrated that there is a physical connection between two previously unconnected regulators of apoptosis: the Bcl-2 family and the caspases. Caspases are cysteine proteases which recognise tetra-peptide motifs and cleave their substrates on the carboxyl side of an aspartate residue (P1 position) (Kumar 1995; Alnemri et al. 1996; Kumar and Lavin 1996). Individual caspases have distinct substrate specificities because they recognise different amino acids at the three residues directly N-terminal of the aspartate (P2, P3 and P4 positions) (Nicholson et al. 1995; Thornberry et al. 1997). The essential role of caspases in apoptosis first became apparent from genetic studies in *C. elegans*. In this nematode, the fate map of all somatic cells has been established (Sulston and Horvitz 1977) and it is known that 131 of the 1090 cells that are formed undergo programmed cell death during the development of this organism. Mutations in genes controlling cell death (*ced* genes) always have the same characteristic phenotype of missing or excess cells. For a comprehensive review of apoptosis in *C. elegans* the reader is referred to previous papers and reviews (Ellis and Horvitz 1986; Ellis et al. 1991; Hengartner and Horvitz 1994c). One of these *ced* genes, *ced-3*, codes for a cysteine protease, closely related to human caspase-1 [interleukin-1β converting enzyme (ICE)] (Yuan et al. 1993). Experiments in which the general caspase inhibitor, baculovirus p35 (Bump et al. 1995; Bertin et al. 1996), was expressed in mammalian cell lines in vitro or in *C. elegans* and *Drosophila melanogaster* in vivo established that caspases are essential for apoptosis in these organisms (Hay et al. 1994; Sugimoto et al. 1994; Beidler et al. 1995; Grether et al. 1995; Martinou et al. 1995; Xue and Horvitz 1995; White et al. 1996; Datta et al. 1997; Manji et al. 1997).

Caspases are synthesised as inactive precursors, zymogens, and themselves need to be cleaved at aspartate residues to facilitate assembly of the fully active enzyme, which is composed of a hetero-tetramer of two subunits of about 20 kDa plus two subunits of about 10 kDa ($p20_2p10_2$) (Walker et al. 1994; Wilson et al. 1994; Rotonda et al. 1996). Caspases can thus auto-activate and activate each other, producing a signalling cascade which results in the cleavage of substrates characteristic of late-stage apoptosis. The end points of this proteolytic cascade have begun to be described, most strikingly with the discovery of two caspase substrates: DNA fragmenting factor (DFF) and PAK2. DFF is a heterodimeric protein which, upon activation by caspase-3 (CPP32, Yama, apopain), can trigger internucleosomal DNA fragmentation in isolated cell nuclei (Liu et al. 1997). PAK2 is a cytosolic protein kinase whose inhibitory domain is removed by caspases, and this leads to uncontrolled PAK2 activation, which appears to be essential for some of the cytosolic changes of apoptosis and plasma membrane blebbing (Rudel and Bokoch 1997). A more detailed description of caspases and their function can be found in Volume 24 of the Results and Problems in Cell Differentiation.

The Bcl-2 family had previously only been genetically and functionally linked to the caspases, and it was feared that the biochemical pathways linking them might be quite complex. New data have gone a long way to allaying these concerns. Ced-4, like Ced-3, is absolutely required for programmed death of somatic cells in *C. elegans* (Ellis and Horvitz 1986; Ellis et al. 1991; Yuan and Horvitz 1992), but, other than this genetic evidence, nothing was known about the function of Ced-4. Ced-9, the *C. elegans* homologue of Bcl-2, has now been shown to bind to Ced-4 (Chinnaiyan et al. 1997; Spector et al. 1997; Wu et al.

1997). Further, Ced-4 was shown to also bind to Ced-3 and to be able to stimulate conversion of Ced-3 zymogens into the active enzyme complexes in insect cells (Chinnaiyan et al. 1997; Seshagiri and Miller 1997). Bcl-x$_L$ and caspase-8 (FLICE, Mach-1) have been shown to be capable of substituting for Ced-9 and Ced-3 in biochemical studies of protein interaction (Chinnaiyan et al. 1997). These analyses provided the first physical link between the Bcl-2 family, caspases and Ced-4, and added further weight to the idea that Ced-4 homologues exist in mammalian cells. The implications of these results for our understanding of the biochemistry of apoptosis are discussed below in Section 6.

4.3
Functional Implications of Intrafamily Protein Interactions

What the above studies do not tell us is which subgroup of Bcl-2 family members regulates the activity of which. Are the pro-apoptotic members essential for apoptosis in mammalian cells, or do they merely act as antagonists of the anti-apoptotic members? Do the anti-apoptotic Bcl-2 proteins promote cell survival by titrating the pro-apoptotic members or are other proteins (e.g. Ced-4-like proteins) their principal targets? There is also the possibility that the active state of Bcl-2 family members is as members of heterodimers, and that the homodimers serve as reservoirs (Oltvai and Korsmeyer 1994). The discovery of mutants of anti-apoptosis family members which cannot bind to pro-apoptotic members but still retain their survival function in mammalian cells makes the model in which the anti-apoptotic proteins are the crucial regulators of apoptosis the one which fits the facts best, and the one yet to be disproved.

Studies in *C. elegans* have added to the puzzle of which biochemical state is the survival-promoting form of Bcl-2 family members. There is a gain of function allele of *ced*-9 (n1950) whose protein, Ced-9 (G169E), is more potent than the wild-type (wt) protein, Ced-9 wt, in protecting cells against apoptosis in most circumstances (Hengartner and Horvitz 1994a). However, in mutant *C. elegans* with a partial loss of function of Ced-4, Ced-9 (G169E) enhances cell death rather than blocking it (Hengartner and Horvitz 1994a). The mutation corresponding to Ced-9 (G169E) in Bcl-2 is Bcl-2 (G145E) and this abolishes both anti-apoptotic activity and binding to Bax (Yin et al. 1994). Our preferred interpretation of these findings is that Ced-9 is a form of Bcl-2 family protein which contains features of both the anti-apoptotic and the pro-apoptotic members and may therefore represent a common ancestor. In this interpretation, conformational alterations to Ced-9 are what antagonise its anti-apoptotic function, and such conformational alterations are similar to the changes produced in Bcl-2 by binding of pro-apoptotic family members. Binding of Ced-9 to Ced-4 appears to be critical for survival function, since mutations in *ced*-9 that block this interaction are loss of function alleles

(Chinnaiyan et al. 1997; Spector et al. 1997; Wu et al. 1997). We speculate that conformational changes induced in Ced-9 by cell death signals or by the G169E mutation alter its interaction with Ced-4 and thus modulate its activity. Before the entire *C. elegans* genome has been sequenced, however, it can not be excluded that BH3 only proteins or other pro-apoptotic Bcl-2 family proteins exist in this organism.

5
Functional Studies of Bcl-2 Family Members

The third key observation drawn from the discovery of new Bcl-2 family members was made when the first such member, Bax, was overexpressed in cultured cell lines (Oltvai et al. 1993). Unlike Bcl-2, Bax enhances cell death triggered by cytotoxic conditions, such as withdrawal of essential growth factors. When Bax was co-expressed with Bcl-2, each protein antagonised the other's effect on apoptosis (Oltvai et al. 1993). All Bcl-2-related proteins presently known have either pro-apoptotic or anti-apoptotic activity and antagonism has been shown to be a general feature of the relationships between members of these two subfamilies (Oltvai and Korsmeyer 1994).

A comprehensive discussion of all the stimuli able to trigger apoptosis is beyond the scope of this chapter. Indeed, it would probably be a smaller task to list those stimuli not reported as being able to cause apoptosis. The reader is referred to several other chapters in this Volume and in Volume 24 of the Results and Problems in Cell Differentiation for details of various apoptotic stimuli and what is known about the signalling pathways that they activate.

5.1
In Vitro

The prototype Bcl-2 family member is Bcl-2 itself; it was the first to be discovered and is presently the best characterised. Functional studies have been carried out in a variety of cell lines, including those of lymphoid, myeloid, erythroid, neuronal, fibroblastoid and epithelioid origin. Overexpression of Bcl-2 protected cell lines that require for their proliferation and/or survival cytokines such as IL-2, IL-3, IL-4, IL-6, IL-7, GM-CSF, G-CSF, EPO or NGF against growth factor withdrawal-induced apoptosis (Vaux et al. 1988; Nunez et al. 1990; Garcia et al. 1992; Baffy et al. 1993; Deng and Podack 1993; Borzillo et al. 1994; Marvel et al. 1994; Chung et al. 1996; Rodel and Link 1996; Silva et al. 1996). In all these analyses, overexpression of Bcl-2 protected cells against apoptosis but was unable to promote cell proliferation. In growth factor-independent cell lines, Bcl-2 antagonised apoptosis induced by exposure to a broad range of cytotoxic conditions, such as γ-radiation, UV-radiation, glucocorticoids, heat shock, cold shock, oxidative stress, calcium ionophores and phorbol ester (Tsujimoto 1989; Miyashita and Reed 1992; Hockenbery et al. 1993; Walton et al. 1993; Strasser et al. 1994b, 1995; Strasser and Anderson 1995). This indicates that independent signal transduction routes to apoptosis converge upon a common cell death effector machinery that is antagonised by Bcl-2.

In those cases in which it has been tested, overexpression of the other anti-apoptotic members of the Bcl-2 family has been shown to be functionally indistinguishable from that of Bcl-2, indicating that these proteins have similar, if not identical biochemical action (Boise et al. 1993; Kozopas et al. 1993; Carrió et al. 1996; Gibson et al. 1996; Lin et al. 1996; Huang et al. 1997). It is noteworthy that there is a dosage effect in these experiments (Huang et al. 1997). This could be interpreted as evidence for the idea of pro-apoptotic and anti-apoptotic Bcl-2 family members titrating each other in the cell, but the dosage effect may equally well be explained by another factor titrating the anti-apoptotic activity of Bcl-2 and its functional homologues.

If Bcl-2 can serve as the prototype for the anti-apoptotic proteins of this family, then Bax can play a similarly representative role for the pro-apoptotic members. It was the first pro-apoptotic protein to be discovered, and is presently the best characterised. Overexpression of Bax results in greatly reduced resistance to cytotoxic stress in many cell lines, and rapid death in others (Oltvai et al. 1993; Strobel et al. 1996; Xiang et al. 1996). The effect of Bax in vitro can be antagonised by simultaneous overexpression of Bcl-2, Bcl-x_L or, importantly for a model of immune evasion by pathogens, also by the adenovirus protein E1B19kDa (Oltvai et al. 1993; Sedlak et al. 1995; Chen et al. 1996; Han et al. 1996a). A more comprehensive discussion of possible mechanisms whereby intracellular organisms can escape immune attack by blocking the apoptotic machinery in their host cells can be found below in Section 7 and in a previous review from our laboratory (Vaux et al. 1994).

5.2
In Vivo

Since bcl-2 was originally identified as a proto-oncogene in a human lymphoid malignancy, the first whole-animal experiments involving the overexpression of Bcl-2 were performed in lymphoid tissues (McDonnell et al. 1989, 1990; Strasser et al. 1990b, 1991a,b; Sentman et al. 1991; Katsumata et al. 1992; Siegel et al. 1992). Subsequently, transgenic mice have been generated that overexpress Bcl-2 in myeloid cells (Lagasse and Weissman 1994), neuronal cells (Martinou et al. 1994; Farlie et al. 1995), hepatocytes (Lacronique et al. 1996; Rodriguez et al. 1996), spermatogonia (Furuchi et al. 1996) or a relatively wide range of cell types (Rodriguez et al. 1997). Three transgenic animal studies of Bcl-x_L were performed in T lymphoid cells (Chao et al. 1995; Grillot et al. 1995; Takahashi et al. 1997) and one in pancreatic β cells (Naik et al. 1996). As expected, these transgenic mice exhibit increased in vitro and in vivo survival of cells in which the anti-apoptotic protein is expressed (McDonnell et al. 1989, 1990; Strasser et al. 1990b, 1991a,b; Sentman et al. 1991; Katsumata et al. 1992; Siegel et al. 1992; Lagasse and Weissman, 1994; Martinou et al. 1994; Smith et al. 1994; Chao et al. 1995; Farlie et al. 1995; Grillot et al. 1995; Lacronique et al. 1996; Naik et al. 1996; Rodriguez et al. 1996, 1997; Takahashi et al. 1997; O'Reilly et al. 1997a,b). Transgene-expressing cells from these animals were resistant to a broad range of cytotoxic conditions, including cytokine deprivation or treatment with γ-radiation, glucocorticoids, calcium ionophores and phorbol ester. This implies a common cell death checkpoint, controlled by Bcl-2 family members, for insults as diverse as treatment with steroid hormone or ionising radiation. No functional difference could be identified between Bcl-2 and Bcl-x$_L$ in transgenic mouse studies and expression of a bcl-x$_L$ transgene was capable of rescuing the survival defect in lymphocytes of bcl-2$^{-/-}$ mice (Chao et al. 1995; see also below). This provides further evidence for the notion that Bcl-2 and Bcl-x$_L$ have similar, if not identical, biochemical function.

The ability of Bcl-2 to protect cells from death induced by withdrawal of growth stimuli has made bcl-2 transgenic mice important research tools in their own right. This has enabled researchers to investigate which physiologically induced cell losses occur by a Bcl-2-inhibitable death programme and has facilitated further study of populations of cells which would otherwise be lost to apoptosis. A few examples of this type of analysis are listed below. Female mice expressing a bcl-2 transgene in a broad range of cell types are infertile because vaginal opening does not occur (Rodriguez et al. 1997), demonstrating for the first time that this process requires cell death. Overexpression of Bcl-2 in the myeloid lineage of mutant op/op mice, which lack functional macrophage colony stimulating factor (M-CSF, also called CSF-1), allowed cells, which otherwise would have died from factor withdrawal, to survive and enabled fundamental questions about the stochastic and instructive models of haemopoietic differentiation to be addressed (Lagasse and Weissman 1997). Overexpression of Bcl-2 rescued T cell development in mice lacking IL-7 receptors (Akashi et al. 1997; Maraskovsky et al. 1997), promoted accumulation of B cells that are unable to express surface immunoglobulin in mutant scid, rag-1$^{-/-}$ and rag-2$^{-/-}$ mice (Strasser et al. 1994a; Young et al. 1997; O'Reilly et al. 1997b; Tarlinton et al. 1997) and enhanced the survival of those T cells which express antigen receptors that are unable to interact with major histocompatibility complex molecules (inability to undergo positive selection) (Linette et al. 1994; Strasser et al. 1994c). These results demonstrate that the absence of positive signals from neighbouring cells triggers in lymphoid and myeloid cells at several developmental checkpoints an intrinsic death programme that can be inhibited by Bcl-2. Interestingly, there are also physiologically induced pathways to apoptosis in haemopoietic cells that cannot be inhibited by Bcl-2 or Bcl-x$_L$ (see Sect. 5.3). Generation of transgenic mice expressing bcl-2 in other tissues and crosses of these animals with mutant mice lacking growth factors, surface receptors or signal transduction molecules is expected to provide important insight into the control of cell production and cell loss during embryogenesis and in adult life. For example, it will be interesting to determine if expression of a *bcl*-2 transgene in erythroid cells can rescue the defect in this lineage that is caused by the absence of EPO, EPO receptor or the GATA-1 transcription factor (Pevny et al. 1991; Orkin 1996).

Bax is so far the only pro-apoptotic member of the Bcl-2 family whose function has been investigated in transgenic mouse studies. Overexpression of Bax did not reduce the number of T lymphoid cells in the whole animal but could accelerate thymocyte apoptosis under conditions of cytotoxic stress (Brady et al. 1996b). We speculate that this may be due to the fact that Bax inactivates anti-apoptotic Bcl-2 family proteins in these cells [e.g. Bcl-x$_L$ (Boise et al. 1995)] and this is consistent with a model in which pro-apoptotic Bcl-2 family members function as attenuators of the anti-apoptotic members. The tumour suppressor gene *p53* is essential for DNA damage-induced apoptosis in thymocytes, pre-B cells, quiescent lymphocytes and some intestinal cells (Clarke et al. 1993, 1994; Lowe et al. 1993; Strasser et al. 1994b). Upon DNA damage p53 activates transcription of the *bax* gene, and it has been speculated that this is essential for the induction of apoptosis (Miyashita et al. 1994). However, γ-radiation-induced cell death occurs normally in *bax*$^{-/-}$ mice (Knudson et al. 1995; see below) and *p53*$^{-/-}$ thymocytes expressing a *bax* transgene are as resistant to γ-radiation as those from *p53*$^{-/-}$ mice (Brady et al. 1996b). This demonstrates that Bax is neither essential nor sufficient for DNA damage-induced apoptosis. It remains, however, possible that the expression of one or more pro-apoptotic Bcl-2 family members is essential for this process.

Gene knock-out technology has been used to determine the essential functions of Bcl-2 family proteins. Mice deficient in Bcl-2 show increased cell death during nephrogenesis and premature demise of their mature lymphocytes (Nakayama et al. 1993, 1994; Veis et al. 1993b; Matsuzaki et al. 1997). These results were expected, given that Bcl-2 is normally expressed at high levels in these tissues (Hockenbery et al. 1991; Veis et al. 1993a; Gratiot-Deans et al. 1994; Merino et al. 1994). Unexpectedly, there was no detectable increase in apoptosis in other tissues in which Bcl-2 is known to be highly expressed, such as the nervous system, intestines and skin (Hockenbery et al. 1991; Merry et al. 1994). These results may imply functional redundancy among the anti-apoptotic Bcl-2 family members, since Bcl-x$_L$, for instance, is also known to be highly expressed in some of the tissues that express Bcl-2 (Gonzalez-Garcia et al. 1994; Krajewski et al. 1994b). The unexpected absence of detectable phenotypic abnormalities in certain tissues may also be explained by the lack of fine discrimination in the available protein expression data. In the case of complex organs, such as the central nervous system, we may not exactly know in which cells and at what stage of their development Bcl-2 is expressed, and therefore do not know how to reveal the consequences of its absence. Mice deficient in *bcl*-x die at embryonic day 13 and show abnormally increased apoptosis in the central nervous system and in fetal liver (Motoyama et al. 1995). We believe that differences in the phenotypes between $bcl\text{-}2^{-/-}$ mice and $bcl\text{-}x^{-/-}$ mice are a reflection of their distinct expression patterns and not a result of differences in biochemical function of the encoded proteins.

Bax is the only pro-apoptotic Bcl-2 family member that has been studied thus far by gene knock-out analysis. Cell production and cell death appear to be normal in most somatic tissues of $bax^{-/-}$ mice, with the exception of mild T cell hyperplasia in the thymus and in peripheral lymphoid organs (Knudson et al. 1995; Knudson and Korsmeyer 1997). Surprisingly, $bax^{-/-}$ male mice exhibit marked hypoplasia in testicular cells and are sterile (Knudson et al. 1995). This phenotypic abnormality is similar to that of mice overexpressing Bcl-2 in spermatogonia (Furuchi et al. 1996), and probably reflects the dependence of spermatid production on apoptosis of neighbouring cells, rather than any fundamental difference in the apoptosis process in germ cells. The finding that most physiologically induced cell deaths do occur normally in $bax^{-/-}$ mice is consistent with the notion that Bcl-2/Bax heterodimers are dispensable for both apoptosis and cell survival. It is anticipated that phenotypic analyses of mice lacking other pro-apoptotic Bcl-2 family members (and crosses between such animals) will reveal whether or not this class of proteins plays an essential role in cell death control.

5.3
Pathways to Apoptosis that are Insensitive to Bcl-2 and Bcl-x$_L$

A gain of function mutation of Ced-9 blocks all programmed deaths of somatic cells in *C. elegans* (Ellis and Horvitz 1986; Ellis et al. 1991; Hengartner et al. 1992). In contrast, Bcl-2 and Bcl-x$_L$ are unable to block all physiologically induced pathways to apoptosis in mammalian cells (Strasser 1995b). For example, Bcl-2 does not prolong the survival of mutant *scid* and $rag\text{-}1^{-/-}$ thymocytes which are unable to express T cell antigen receptor (TCR) chains (Strasser et al. 1994a; Maraskovsky et al. 1997), and neither Bcl-2 nor Bcl-x$_L$ can block deletion of self-antigen-specific immature B cells (Hartley et al. 1993; Nisitani et al. 1993) and T cells (Sentman et al. 1991; Strasser et al. 1991a, 1994c; Tao et al. 1994; Chao et al. 1995; Grillot et al. 1995).

Members of the tumour necrosis factor receptor (TNF-R) family play critical roles in apoptosis of lymphocytes that have been chronically activated via their antigen receptors (Strasser 1995a,b; Golstein 1997; Nagata 1997). Ligation of the TCR/CD3 complex on proliferating normal T cells or T lymphoma lines induces expression of CD95 ligand (also called FasL or APO-1L) and its receptor CD95 (also called Fas or APO-1). This triggers apoptosis by autocrine or paracrine CD95 stimulation (Russell et al. 1993; Russell and Wang 1993; Alderson et al. 1995; Brunner et al. 1995; Dhein et al. 1995; Ju et al. 1995; Strasser 1995a) and this death cannot be blocked by Bcl-2 (Strasser et al. 1995). Experiments in some transformed cell lines (e.g. L929, WEHI 164, NIH 3T3 or

293T) demonstrated that none of the anti-apoptotic proteins Bcl-2, Bcl-x$_L$, Bcl-w and adenovirus protein E1B19kDa afforded any protection against apoptosis triggered by ligation of p55 TNF-RI, DR3 and possibly other members of the TNF-R family (Vanhaesebroeck et al. 1993; Memon et al. 1995; Strasser et al. 1995; Chinnaiyan et al. 1996a; Bodmer et al. 1997; Huang et al. 1997). These receptors all contain a region called the death domain, so called because it is essential for triggering apoptosis (Itoh and Nagata 1993; Tartaglia et al. 1993). Upon receptor aggregation, this region binds to related death domains in cytosolic adaptor molecules, such as FADD (also called Mort-1) and TRADD (Boldin et al. 1995; Chinnaiyan et al. 1995; Hsu et al. 1995). FADD/Mort-1 binds directly to some receptors and indirectly via TRADD to others. The so-called death effector domain in FADD/Mort-1 serves to recruit death effector domain bearing caspases (caspase-8/FLICE-1/Mach-1 or caspase-10b/FLICE-2) via homotypic interaction (Boldin et al. 1996; Muzio et al. 1996). It is believed, but not yet proven, that receptor-mediated aggregation leads to activation of caspase-8 and/or caspase-10b zymogens. Consistent with the finding that Bcl-2 and its homologues cannot block CD95- or p55 TNF-RI-transduced apoptosis, overexpression of the downstream signalling molecules FADD/Mort-1 or TRADD was shown to kill 293T cells in a manner that is insensitive to Bcl-2, Bcl-x$_L$ and adenovirus protein E1B19kDa (Hsu et al. 1995; Chinnaiyan et al. 1996b; Hsu et al. 1996). Bcl-2 family proteins are not localised to the plasma membrane (see Sect. 3.1), the site where caspase activation occurs after ligation of CD95 (and possibly also after crosslinking of related receptors) (Kischkel et al. 1995; Medema et al. 1997). It therefore makes sense that Bcl-2 and its homologues cannot block this pathway to apoptosis.

The cowpox virus serpin CrmA (cytokine response modifier A) (Ray et al. 1992) is a potent inhibitor of caspase-8, caspase-10b and caspase-1, but is a poor antagonist of other caspases (Komiyama et al. 1994; Srinivasula et al. 1996; Zhou et al. 1997). Expression of CrmA in cultured cell lines or thymocytes of transgenic mice blocks CD95- and TNF-RI-transduced apoptosis but has no impact on those pathways to apoptosis (e.g. growth factor deprivation) that can be inhibited by Bcl-2 or its close homologues (Memon et al. 1995; Miura et al. 1995; Strasser et al. 1995; Talley et al. 1995; Tewari and Dixit 1995; Chinnaiyan et al. 1996a; Smith et al. 1996; Srinivasula et al. 1996). All of these cell death pathways can be blocked by p35 (Datta et al. 1997), a potent inhibitor of all known caspases (Bump et al. 1995; Bertin et al. 1996). This indicates that there are two distinct physiological pathways to apoptosis: 1) absence of growth factors or metabolic disturbance trigger un-known mechanisms which kill cells by activating CrmA-insensitve (p35-sensi-tive) caspases by a Bcl-2-inhibitable mechanism; and 2) chronic antigen receptor crosslinking leads to autocrine and/or paracrine stimulation of CD95 (and probably also related receptors) and this triggers activation of CrmA-sensitive (p35-sensitive) caspases by a mechanism that cannot be blocked by Bcl-2.

In some cell types, such as the MCF-7 breast cancer-derived cell line and normal hepatocytes, Bcl-2, Bcl-x$_L$ and E1B19kDa protein can delay CD95-transduced apoptosis (White et al. 1992; Jäättelä et al. 1995; Lacronique et al. 1996; Rodriguez et al. 1996; Huang et al. 1997), indicating that this receptor may be able to trigger additional apoptotic pathways (besides FADD/Mort-1-caspase-8). Indeed, it has been found that CD95 can bind via its death domain to Daxx, a cytosolic adaptor which does not contain a death domain. Recruitment of Daxx to oligomerised CD95 receptors stimulates, by an unknown mechanism, Jun kinase activation, which ultimately leads to apoptosis. This pathway to cell death, unlike the FADD/Mort-1-caspase-8 pathway, can be blocked by Bcl-2. It is possible that other members of the TNF-R family can also trigger multiple pathways to apoptosis.

In many cases of cell loss, particularly those that occur in embryogenesis, it is unknown whether they occur by apoptosis and what the mechanism of cell death might be. Usage of transgenes encoding apoptosis inhibitors, such as Bcl-2, caspase inhibitors (CrmA and p35) or dominant interfering mutants of signalling molecules (e.g. FADD/Mort-1 and Daxx), is expected to provide insight into these processes.

5.4
The Bcl-2 Family has an Effect on Cell Cycle Control

Until recently, the only function ascribed to Bcl-2 family members was that of controlling apoptosis. Bcl-2, Bcl-x$_L$ and adenovirus protein E1B19kD have now been shown to have an effect on the cell cycle, independent of their anti-apoptotic function. Bcl-2 overexpression inhibits cell cycle entry of quiescent cells in cultured cell lines and in lymphocytes from transgenic mice (Linette et al. 1996; Mazel et al. 1996; O'Reilly et al. 1996, 1997a,b; Huang et al. 1997). Bcl-2 has also been reported to hasten exit from the cell cycle of cells exposed to conditions of cytotoxic stress (Vairo et al. 1996). *Caenorhabditis elegans* Ced-9 may also have an effect on the cell cycle, since *ced*-9 loss of function mutant embryos not only show abnormally increased cell death but also exhibit abnormal mitoses (Hengartner et al. 1992). These results are reminiscent of several observations which suggested a link between the Bcl-2 family and the cell cycle. Rb is a tumour suppressor which controls entry into the S-phase of the cell cycle. Interesting similarities in phenotype of *bcl-x*$^{-/-}$ mice (Motoyama et al. 1995) and *Rb*$^{-/-}$ mice (Jacks et al. 1992; Macleod et al. 1996) have been noted. Furthermore, proliferating cells lacking the tumour suppressor p53 arrest at the G$_2$/M boundary of the cell cycle upon DNA damage, but when these cells also express Bcl-2, a proportion of them arrests in G$_0$/G$_1$ (Strasser et al. 1994b).

Deletion of the BH4 region and point mutations in BH1 (G145E) or BH2 (W188A) within Bcl-2 abrogate both its anti-apoptotic activity and its effect on the cell cycle (O'Reilly et al. 1996). In addition, co-expression of Bax reduces the inhibitory effect of Bcl-2 on cell cycle entry in NIH 3T3 fibroblasts (O'Reilly et al. 1996) and in T cells of doubly transgenic mice (Brady et al. 1996a). This indicates that the cell cycle effect of Bcl-2 may be inseparable from its anti-apoptotic activity and demonstrates that Bax can function as an attenuator of both of these functions. The discovery of mutant forms of Bcl-2 (Y28A, Y28S and Y28F) which lack the inhibitory effect on cell cycle entry but have wild-type anti-apoptotic activity genetically separates these two functions (Huang et al. 1997). We are intrigued by the possibility that the physical interaction between Bcl-2 and calcineurin (Shibasaki et al. 1997), a phosphatase that regulates the activity of the NF-AT transcription factor family (Beals et al. 1997), may play a role in the effect of Bcl-2 on the cell cycle.

6
How Do Bcl-2 Family Members Regulate Apoptosis?

Members of the Bcl-2 family have at various times been suggested to function through a number of biochemical mechanisms. Many of these models have since been questioned in the light of new data, but aspects of some of them still persist in the ideas of those not directly involved in this field, so it may be useful to lay them to rest here. Bcl-2 has been described as being able to bind GTP, implicating the G-protein signal transduction system in apoptosis (Haldar et al. 1989). This has been contradicted; directly, with the report of an inability to reproduce these interactions (Monica et al. 1990), and indirectly, with the suggestion that the interaction is mediated through p23 R-ras (Fernandez-Sarabia and Bischoff 1993). Mistaken localisation of Bcl-2 to the inner (rather than outer) membranes of mitochondria (Hockenbery et al. 1990) led to the suggestion that Bcl-2 might function to regulate cellular energy levels in the form of ATP. The discovery that Bcl-2 can function in ρ^0 cells (which lack mitochondrial DNA) and elucidation of the subcellular localisation of Bcl-2 by more sensitive techniques (see Sect. 3.1) disproved this hypothesis (Jacobson et al. 1993). Reactive oxygen species have been proposed to be key effectors of apoptosis (Hockenbery et al. 1993; Kane et al. 1993), but apoptosis has subsequently been demonstrated under hypoxic conditions, and since Bcl-2 and Bcl-x_L could antagonise cell death under these conditions (Jacobson and Raff 1995; Shimizu et al. 1995), this hypothesis should also be treated with caution.

We propose here the simplest model for the function of the Bcl-2 family that we can conceive and which fits all the available data (Fig. 2). In this model, the putative homologue of Ced-4 (and this does not preclude there being more than one) is a key activator of the caspase cascade. This activity of Ced-4 proteins can be antagonised by anti-apoptotic Bcl-2 family members, by direct physical interaction. Whether this inactivation takes the form of direct competition between anti-apoptotic Bcl-2 family members and caspases for binding to the same region in Ced-4 is not clear. The role of pro-apoptotic Bcl-2 family

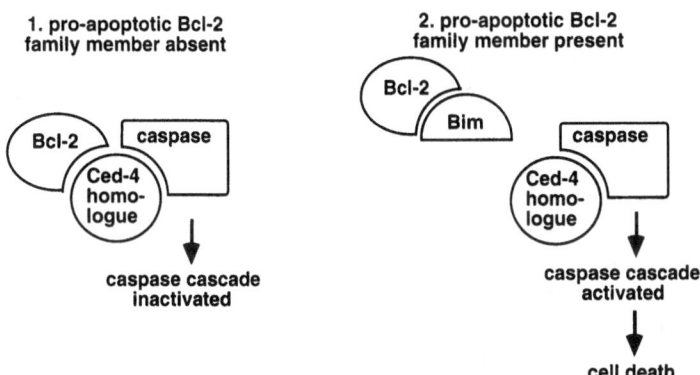

Fig. 2. A proposed system for the regulation of apoptosis by the Bcl-2 family

members in this system is as inactivators of the anti-apoptotic members, although again it is not clear whether this inactivation is by direct competition for a common binding site or by steric hindrance.

This model predicts that caspase activation is the critical step and constitutes the point of no return in apoptosis. We speculate that zymogens possess some low level enzymatic activity and that conversion into fully active caspases is brought about by increasing their local concentration. Caspases containing death effector domains are recruited by FADD/Mort-1 (and perhaps also by related molecules) to receptor complexes at the plasma membrane from where the caspase cascade is initiated (Kischkel et al. 1995; Boldin et al. 1996; Muzio et al. 1996; Medema et al. 1997). We believe that Bcl-2 family-inhibitable pathways to apoptosis are initiated at the subcellular localisations where these proteins reside, the nuclear envelope, endoplasmic reticulum and the outer mitochondrial membrane. Metabolic disturbances in these organelles, such as collapse of the mitochondrial transmembrane potential and release of cytochrome c into the cytosol (Liu et al. 1996; Marchetti et al. 1996; Kharbanda et al. 1997; Kluck et al. 1997; Yang et al. 1997), may be the initial trigger for apoptosis. This may then lead to Ced-4 protein mediated aggregation and consequent activation of caspase zymogens. Analogously to the interactions of death effector domains between caspase-8 and FADD/Mort-1, caspases with long pro-domains, such as caspase-2 (Nedd2/Ich-1) (Kumar et al. 1994; Wang et al. 1994), are likely to be the ones that interact with Ced-4. Caspases with short pro-domains, such as caspase-3, may serve as downstream effectors in the proteolytic cascade. It appears that mammals (and possibly also other species) have evolved multiple caspases with specialised functions, multiple Bcl-2 proteins and distinct pathways to apoptosis (Bcl-2-sensitive and Bcl-2-insensitive) to allow for the subtle control of cell death that is required for the development of complex multi-cellular organisms.

7
The Bcl-2 Protein Family and Disease

That abnormalities in the Bcl-2 protein family can play a role in disease processes is now beyond dispute. As mentioned in Section 2, *bcl*-2 itself was originally isolated as a proto-oncogene (Tsujimoto et al. 1984; Bakhshi et al. 1985;
Cleary et al. 1986), and transgenic mice overexpressing Bcl-2 in lymphocytes
are clearly more prone to tumourigenesis than control littermates (Strasser et
al. 1990a; McDonnell and Korsmeyer 1991; Strasser et al. 1993; Linette et al.
1995). On its own *bcl*-2 is a relatively weak transforming oncogene, but it
synergises potently in lymphomagenesis with growth promoting oncogenes,
such as *c-myc* and *pim*-1, in doubly transgenic mice (Strasser et al. 1990a;
Acton et al. 1992). This has led to the conclusion that Bcl-2 overexpression
functions in neoplastic transformation by extending the lifespan of cells,
thereby facilitating acquisition of further oncogenic mutations.

Recent experiments have provided evidence that the pro-apoptotic members of the Bcl-2 family can act as tumour suppressors. Somatic frameshift
mutations in the *bax* gene have been found in some cases of colon cancer with
the microsatellite mutator phenotype (Rampino et al. 1997). Even more con-
vincingly, it was shown that transformation of choroid plexus epithelial cells
by transgenic expression of a truncated version of the SV40 large T antigen
(which inactivates Rb but not p53) is accelerated in $bax^{-/-}$ mice (Yin et al.
1997).

Chemotherapeutic anti-cancer drugs and radiation induce apoptosis in tumour cells (Kerr et al. 1972; Wyllie et al. 1980). Oncogenes and tumour suppressor genes which regulate cell death influence the response of tumour cells
to anti-cancer therapy. Over-expression of Bcl-2 or inactivation of Bax not
only provide short-term protection against apoptosis, but also can significantly increase long-term survival, with retention of clonogenicity, in tumour
cells that have been treated with anti-cancer drugs or γ-radiation (Strasser et
al. 1994b; McCurrach et al. 1997). Thus, the response of cancer cells to therapy
is determined by at least two processes: the propensity to undergo "mitotic
death" and the sensitivity to apoptotic stimuli.

Inappropriate survival of cells does not have consequences only for
tumourigenesis. The immune response is characterised by rapid cell proliferation in response to pathogens and equally rapid apoptotic elimination of
responder cells after the pathogen has been eliminated (Golstein et al. 1991;
Strasser 1995a,b). Abnormally prolonged survival of activated lymphocytes,
which produce effector molecules that may be damaging to the host, can have
dire consequences. Over-expression of Bcl-2 in B lymphocytes of transgenic
mice results in prolonged humoral immune response and pathological accumulation of plasma cells, which can eventually lead to systemic lupus
erythematosus-like fatal autoimmune disease (Strasser et al. 1991b, 1992). The
discovery that mutations in CD95 or its ligand cause lymphadenopathy and

autoimmunity in mice (Watanabe-Fukunaga et al. 1992; Lynch et al. 1994; Takahashi et al. 1994; Adachi et al. 1995) and humans (Fisher et al. 1995; Rieux-Laucat et al. 1995) provided additional support for the notion that the normally short lifespan of effector lymphocytes constitutes a vital barrier against autoimmune attack.

The Bcl-2 protein family also plays a role in viral infection. Apoptosis is used by multi-cellular organisms as a mechanism to reduce viral spread (Haecker and Vaux 1994; Vaux et al. 1994). It has been postulated that host cells have stress sensors which detect metabolic disturbances caused by the virus and trigger the apoptotic machinery. As a counter-attack, several viruses have co-evolved Bcl-2 homologues (see Sect. 3) and presumably utilise them to enhance virus production by blocking the apoptotic response in infected host cells (White et al. 1992; Henderson et al. 1993; Neilan et al. 1993; Vaux et al. 1994; Cheng et al. 1997; Nava et al. 1997; Sarid et al. 1997).

8
Future Directions

New members of the Bcl-2 family are still being discovered, albeit less frequently, and the functional analysis of this protein family is by no means complete. Detailed functional data, including mutagenesis, generation of transgenic mice and gene knock-out mice, are yet to be published for the more recently discovered members. The possibility that pro-apoptotic Bcl-2 family proteins can act as tumour suppressors deserves further attention. Quantitative biochemical data on the nature of intrafamily interactions are largely unavailable and are required for better understanding of their activity. The discovery of the interaction between 14-3-3 proteins and Bad is exciting and we anticipate that other mechanisms for regulating the activity of pro-apoptotic Bcl-2 family members will soon emerge. The recent linking of the Bcl-2 family to the caspases via Ced-4 will also act as a springboard for new investigation. The discovery of mammalian Ced-4 homologues must be imminent and will undoubtedly lead to new insight into the function of the Bcl-2 family, and thereby bring us closer to a true understanding of apoptosis.

Acknowledgements. Current work by the authors is supported by the National Health and Medical Research Council (Canberra), a Clinical Investigator Award from the Cancer Research Institute (New York), the Anti-Cancer Council of Victoria, the Hazel and Pip Appel Foundation, the Australian and New Zealand Bank trust, a grant from the US National Cancer Institute (#CA43540; principal investigator S. Cory) and the Howard Hughes Medical Institute (#75193-531101; principal investigator S. Cory). L.O'C. is supported by an Edith Moffat Ph.D. scholarship from the Walter and Eliza Hall Institute of Medical Research. A. S. has been supported by fellowships from the Leukemia Society of America, the Swiss National Science Foundation and the L. and Th. La Roche Foundation and is currently a Scholar of the Leukemia Society of America. We thank all our past and present collaborators, particularly Drs. J. Adams, S. Cory, D.

Huang, D. Vaux, L. O'Reilly and A. Harris, for stimulating discussions and for their input into our work and J. Tyers and E. Bonnici for editorial assistance. We apologise to those scientists in the field whose work was mentioned in the references only indirectly through reviews.

References

Acton D, Domen J, Jacobs H, Vlaar M, Korsmeyer S, Berns A (1992) Collaboration of PIM-1 and BCL-2 in lymphomagenesis. Curr Top Microbiol Immunol 182:293–298

Adachi M, Suematsu S, Kondo T, Ogasawara J, Tanaka T, Yoshida N, Nagata S (1995) Targeted mutation in the *Fas* gene causes hyperplasia in peripheral lymphoid organs and liver. Nat Genet 11:294–300

Ahmad M, Srinivasula SM, Wang LJ, Litwack G, Fernandes-Alnemri T, Alnemri ES (1997) *Spodoptera frugiperda* caspase-1, a novel insect death protease that cleaves the nuclear immunophilin fkbp46, is the target of the baculovirus antiapoptotic protein p35. J Biol Chem 272:1421–1424

Aitken A (1995) 14-3-3 proteins on the MAP TIBS 20:95–97

Akashi K, Kondo M, von Freeden-Jeffry U, Murray R, Weissman IL (1997) Bcl-2 rescues T lymphopoiesis in interleukin-7 receptor-deficient mice. Cell 89:1033–1041

Alderson MR, Tough TW, Davis-Smith T, Braddy S, Falk B, Schooley KA, Goodwin RG, Smith CA, Ramsdell F, Lynch DH (1995) Fas ligand mediates activation-induced cell death in human T lymphocytes. J Exp Med 181:71–77

Alnemri ES, Livingston DJ, Nicholson DW, Salvesen G, Thornberry NA, Wong WW, Yuan J (1996) Human ICE/CED-3 protease nomenclature. Cell 87:171

Ameisen JC, Estaquier J, Idziorek T (1994) From AIDS to parasite infection: pathogen-mediated subversion of programmed cell death as a mechanism for immune dysregulation. Immunol Rev 142:9–51

Baffy G, Miyashita T, Williamson JR, Reed JC (1993) Apoptosis induced by withdrawal of interleukin-3 from an IL-3-dependent hematopoietic cell line is associated with repartitioning of intracellular calcium and is blocked by enforced Bcl-2 oncoprotein production. J Biol Chem 268:6511–6519

Bakhshi A, Jensen JP, Goldman P, Wright JJ, McBride OW, Epstein AL, Korsmeyer SJ (1985) Cloning the chromosomal breakpoint of t(14;18) human lymphomas: clustering around J_H on chromosome 14 and near a transcriptional unit on 18. Cell 41:899–906

Beals CR, Clipstone NA, Ho SN, Crabtree GR (1997) Nuclear localization of NF-ATc by a calcineurin-dependent, cyclosporin-sensitive intramolecular interaction. Genes Dev 11:824–834

Beidler DR, Tewari M, Friesen PD, Poirier G, Dixit VM (1995) The baculovirus p35 protein inhibits Fas- and tumor necrosis factor-induced apoptosis. J Biol Chem 270:16526–16528

Bertin J, Mendrysa SM, LaCount DJ, Gaur S, Krebs JF, Armstrong RC, Tomaselli KJ, Friesen PD (1996) Apoptotic suppression by baculovirus p35 involves cleavage by and inhibition of a virus-induced CED-3/ICE-like protease. J Virol 70:6251–6259

Bodmer J-L, Burns K, Schneider P, Hofmann K, Steiner V, Thome M, Bornand T, Hahne M, Schröter M, Becker K, Wilson A, French LE, Browning JL, MacDonald HR, Tschopp J (1997) TRAMP, a novel apoptosis-mediating receptor with sequence homology to tumor ecrosis factor receptor 1 and Fas (Apo-1/CD95). Immunity 6:79–88

Boise LH, Gonzalez-Garcia M, Postema CE, Ding L, Lindsten T, Turka LA, Mao X, Nunez G, Thompson CB (1993) *bcl-x*, a *bcl*-2-related gene that functions as a dominant regulator of apoptotic cell death. Cell 74:597–608

Boise LH, Minn AJ, June CH, Lindsten T, Thompson CB (1995) Growth factors can enhance lymphocyte survival without committing the cell to undergo cell division. Proc Natl Acad Sci USA 92:5491–5495

Boldin MP, Varfolomeev EE, Pancer Z, Mett IL, Camonis JH, Wallach D (1995) A novel protein that interacts with the death domain of Fas/APO1 contains a sequence motif related to the death domain. J Biol Chem 270:7795–7798

Boldin MP, Goncharov TM, Goltsev YV, Wallach D (1996) Involvement of MACH, a novel MORT1/FADD-interacting protease, in Fas/APO-1- and TNF receptor-induced cell death. Cell 85:803–815

Borner C, Martinou I, Mattmann C, Irmler M, Schaerer E, Martinou J-C, Tschopp J (1994) The protein bcl-2α does not require membrane attachment, but two conserved domains to suppress apoptosis. J Cell Biol 126:1059–1068

Borzillo GV, Endo K, Tsujimoto Y (1994) bcl-2 confers growth and survival advantage to interleukin 7-dependent early pre-B cells which become factor independent by a multistep process in culture. Oncogene 7:869–876

Boyd JM, Malstrom S, Subramanian T, Venkatesh LK, Schaeper U, Elangovan B, D'Sa-Eipper C, Chinnadurai G (1994) Adenovirus E1B 19 kDa and Bcl-2 proteins interact with a common set of cellular proteins. Cell 79:341–351

Boyd JM, Gallo GJ, Elangovan B, Houghton AB, Malstrom S, Avery BJ, Ebb RG, Subramanian T, Chittenden T, Lutz RJ, Chinnadurai G (1995) Bik, a novel death-inducing protein shares a distinct sequence motif with Bcl-2 family proteins and interacts with viral and cellular survival-promoting proteins. Oncogene 11:1921–1928

Brady HJM, Gil-Gómez G, Kirberg J, Berns AJM (1996a) Baxα perturbs T cell development and affects cell cycle entry of T cells. EMBO J 15:6991–7001

Brady HJM, Salomns GS, Bobeldijk RC, Berns AJM (1996b) T cells from baxα transgenic mice show accelerated apoptosis in response to stimuli but do not show restored DNA damage-induced cell death in the absence of p53. EMBO J 15:1221–1230

Brunner T, Mogil RJ, LaFace D, Yoo NJ, Mahboubi A, Echeverri F, Martin SJ, Force WR, Lynch DH, Ware CF, Green DR (1995) Cell-autonomous Fas (CD95)/Fas-ligand interaction mediates activation-induced apoptosis in T-cell hybridomas. Nature 373:441–444

Bump NJ, Hackett M, Hugunin M, Seshagiri S, Brady K, Chen P, Ferenz C, Franklin S, Ghayur T, Li P, Licari P, Mankovich J, Shi L, Greenberg AH, Miller LK, Wong WW (1995) Inhibition of ICE family proteases by baculovirus antiapoptotic protein p35. Science 269:1885–1888

Carrió R, López-Hoyos M, Jimeno J, Benedict MA, Merino R, Benito MA, Fernández-Luna JL, Núñez G, García-Porrero JA, Merino J (1996) A1 demonstrates restricted tissue distribution during embryonic development and functions to protect against cell death. Am J Pathol 149:2133–2142

Casciola-Rosen L, Nicholson DW, Chong T, Rowan KR, Thornberry NA, Miller DK, Rosen A (1996) Apopain/CPP32 cleaves proteins that are essential for cellular repair: a fundamental principle of apoptotic death. J Exp Med 183:1957–1964

Chang BS, Minn AJ, Muchmore SW, Fesik SW, Thompson CB (1997) Identification of a novel regulatory domain in Bcl-x_L and Bcl-2. EMBO J 16:968–977

Chao DT, Linette GP, Boise LH, White LS, Thompson CB, Korsmeyer SJ (1995) Bcl-x(L) and Bcl-2 repress a common pathway of cell death. J Exp Med 182:821–828

Chen G, Branton PE, Yang E, Korsmeyer SJ, Shore GC (1996) Adenovirus E1B 19-kDa death suppressor protein interacts with Bax but not with Bad. J Biol Chem 271:24221–24225

Chen-Levy Z, Cleary ML (1990) Membrane topology of the Bcl-2 proto-oncogenic protein demonstrated in vitro. J Biol Chem 265:4929–4933

Chen-Levy Z, Nourse J, Cleary ML (1989) The bcl-2 candidate proto-oncogene product is a 24-kilodalton integral-membrane protein highly expressed in lymphoid cell lines and lymphomas carrying the t(14;18) translocation. Mol Cell Biol 9:701–710

Cheng EH-Y, Levine B, Boise LH, Thompson CG, Hardwick JM (1996) Bax-independent inhibition of apoptosis by Bcl-x_L. Nature 379:554–556

Cheng EH-Y, Nicholas J, Bellows DS, Hayward GS, Guo H-G, Reitz MS, Hardwick JM (1997) A Bcl-2 homolog encoded by Kaposi sarcoma-associated virus, human herpesvirus 8, inhibits apoptosis but does not heterodimerize with Bax or Bak. Proc Natl Acad Sci USA 94:690–694

Chinnaiyan AM, O'Rourke K, Tewari M, Dixit VM (1995) FADD, a novel death domain-containing protein, interacts with the death domain of Fas and initiates apoptosis. Cell 81:505–512

Chinnaiyan AM, Orth K, O'Rourke K, Duan H, Poirier GG, Dixit VM (1996a) Molecular ordering of the cell death pathway. J Biol Chem 271:4573–4576

Chinnaiyan AM, Tepper CG, Seldin MF, O'Rourke K, Kischkel FC, Hellbardt S, Krammer PH, Peter ME, Dixit VM (1996b) FADD/MORT1 is a common mediator of CD95 (Fas/APO-1) and tumor necrosis factor receptor-induced apoptosis. J Biol Chem 271:4961–4965

Chinnaiyan AM, O'Rourke K, Lane BR, Dixit VM (1997) Interaction of CED-4 with CED-3 and CED-9: a molecular framework for cell death. Science 275:1122–1126

Chittenden T, Flemington C, Houghton AB, Ebb RG, Gallo GJ, Elangovan B, Chinnadurai G, Lutz RJ (1995a) A conserved domain in Bak, distinct from BH1 and BH2, mediates cell death and protein binding functions. EMBO J 14:5589–5596

Chittenden T, Harrington EA, O'Connor R, Flemington C, Lutz RJ, Evan GI, Guild BC (1995b) Induction of apoptosis by the Bcl-2 homologue Bak. Nature 374:733–736

Chung J, Deutsch HH, Kalthoff FS (1996) IL-4-dependent proliferation of BA/F3 cells expressing a growth-negative mutant of the human IL-4 receptor is restored by enforced expression of Bcl-2. J Leuk Biol 59:586–590

Clarke AR, Purdie CA, Harrison DJ, Morris RG, Bird CC, Hooper ML, Wyllie AH (1993) Thymocyte apoptosis induced by p53-dependent and independent pathways. Nature 362:849–852

Clarke AR, Gledhill S, Hooper ML, Bird CC, Wyllie AH (1994) p53 dependence of early apoptotic and proliferative responses within the mouse intestinal epithelium following γ-irradiation. Oncogene 9:1767–1773

Cleary ML, Sklar J (1987) Formation of a hybrid *bcl*-2/immunoglobulin transcript as a result of t(14;18) chromosomal translocation. Hämatol Bluttransfus 31:314–319

Cleary ML, Smith SD, Sklar J (1986) Cloning and structural analysis of cDNAs for *bcl*-2 and a hybrid *bcl*-2/immunoglobulin transcript resulting from the t(14;18) translocation. Cell 47:19–28

Cohen JJ, Duke RC, Fadok VA, Sellins KS (1992) Apoptosis and programmed cell death in immunity. Annu Rev Immunol 10:267–293

Cory S (1995) Regulation of lymphocyte survival by the *BCL*-2 gene family. Annu Rev Immunol 13:513–543

Cory S, Harris AW, Strasser A (1994) Insights from transgenic mice regarding the role of *bcl*-2 in normal and neoplastic lymphoid cells. Proc R Soc Lond 345:289–295

Datta R, Kojima H, Banach D, Bump NJ, Talanian RV, Alnemri ES, Weichselbaum RR, Wong WW, Kufe DW (1997) Activation of a CrmA-insensitive, p35-sensitive pathway in ionizing radiation-induced apoptosis. J Biol Chem 272:1965–1969

de Jong D, Prins F, van Krieken HHJM, Mason DY, van Ommen G, Kluin PM (1992) Subcellular localization of *bcl*-2 protein. Curr Top Microbiol Immunol 182:287–292

de Jong D, Prins FA, Mason DY, Reed JC, van Ommen GB, Kluin PM (1994) Subcellular localization of the bcl-2 protein in malignant and normal lymphoid cells. Cancer Res 54:256–260

Deng G, Podack ER (1993) Suppression of apoptosis in a cytotoxic T-cell line by interleukin-2-mediated gene transcription and deregulated expression of the protooncogene *bcl*-2. Proc Natl Acad Sci USA 90:2189–2193

Dhein J, Walczak H, Baumler C, Debatin K-M, Krammer PH (1995) Autocrine T-cell suicide mediated by APO-1/(Fas/CD95). Nature 373:438–441

Diaz J-L, Oltersdorf T, Horne W, McConnell M, Wilson G, Weeks S, Garcia T, Fritz LC (1997) A common binding site mediates heterodimerization and homodimerization of Bcl-2 family members. J Biol Chem 272:11350–11355

Ellis HM, Horvitz HR (1986) Genetic control of programmed cell death in the nematode *C. elegans*. Cell 44:817–829

Ellis RE, Yuan J, Horvitz HR (1991) Mechanisms and functions of cell death. Annu Rev Cell Biol 7:663–698

Farlie PG, Dringen R, Rees SM, Kannourakis G, Bernard O (1995) *bcl-2* transgene expression can protect neurons against developmental and induced cell death. Proc Natl Acad Sci USA 92:4397–4401

Farrow SN, White JHM, Martinou I, Raven T, Pun K-T, Grinham CJ, Martinou J-C, Brown R (1995) Cloning of a *bcl-2* homologue by interaction with adenovirus E1B 19K. Nature 374:731–733

Fernandez-Sarabia MJ, Bischoff JR (1993) Bcl-2 associates with the *ras*-related protein R-*ras* p23. Nature 366:274–275

Fisher GH, Rosenberg FJ, Straus SE, Dale JK, Middelton LA, Lin AY, Strober W, Lenardo MJ, Puck JM (1995) Dominant interfering Fas gene mutations impair apoptosis in a human autoimmune lymphoproliferative syndrome. Cell 81:935–946

Fraser AG, Evan GI (1997) Identification of a *Drosophila melanogaster* ICE/CED-3-related protease, drICE. EMBO J 16:2805–2813

Furuchi T, Masuko K, Nishimune Y, Obinata M, Matsui Y (1996) Inhibition of testicular germ cell apoptosis and differentiation in mice misexpressing Bcl-2 in spermatogonia. Development 122:1703–1709

Garcia I, Martinou I, Tsujimoto Y, Martinou J-C (1992) Prevention of programmed cell death of sympathetic neurons by the *bcl-2* proto-oncogene. Science 258:302–304

Gibson L, Holmgreen S, Huang DCS, Bernard O, Copeland NG, Jenkins NA, Sutherland GR, Baker E, Adams JM, Cory S (1996) *bcl-w*, a novel member of the *bcl-2* family, promotes cell survival. Oncogene 13:665–675

Golstein P (1997) Controlling cell death. Science 275:1081–1082

Golstein P, Ojcius DM, Young D-E (1991) Cell death mechanisms and the immune system. Immunol Rev 121:29–65

Gonzalez-Garcia M, Perez-Ballestero R, Ding L, Duan L, Boise LH, Thompson CB, Nunez G (1994) Bcl-X$_L$ is the major bcl-x mRNA form expressed during murine development and its product localizes to mitochondria. Development 120:3033–3042

Graninger WB, Seto M, Boutain B, Goldman P, Korsmeyer SJ (1987) Expression of *Bcl-2* and *Bcl-2*-Ig fusion transcripts in normal and neoplastic cells. J Clin Invest 80:1512–1515

Gratiot-Deans J, Merino R, Nunez G, Turka LA (1994) Bcl-2 expression during T-cell development: early loss and late return occur at specific stages of commitment to differentiation and survival. Proc Natl Acad Sci USA 91:10685–10689

Grether ME, Abrams JM, Agapite J, White K, Steller H (1995) The *head involution defective* gene of *Drosophila melanogaster* functions in programmed cell death. Genes Dev 9:1694–1708

Grillot DAM, Merino R, Nunez G (1995) Bcl-x$_L$ displays restricted distribution during T cell development and inhibits multiple forms of apoptosis but not clonal deletion in transgenic mice. J Exp Med 182:1973–1983

Haecker G, Vaux DL (1994) Viral, worm and radical implications for apoptosis. Trends Biochem Sci 19:99–100

Haldar S, Beatty C, Tsujimoto Y, Croce CM (1989) The *bcl-2* gene encodes a novel G protein. Nature 342:195–198

Han J, Sabbatini P, Perez D, Rao L, Modha D, White E (1996a) The E1B 19K protein blocks apoptosis by interacting with and inhibiting the p53-inducible and death-promoting Bax protein. Genes Dev 10:461–477

Han J, Sabbatini P, White E (1996b) Induction of apoptosis by human Nbk/Bik, a BH3-containing protein that interacts with E1B 19K.Mol Cell Biol 16:5857–5864

Hanada M, Aime-Sempe C, Sato T, Reed JC (1995) Structure-function analysis of Bcl-2 protein. Identification of conserved domains important for homodimerization with Bcl-2 and heterodimerization with Bax. J Biol Chem 270:11962–11969

Hartley SB, Cooke MP, Fulcher DA, Harris AW, Cory S, Basten A, Goodnow CC (1993) Elimination of self-reactive B lymphocytes proceeds in two stages: arrested development and cell death. Cell 72:325–335

Hay BA, Wolff T, Rubin GM (1994) Expression of baculovirus p35 prevents cell death in *Drosophila*. Development 120:2121–2129

Henderson S, Huen D, Rowe M, Dawson C, Johnson G, Rickinson A (1993) Epstein virus-coded BHRF 1 protein, a viral homologue of Bcl-2 protects human B cells from programmed cell death. Proc Natl Acad Sci USA 90:8479–8483

Hengartner MO, Horvitz HR (1994a) Activation of *C. elegans* cell death protein CED-9 by an amino-acid substitution in a domain conserved in Bcl-2. Nature 369:318–320

Hengartner MO, Horvitz HR (1994b) *C. elegans* cell survival gene *ced*-9 encodes a functional homolog of the mammalian proto-oncogene *bcl*-2. Cell 76:665–676

Hengartner MO, Horvitz HR (1994c) Programmed cell death in *Caenorhabditis elegans*. Curr Opin Genet Dev 4:581–586

Hengartner MO, Ellis RE, Horvitz HR (1992) *Caenorhabditis elegans* gene *ced*-9 protects cells from programmed cell death. Nature 356:494–499

Hockenbery D, Nunez G, Milliman C, Schreiber RD, Korsmeyer SJ (1990) *Bcl*-2 is an inner mitochondrial membrane protein that blocks programmed cell death. Nature 348:334–336

Hockenbery DM, Zutter M, Hickey W, Nahm M, Korsmeyer S (1991) BCL2 protein is topographically restricted in tissues characterized by apoptotic cell death. Proc Natl Acad Sci USA 88:6961–6965

Hockenbery DM, Oltvai ZN, Yin X-M, Milliman CL, Korsmeyer S (1993) Bcl-2 functions in an antioxidant pathway to prevent apoptosis. Cell 74:241–251

Hsu H, Xiong J, Goeddel DV (1995) The TNF receptor 1-associated protein TRADD signals cell death and NF-$\varkappa$B activation. Cell 81:495–504

Hsu H, Shu H-B, Pan M-G, Goeddel DV (1996) TRADD-TRAF2 and TRADD-FADD interactions define two distinct TNF receptor 1 signal transduction pathways. Cell 84:299–308

Huang DCS, Cory S, Strasser A (1997) Bcl-2 Bcl-x$_L$ and adenovirus protein E1B19kD are functionally equivalent in their ability to inhibit cell death. Oncogene 14:405–414

Huang DCS, O'Reilly LA, Strasser A, Cory S (1997) The anti-apoptosis function of Bcl-2 can be genetically separated from its inhibitory effect on cell cycle entry. EMBO J 16:4638–4648

Inohara N, Ding L, Chen S, Núñez G (1997) *harakiri*, a novel regulator of cell death, encodes a protein that activates apoptosis and interacts selectively with survival-promoting proteins Bcl-2 and Bcl-X$_L$. EMBO J 16:1686–1694

Ito T, Deng X, Carr B, May WS (1997) Bcl-2 phosphorylation required for anti-apoptosis function. J Biol Chem 272:11671–11673

Itoh N, Nagata S (1993) A novel protein domain required for apoptosis. J Biol Chem 268:10932–10937

Jäättelä M, Benedict M, Tewari M, Shayman JA, Dixit VM (1995) Bcl-x and Bcl-2 inhibit TNF and Fas-induced apoptosis and activation of phospholipase A$_2$ in breast carcinoma cells. Oncogene 10:2297–2305

Jacks T, Fazeli A, Schmitt EM, Bronson RT, Goodell MA, Weinberg RA (1992) Effects of an Rb mutation in the mouse. Nature 359:295–300

Jacobson MD, Raff MC (1995) Programmed cell death and Bcl-2 protection in very low oxygen. Nature 374:814–816

Jacobson MD, Burne JF, King MP, Miyashita T, Reed JC, Raff MC (1993) Bcl-2 blocks apoptosis in cells lacking mitochondrial DNA. Nature 361:365–369

Jacobson MD, Weil M, Raff MC (1997) Programmed cell death in animal development. Cell 88:347–354

Ju S-T, Panka DJ, Cui H, Ettinger R, El-Khatib M, Sherr DH, Stanger BZ, Marshak-Rothstein A (1995) Fas(CD95)/FasL interactions required for programmed cell death after T-cell activation. Nature 373:444–448

Kane DJ, Sarafian TA, Anton R, Hahn H, Gralla EB, Valentine JS, Ord T, Bredesen DE (1993) Bcl-2 inhibition of neural death: decreased generation of reactive oxygen species. Science 262:1274–1277

Katsumata M, Siegel RM, Louie DC, Miyashita T, Tsujimoto Y, Nowell PC, Greene MI, Reed JC (1992) Differential effects of Bcl-2 on T and B cells in transgenic mice. Proc Natl Acad Sci USA 89:11376–11380

Kerr JFR, Wyllie AH, Currie AR (1972) Apoptosis: a basic biological phenomenon with wide-ranging implications in tissue kinetics. Br J Cancer 26:239–257

Kharbanda S, Pandey P, Schofield L, Israels S, Roncinske R, Yoshida K, Bharti A, Yuan Z-M, Saxena S, Weichselbaum R, Nalin C, Kufe D (1997) Role for Bcl-x$_L$ as an inhibitor of cytosolic cytochrome c accumulation in DNA damage-induced apoptosis. Proc Natl Acad Sci USA 94:6939–6942

Kiefer MC, Brauer MJ, Powers VC, Wu JJ, Umansky SR, Tomei LD, Barr PJ (1995) Modulation of apoptosis by the widely distributed Bcl-2 homologue Bak. Nature 374:736–739

Kischkel FC, Hellbardt S, Behrmann I, Germer M, Pawlita M, Krammer PH, Peter ME (1995) Cytotoxicity-dependent APO-1 (Fas/CD95) – associated proteins form a death-inducing signaling complex (DISC) with the receptor. EMBO J 14:5579–5588

Kluck RM, Bossy Wetzel E, Green DR, Newmeyer DD (1997) The release of cytochrome c from mitochondria: a primary site for Bcl-2 regulation of apoptosis. Science 275:1132–1136

Knudson CM, Korsmeyer SJ (1997) Bcl-2 and Bax function independently to regulate cell death. Nat Genet 16:358–363

Knudson CM, Tung KSK, Tourtellotte WG, Brown GAJ, Korsmeyer SJ (1995) Bax-deficient mice with lymphoid hyperplasia and male germ cell death. Science 270:96–99

Komiyama T, Ray CA, Pickup DJ, Howard AD, Thornberry NA, Peterson EP, Salvesen G (1994) Inhibition of interleukin-1 beta converting enzyme by the cowpox virus serpin CrmA. An example of cross-class inhibition. J Biol Chem 269:19331–19337

Korsmeyer SJ (1992) Bcl-2: a repressor of lymphocyte death. Immunol Today 13:285–288

Korsmeyer SJ (1995) Regulators of cell death. Trends Genet 11:101–105

Kozopas KM, Yang T, Buchan HL, Zhou P, Craig RW (1993) MCL1, a gene expressed in programmed myeloid cell differentiation, has sequence similarity to bcl-2. Proc Natl Acad Sci USA 90:3516–3520

Krajewski S, Tanaka S, Takayama S, Schibler MJ, Fenton W, Reed JC (1993) Investigation of the subcellular distribution of the bcl-2 oncoprotein: residence in the nuclear envelope, endoplasmic reticulum, and outer mitochondrial membranes. Cancer Res 53:4701–4714

Krajewski S, Krajewska M, Shabaik A, Miyashita T, Wang HG, Reed JC (1994a) Immunohistochemical determination of in vivo distribution of Bax, a dominant inhibitor of Bcl-2. Am J Pathol 145:1323–1336

Krajewski S, Krajewska M, Shabaik A, Wang HG, Irie S, Fong L, Reed JC (1994b) Immunohistochemical analysis of in vivo patterns of Bcl-X expression. Cancer Res 54:5501–5507

Krajewski S, Bodrug S, Krajewska M, Shabaik A, Gascoyne R, Berean K, Reed JC (1995) Immunohistochemical analysis of Mcl-1 protein in human tissues. Am J Pathol 146:1309–1319

Krajewski S, Krajewska M, Reed JC (1996) Immunohistochemical analysis of in vivo patterns of bak expression, a proapoptotic member of the Bcl-2 protein family. Cancer Res 56:2849–2855

Kumar S (1995) ICE-like proteases in apoptosis. Trends Biochem Sci 20:198–202

Kumar S, Lavin MF (1996) The ICE family of cysteine proteases as effectors of cell death. Cell Death Differ 3:255–267

Kumar S, Kinoshita M, Noda M, Copeland NG, Jenkins NA (1994) Induction of apoptosis by the mouse Nedd2 gene, which encodes a protein similar to the product of the Caenorhabditis elegans cell death gene ced-3 and the mammalian IL-1β-converting enzyme. Genes Dev 8:1613–1626

Lacronique V, Mignon A, Fabre M, Viollet B, Rouquet N, Molina T, Porteu A, Henrion A, Bouscary D, Varlet P, Joulin V, Kahn A (1996) Bcl-2 protects from lethal hepatic apoptosis induced by an anti-Fas antibody in mice. Nat Med 2:80–86

Lagasse E, Weissman IL (1994) Bcl-2 inhibits apoptosis of neutrophils but not their engulfment by macrophages. J Exp Med 179:1047–1052

Lagasse E, Weissman IL (1997) Enforced expression of Bcl-2 in monocytes rescues macrophages and partially reverses osteopetrosis in *op/op* mice. Cell 89:1021–1031

Lin EY, Orlofsky A, Berger MS, Prystowsky MB (1993) Characterization of A1, a novel hemopoietic-specific early-response gene with sequence similarity to *bcl-2*. J Immunol 151:1979–1988

Lin EY, Orlofsky A, Wang H-G, Reed JC, Prystowsky MB (1996) A1, a *bcl-2* family member, prolongs cell survival and permits myeloid differentiation. Blood 87:983–992

Linette GP, Grusby MJ, Hedrick SM, Hansen TH, Glimcher LH, Korsmeyer SJ (1994) Bcl-2 is upregulated at the $CD4^+CD8^+$ stage during positive selection and promotes thymocyte differentiation at several control points. Immunity 1:197–205

Linette GP, Hess JL, Sentman CL, Korsmeyer SJ (1995) Peripheral T-cell lymphoma in lckpr-bcl-2 transgenic mice. Blood 86:1255–1260

Linette GP, Li Y, Roth K, Korsmeyer SJ (1996) Cross talk between cell death and cell cycle progression: BCL-2 regulates NFAT-mediated activation. Proc Natl Acad Sci USA 93:9545–9552

Lithgow T, van Driel R, Bertram JF, Strasser A (1994) The protein product of the oncogene *bcl-2* is a component of the nuclear envelope, the endoplasmic reticulum and the outer mitochondrial membrane. Cell Growth Differ 5:411–417

Liu X, Kim CN, Yang J, Jemmerson R, Wang X (1996) Induction of apoptotic program in cell-free extracts: requirement for dATP and cytochrome c. Cell 86:147–157

Liu X, Zou H, Slaughter C, Wang X (1997) DFF, a heterodimeric protein that functions downstream of caspase-3 to trigger DNA fragmentation during apoptosis. Cell 89:175–184

Lowe SW, Schmitt EM, Smith SW, Osborne BA, Jacks T (1993) p53 is required for radiation-induced apoptosis in mouse thymocytes. Nature 362:847–849

Lynch DH, Watson ML, Alderson MR, Baum PR, Miller RE, Tough T, Gibson M, Davis-Smith T, Smith CA, Hunter K, Bhat D, Din W, Goodwin RG, Seldin MF (1994) The mouse fas-ligand gene is mutated in *gld* mice and is part of a TNF family gene cluster. Immunity 1:131–136

Macleod KF, Hu Y, Jacks T (1996) Loss of *Rb* activates both *p53*-dependent and independent cell death pathways in the developing mouse nervous system. EMBO J 15:6178–6188

Manji GA, Hozak RR, LaCount DJ, Friesen PD (1997) Baculovirus inhibitor of apoptosis functions at or upstream of the apoptotic suppressor p35 to prevent programmed cell death. J Virol 71:4509–4516

Maraskovsky E, O'Reilly LA, Teepe M, Corcoran LM, Peschon JJ, Strasser A (1997) Bcl-2 can rescue T lymphocyte development in interleukin-7 receptor-deficient mice but not in mutant *rag-1*$^{-/-}$ mice. Cell 89:1011–1019

Marchetti P, Castedo M, Susin SA, Zamzami N, Hirsch T, Macho A, Haeffner A, Hirsch F, Geuskens M, Kroemer G (1996) Mitochondrial permeability transition is a central coordinating event of apoptosis. J Exp Med 184:1155–1160

Martinou I, Fernandez PA, Missotten M, White E, Allet B, Sadoul R, Martinou JC (1995) Viral proteins E1B19K and p35 protect sympathetic neurons from cell death induced by NGF deprivation. J Cell Biol 128:201–208

Martinou J-C, Dubois-Dauphin M, Staple JK, Rodriguez I, Frankowski H, Missotten M, Albertini P, Talabot D, Catsicas S, Pietra C, Huarte J (1994) Overexpression of Bcl-2 in transgenic mice protects neurons from naturally occurring cell death and experimental ischemia. Neuron 13:1017–1030

Marvel J, Perkins GR, Lopez-Rivas A, Collins MKL (1994) Growth factor starvation of bcl-2 overexpressing murine bone marrow cells induced refractoriness of IL-3 stimulation of proliferation. Oncogene 9:1117–1122

Matsuzaki Y, Nakayama K-I, Nakayama K, Tomita T, Isoda M, Loh DY, Nakauchi H (1997) Role of bcl-2 in the development of lymphoid cells from the hematopoietic stem cell. Blood 89:853–862

Mazel S, Burtrum D, Petrie HT (1996) Regulation of cell division cycle progression by *bcl-2* expression: a potential mechanism for inhibition of programmed cell death. J Exp Med 183:2219–2226

McCurrach ME, Connor TMF, Knudson CM, Korsmeyer SJ, Lowe SW (1997) *bax*-deficiency promotes drug resistance and oncogenic transformation by attenuating p53-dependent apoptosis. Proc Natl Acad Sci USA 94:2345–2349

McDonnell TJ, Korsmeyer SJ (1991) Progression from lymphoid hyperplasia to high-grade malignant lymphoma in mice transgenic for the t(14;18) Nature 349:254–256

McDonnell TJ, Deane N, Platt FM, Nunez G, Jaeger U, McKearn JP, Korsmeyer SJ (1989) *bcl-2*-immunoglobulin transgenic mice demonstrate extended B cell survival and follicular lymphoproliferation. Cell 57:79–88

McDonnell TJ, Nunez G, Platt FM, Hockenbery D, London L, McKearn JP, Korsmeyer SJ (1990) Deregulated *Bcl-2*-immunoglobulin transgene expands a resting but responsive immunoglobulin M and D-expressing B-cell population. Mol Cell Biol 10:1901–1907

Medema JP, Scaffidi C, Kischkel FC, Shevchenko A, Mann M, Krammer PH, Peter ME (1997) FLICE is activated by association with the CD95 death-inducing signaling complex (DISC) EMBO J 16:2794–2804

Memon SA, Moreno MB, Petrak D, Zacharchuk CM (1995) Bcl-2 blocks glucocorticoid – but not Fas – or activation-induced apoptosis in a T cell hybridoma. J Immunol 115:4644–4652

Merino R, Ding L, Veis DJ, Korsmeyer SJ, Nuñez G (1994) Developmental regulation of the Bcl-2 protein and susceptibility to cell death in B lymphocytes. EMBO J 13:683–691

Merry DE, Veis DJ, Hickey WF, Korsmeyer SJ (1994) Bcl-2 protein expression is widespread in the developing nervous system and retained in the adult PNS. Development 120:301–311

Minn AJ, Vélez P, Schendel SL, Liang H, Muchmore SW, Fesik SW, Fill M, Thompson CB (1997) Bcl-x$_L$ forms an ion channel in synthetic lipid membranes. Nature 385:353–357

Miura M, Friedlander RM, Yuan J (1995) Tumor necrosis factor-induced apoptosis is mediated by a CrmA-sensitive cell death pathway. Proc Natl Acad Sci USA 92:8318–8322

Miyashita T, Reed JC (1992) *bcl-2* gene transfer increases relative resistance of S49.1 and WEHI17.2 lymphoid cells to cell death and DNA fragmentation induced by glucocorticoids and multiple chemotherapeutic drugs. Cancer Res 52:5407–5411

Miyashita T, Krajewski S, Krajewska M, Wang HG, Lin HK, Liebermann DA, Hoffman B, Reed JC (1994) Tumor suppressor p53 is a regulator of *bcl-2* and *bax* gene expression in vitro and in vivo. Oncogene 9:1799–1805

Monaghan P, Robertson D, Amos TAS, Dyer MJS, Mason DY, Greaves MF (1992) Ultrastructural localization of BCL-2 protein. J Histochem. Cytochem. 40:1819–1825

Monica K, Chen Levy Z, Cleary ML (1990) Small G proteins are expressed ubiquitously in lymphoid cells and do not correspond to *Bcl-2*. Nature 346:189–191

Motoyama N, Wang FP, Roth KA, Sawa H, Nakayama K, Nakayama K, Negishi I, Senju S, Zhang Q, Fujii S, Loh DY (1995) Massive cell death of immature hematopoietic cells and neurons in Bcl-x deficient mice. Science 267:1506–1510

Muchmore SW, Sattler M, Liang H, Meadows RP, Harlan JE, Yoon HS, Nettesheim D, Chang BS, Thompson CB, Wong S-L, Ng S-C, Fesik SW (1996) X-ray and NMR structure of human Bcl-x$_L$, an inhibitor of programmed cell death. Nature 381:335–341

Muzio M, Chinnaiyan AM, Kischkel FC, O'Rourke K, Shevchenko A, Ni J, Scaffidi C, Bretz JD, Zhang M, Gentz R, Mann M, Krammer PH, Peter ME, Dixit VM (1996) FLICE, a novel FADD homologous ICE/CED-3-like protease, is recruited to the CD95 (Fas/Apo-1) death-inducing signaling complex. Cell 85:817–827

Nagata S (1997) Apoptosis by death factor. Cell 88:355–365

Naik P, Karrim J, Hanahan D (1996) The rise and fall of apoptosis during multistage tumorigenesis: down-modulation contributes to tumor progression from angiogenic progenitors. Genes Dev 10:2105–2116

Nakayama K, Nakayama K-I, Negishi I, Kuida K, Sawa H, Loh DY (1994) Targeted disruption of bcl-2αβ in mice: occurrence of gray hair, polycystic kidney disease, and lymphocytopenia. Proc Natl Acad Sci USA 91:3700–3704

Nakayama K-I, Nakayama K, Izumi N, Kulda K, Shinkai Y, Louie MC, Fields LE, Lucas PJ, Stewart V, Alt FW, Loh DY (1993) Disappearance of the lymphoid system in Bcl-2 homozygous mutant chimeric mice. Science 261:1884–1888

Naumovski L, Cleary ML (1996) The p53-binding protein 53BP2 also interacts with Bcl2 and impedes cell cycle progression at G_2/M. Mol Cell Biol 16:3884–3892

Nava VE, Cheng EH-Y, Veliuona M, Zou S, Clem RJ, Mayer ML, Hardwick JM (1997) Herpesvirus saimiri encodes a functional homolog of the human bcl-2 oncogene. J Virol 71:4118–4122

Neilan JG, Lu Z, Alfonso CL, Kutish GF, Sussman MD, Rock DL (1993) An African swine fever virus gene with similarity to the proto-oncogene bcl-2 and the Epstein–Barr virus gene BHRF1. J Virol 67:4391–4394

Nicholson DW, Ali A, Thornberry NA, Vaillancourt JP, Ding CK, Gallant M, Gareau Y, Griffin PR, Labelle M, Lazebnik YA, Munday NA, Raju SM, Smulson ME, Yamin T-T, Yu VL, Miller DK (1995) Identification and inhibition of the ICE/CED-3 protease necessary for mammalian apoptosis. Nature 376:37–43

Nisitani S, Tsubata T, Murakami M, Okamoto M, Honjo T (1993) The bcl-2 gene product inhibits clonal deletion of self-reactive B lymphocytes in the periphery but not in the bone marrow. J Exp Med 178:1247–1254

Nunez G, London L, Hockenbery D, Alexander M, McKearn JP, Korsmeyer SJ (1990) Deregulated Bcl-2 gene expression selectively prolongs survival of growth factor-deprived hemopoietic cell lines. J Immunol 144:3602–3610

O'Connor L, Strassar A, O'Reilly LA, Hausmann G, Adams J, Cory S, Huang DCS (1998) Bim: a novel member of the Bcl-2 family that promotes apoptosis. EMBO J 17:384–395

Oltvai ZN, Korsmeyer SJ (1994) Checkpoints of dueling dimers foil death wishes. Cell 79:189–192

Oltvai ZN, Milliman CL, Korsmeyer SJ (1993) Bcl-2 heterodimerizes in vivo with a conserved homolog, Bax, that accelerates programmed cell death. Cell 74:609–619

O'Reilly L, Huang DCS, Strasser A (1996) The cell death inhibitor Bcl-2 and its homologues influence control of cell cycle entry. EMBO J 15:6979–6990

O'Reilly LA, Harris AW, Strasser A (1997a) Bcl-2 transgene expression promotes survival and reduces proliferation of CD3⁻4⁻8⁻ T cell progenitors. Int Immunol 9:1291–1301

O'Reilly LA, Harris AW, Tarlinton DM, Corcoran LM, Strasser A (1997b) Expression of a bcl-2 transgene reduces proliferation and slows turnover of developing B cells in vivo. J Immunol 159:2301–2311

Orkin SH (1996) Development of the hematopoietic system. Curr Opin Genet Dev 6:597–602

Parker MW, Pattus F (1993) Rendering a membrane protein soluble in water: a common packing motif in bacterial protein toxins. TIBS 18:391–395

Pevny L, Simon MC, Robertson E, Klein WH, Tsai SF, D Agati V, Orkin SH, Costantini F (1991) Erythroid differentiation in chimeric mice blocked by a targeted mutation in the gene for transcription factor GATA-1. Nature 349:257–260

Raff MC (1992) Social controls on cell survival and cell death. Nature 356:397–400

Rampino N, Yamamoto H, Ionov Y, Li Y, Sawai H, Reed JC, Perucho M (1997) Somatic frameshift mutations in the BAX gene in colon cancers of the microsatellite mutator phenotype. Science 275:967–969

Ray CA, Black RA, Kronheim SR, Greenstreet GS, Pickup DJ (1992) Viral inhibition of inflammation: cowpox virus encodes an inhibitor of the interleukin-1β converting enzyme. Cell 69:597–604

Reed JC (1994) Bcl-2 and the regulation of programmed cell death. J Cell Biol 124:1–6

Reed JC (1996) Mechanisms of Bcl-2 family protein function and dysfunction in health and disease. Behring Inst Mitt 97:72–100

Reed JC (1997) Double identity for proteins of the Bcl-2 family. Nature 387:773–776

Reed JC, Miyashita T, Takayama S, Wang H-G, Sato T, Krajewski S, Aimé-Sempé C, Bodrug S, Kitada S, Hanada M (1996) BCL-2 family proteins: regulators of cell death involved in the pathogenesis of cancer and resistance to therapy. J Cell Biochem 60:23–32

Rieux-Laucat F, Le Deist F, Hivroz C, Roberts IAG, Debatin KM, Fischer A, de Villartay JP (1995) Mutations in Fas associated with human lymphoproliferative syndrome and autoimmunity. Science 268:1347–1349

Rodel JE, Link DC (1996) Suppression of apoptosis during cytokine deprivation of 32D cells is not sufficient to induce complete granulocytic differentiation. Blood 87:858–864

Rodriguez I, Matsuura K, Khatib K, Reed JC, Nagata S, Vassalli P (1996) A bcl-2 transgene expressed in hepatocytes protects mice from fulminant liver destruction but not from rapid death induced by anti-Fas antibody injection. J Exp Med 183:1031–1036

Rodriguez I, Araki K, Khatib K, Martinou J-C, Vassalli P (1997) Mouse vaginal opening is an apoptosis-dependent process which can be prevented by the overexpression of Bcl2. Dev Biol 184:115–121

Rotonda J, Nicholson DW, Fazil KM, Gallant M, Gareau Y, Labelle M, Peterson EP, Rasper DM, Ruel R, Vaillancourt JP, Thornberry NA, Becker JW (1996) The three-dimensional structure of apopain/cpp32, a key mediator of apoptosis. Nat Struct Biol 3:619–625

Rudel T, Bokoch GM (1997) Membrane and morphological changes in apoptotic cells regulated by caspase-mediated activation of PAK2. Science 276:1571–1574

Russell JH, Wang R (1993) Autoimmune *gld* mutation uncouples suicide and cytokine/proliferation pathways in activated, mature T cells. Eur J Immunol 23:2379–2382

Russell JH, Rush B, Weaver C, Wang R (1993) Mature T cells of autoimmune *lpr/lpr* mice have a defect in antigen-stimulated suicide. Proc Natl Acad Sci USA 90:4409–4413

Sarid R, Sato T, Bohenzky RA, Russo JJ, Chang Y (1997) Kaposi's sarcoma-associated herpesvirus encodes a functional Bcl-2 homologue. Nat Med 3:293–298

Sattler M, Liang H, Nettesheim D, Meadows RP, Harlan JE, Eberstadt M, Yoon HS, Shuker SB, Chang BS, Minn AJ, Thompson CB, Fesik SW (1997) Structure of Bcl-x$_L$-Bak peptide complex: recognition between regulators of apoptosis. Science 275:983–986

Savill J, Fadok V, Henson P, Haslett C (1993) Phagocyte recognition of cells undergoing apoptosis. Immunol Today 14:131–136

Schendel SL, Xie Z, Oblatt Montal M, Matsuyama S, Montal M, Reed JC (1997) Channel formation by antiapoptotic protein Bcl-2. Proc Natl Acad Sci USA 94:5113–5118

Sedlak TW, Oltvai ZN, Yang E, Wang K, Boise LH, Thompson CB, Korsmeyer SJ (1995) Multiple Bcl-2 family members demonstrate selective dimerizations with Bax. Proc Natl Acad Sci USA 92:7834–7838

Sentman CL, Shutter JR, Hockenbery D, Kanagawa O, Korsmeyer SJ (1991) *bcl*-2 inhibits multiple forms of apoptosis but not negative selection in thymocytes. Cell 67:879–888

Seshagiri S, Miller LK (1997) *Caenorhabditis elegans* CED-4 stimulates CED-3 processing and CED-3-induced apoptosis. Curr Biol 7:455–460

Shibasaki F, Kondo E, Akagi T, McKeon F (1997) Suppression of signalling through transcription factor NF-AT by interactions between calcineurin and Bcl-2. Nature 386:728–731

Shimizu S, Eguchi Y, Kosaka H, Kamiike W, Matsuda H, Tsujimoto Y (1995) Prevention of hypoxia-induced cell death by Bcl-2 and Bcl-x$_L$. Nature 374:811–813

Siegel RM, Katsumata M, Miyashita T, Louie DC, Greene MI, Reed JC (1992) Inhibition of thymocyte apoptosis and negative antigenic selection in *bcl*-2 transgenic mice. Proc Natl Acad Sci USA 89:7003–7007

Silva M, Grillot D, Benito A, Richard C, Nuñez G, Fernández-Luna JL (1996) Erythropoietin can promote erythroid progenitor survival by repressing apoptosis through Bcl-X$_L$ and Bcl-2. Blood 88:1576–1582

Simonian PL, Grillot DAM, Andrews DW, Leber B, Nuñez G (1996a) Bax homodimerization is not required for Bax to accelerate chemotherapy-induced cell death. J Biol Chem 271:32073–32077

Simonian PL, Grillot DAM, Merino R, Nuñez G (1996b) Bax can antagonize Bcl-X$_L$ during etoposide and cisplatin-induced cell death independently of its heterodimerization wtih Bcl-X$_L$. J Biol Chem 271:22764–22772

Smith KGC, Weiss U, Rajewsky K, Nossal GJV, Tarlinton DM (1994) BCL-2 increases memory B cell recruitment but does not perturb selection in germinal centers. Immunity 1:808–813

Smith KGC, Strasser A, Vaux DL (1996) CrmA expression in T lymphocytes of transgenic mice inhibits CD95 (Fas/APO-1)-transduced apoptosis, but does not cause lymphadenopathy or autoimmune disease. EMBO J 15:5167–5176

Song Z, McCall K, Steller H (1997) DCP-1 a *Drosophila* cell death protease essential for development. Science 275:536–540

Spector MS, Desnoyers S, Hoeppner DJ, Hengartner MO (1997) Interaction between the *C. elegans* cell-death regulators CED-9 and CED-4. Nature 385:653–656

Srinivasula SM, Ahmad M, Fernandes-Alnemri T, Litwack G, Alnemri ES (1996) Molecular ordering of the Fas-apoptotic pathway: the Fas/APO-1 protease Mch5 is a CrmA-inhibitable protease that activates multiple Ced-3/ICE-like cysteine proteases. Proc Natl Acad Sci USA 93:14486–14491

Steller H (1995) Mechanisms and genes of cellular suicide. Science 267:1445–1449

Strasser A (1995a) Death of a T cell. Nature 373:385–386

Strasser A (1995b) Life and death during lymphocyte development and function: evidence for two distinct killing mechanisms. Curr Opin Immunol 7:228–234

Strasser A, Anderson RL (1995) Bcl-2 and thermotolerance cooperate in cell survival. Cell Growth Differ 6:799–805

Strasser A, Harris AW, Bath ML, Cory S (1990a) Novel primitive lymphoid tumours induced in transgenic mice by cooperation between *myc* and *bcl*-2. Nature 348:331–333

Strasser A, Harris AW, Vaux DL, Webb E, Bath ML, Adams JM, Cory S (1990b) Abnormalities of the immune system induced by dysregulated *bcl*-2 expression in transgenic mice. Curr Top Microbiol Immunol 166:175–181

Strasser A, Harris AW, Cory S (1991a) Bcl-2 transgene inhibits T cell death and perturbs thymic self-censorship. Cell 67:889–899

Strasser A, Whittingham S, Vaux DL, Bath ML, Adams JM, Cory S, Harris AW (1991b) Enforced BCL2 expression in B-lymphoid cells prolongs antibody responses and elicits autoimmune disease. Proc Natl Acad Sci USA 88:8661–8665

Strasser A, Harris AW, Cory S (1992) The role of *bcl*-2 in lymphoid differentiation and transformation. Curr Top Microbiol Immunol 182:299–302

Strasser A, Harris AW, Cory S (1993) Eµ-*bcl*-2 transgene facilitates spontaneous transformation of early pre-B and immunoglobulin-secreting cells but not T cells. Oncogene 8:1–9

Strasser A, Harris AW, Corcoran LM, Cory S (1994a) *bcl*-2 expression promotes B but not T lymphoid development in *scid* mice. Nature 368:457–460

Strasser A, Harris AW, Jacks T, Cory S (1994b) DNA damage can induce apoptosis in proliferating lymphoid cells via p53-independent mechanisms inhibitable by Bcl-2. Cell 79:329–339

Strasser A, Harris AW, Von Boehmer H, Cory S (1994c) Positive and negative selection of T cells in T cell receptor transgenic mice expressing a *bcl*-2 transgene. Proc Natl Acad Sci USA 91:1376–1380

Strasser A, Harris AW, Huang DCS, Krammer PH, Cory S (1995) Bcl-2 and Fas/APO-1 regulate distinct pathways to lymphocyte apoptosis. EMBO J 14:6136–6147

Strasser A, O'Connor L, Huang DCS, O'Reilly LA, Stanley ML, Bath ML, Adams JM, Cory S, Harris AW (1996) Lessons from *bcl*-2 transgenic mice for immunology, cancer biology and cell death research. Behring Inst Mitt 97:101–117

Strasser A, Huang DCS, Vaux DL (1997) The role of the bcl-2/ced-9 gene family in cancer and general implications of defects in cell death control in tumourigenesis and resistance to chemotherapy. Biochim Biophys Acta 1333:F151–F178

Strobel T, Swanson L, Korsmeyer S, Cannistra SA (1996) BAX enhances paclitaxel-induced apoptosis through a p53-independent pathway. Proc Natl Acad Sci USA 93:14094–14099

Sugimoto A, Friesen PD, Rothman JH (1994) Baculovirus p35 prevents developmentally programmed cell death and rescues a ced-9 mutant in the nematode *Caenorhabditis elegans*.
EMBO J 13:2023–2028

Sulston JE, Horvitz HR (1977) Postembryonic cell lineages of the nematode *Caenorhabditis
elegans*. Dev Biol 56:110–156

Takahashi T, Tanaka M, Brannan CI, Jenkins NA, Copeland NG, Suda T, Nagata S (1994) Generalized lymphoproliferative disease in mice, caused by a point mutation in the Fas ligand. Cell
76:969–976

Takahashi T, Honda H, Hirai H, Tsujimoto Y (1997) Overexpressed Bcl-x$_L$ prevents bacterial
superantigen-induced apoptosis of thymocytes in vitro. Cell Death Differ 4:159–165

Takayama S, Sato T, Krajewski S, Kochel K, Irie S, Millan JA, Reed JC (1995) Cloning and
functional analysis of BAG-1: a novel Bcl-2-binding protein with anti-cell death activity. Cell
80:279–284

Talley AK, Dewhurst S, Perry SW, Dollard SC, Gummuluru S, Fine SM, New D, Epstein LG,
Gendelman HE, Gelbard HA (1995) Tumor necrosis factor alpha-induced apoptosis in human
neuronal cells: protection by the antioxidant *N*-acetylcysteine and the genes *bcl-2* and *crmA*.
Mol Cell Biol 15:2359–2366

Tao W, Teh S-J, Melhado I, Jirik F, Korsmeyer SJ, Teh H-S (1994) The T cell receptor repertoire
of CD4⁻8⁺ thymocytes is altered by overexpression of the *BCL-2* protooncogene in the thymus. J Exp Med 179:145–153

Tarlinton DM, Corcoran LM, Strasser A (1997) Cell survival and cell differentiation during B
lymphopoiesis are subject to distinct control. Int Immunol 9:1481–1494

Tartaglia LA, Ayres TM, Wong GHW, Goeddel DV (1993) A novel domain within the 55k TNF
receptor signals cell death. Cell 74:845–853

Tewari M, Dixit VM (1995) Fas- and tumor necrosis factor-induced apoptosis is inhibited by the
poxvirus crmA gene product. J Biol Chem 270:3255–3260

Thornberry NA, Rano TA, Peterson EP, Rasper DM, Timkey T, Garcia-Calvo M, Houtzager VM,
Nordstrom PA, Roy S, Vaillancourt JP, Chapman KT, Nicholson DW (1997) A combinatorial
approach defines specificities of members of the caspase family and granzyme B. J Biol Chem
272:17907–17911

Tsujimoto Y (1989) Stress-resistance conferred by high level of *bcl-2* alpha protein in human B
lymphoblastoid cell. Oncogene 4:1331–1336

Tsujimoto Y, Croce CM (1986) Analysis of the structure, transcripts, and protein products of *bcl-
2*, the gene involved in human follicular lymphoma. Proc Natl Acad Sci USA 83:5214–5218

Tsujimoto Y, Finger LR, Yunis J, Nowell PC, Croce CM (1984) Cloning of the chromosome
breakpoint of neoplastic B cells with the t(14;18) chromosome translocation. Science
226:1097–1099

Tsujimoto Y, Ikegaki N, Croce CM (1987) Characterization of the protein product of *bcl-2*, the
gene involved in human follicular lymphoma. Oncogene 2:3–7

Vairo G, Innes KM, Adams JM (1996) Bcl-2 has a cell cycle inhibitory function separable from its
enhancement of cell survival. Oncogene 13:1511–1519

Vanhaesebroeck B, Reed JC, de Valck D, Grooten J, Miyashita T, Tanaka S, Beyaert R, van Roy F,
Fiers W (1993) Effect of *bcl-2* proto-oncogene expression on cellular sensitivity to tumor
necrosis factor-mediated cytotoxicity. Oncogene 8:1075–1081

Vaux DL, Strasser A (1996) The molecular biology of apoptosis. Proc Natl Acad Sci USA 93:2239–
2244

Vaux DL, Cory S, Adams JM (1988) *Bcl-2* gene promotes haemopoietic cell survival and cooperates with c-*myc* to immortalize pre-B cells. Nature 335:440–442

Vaux DL, Haecker G, Strasser A (1994) An evolutionary perspective on apoptosis. Cell 76:777–779

Vaux DL, Weissman IL, Kim SK (1992) Prevention of programmed cell death in *Caenorhabditis
elegans* by human *bcl-2*. Science 258:1955–1957

Veis DJ, Sentman CL, Bach EA, Korsmeyer SJ (1993a) Expression of the Bcl-2 protein in murine
and human thymocytes and in peripheral T lymphocytes. J Immunol 151:2546–2554

Veis DJ, Sorenson CM, Shutter JR, Korsmeyer SJ (1993b) Bcl-2-deficient mice demonstrate fulminant lymphoid apoptosis, polycystic kidneys, and hypopigmented hair. Cell 75:229–240

Walker NPC, Talanian RV, Brady KD, Dang LC, Bump NJ, Ferenz CR, Franklin S, Ghayur T, Hackett MC, Hammill LD, Herzog L, Hugunin M, Houy W, Mankovich JA, McGuiness L, Orlewicz E, Paskind M, Pratt CA, Reis P, Summani A, Terranova M, Welch JP, Xiong L, Moller A, Tracey DE, Kamen R, Wong WW (1994) Crystal structure of cysteine protease interleukin-1β-converting enzyme: a (p20/p10)₂ homodimer. Cell 78:343–352

Walton MI, Whysong D, O'Connor PM, Hockenbery D, Korsmeyer SJ, Kohn KW (1993) Constitutive expression of human Bcl-2 modulates nitrogen mustard and camptothecin induced apoptosis. Cancer Res 53:1853–1861

Wang H-G, Rapp UR, Reed JC (1996a) Bcl-2 targets the protein kinase Raf-1 to mitochondria. Cell 87:629–638

Wang K, Yin X-M, Chao DT, Milliman CL, Korsmeyer SJ (1996b) BID: a novel BH3 domain-only death agonist. Genes Dev 10:2859–2869

Wang L, Miura M, Bergeron L, Zhu H, Yuan J (1994) Ich-1, an Ice/ced-3-related gene, encodes both positive and negative regulators of programmed cell death. Cell 78:739–750

Watanabe-Fukunaga R, Brannan CI, Copeland NG, Jenkins NA, Nagata S (1992) Lymphoproliferation disorder in mice explained by defects in Fas antigen that mediates apoptosis. Nature 356:314–317

White E (1996) Life, death, and the pursuit of apoptosis. Genes Dev 10:1–15

White E, Sabbatini P, Debbas M, Wold WSM, Kusher DI, Gooding LR (1992) The 19-kilodalton adenovirus E1B transforming protein inhibits programmed cell death and prevents cytolysis by tumor necrosis factor α. Mol Cell Biol 12:2570–2580

White K, Tahaoglu E, Steller H (1996) Cell killing by the Drosophila gene reaper. Science 271:805–807

Wilson KP, Black J-A F, Thomson JA, Kim EE, Griffith JP, Navia MA, Murcko MA, Chambers SP, Aldape RA, Raybuck SA, Livingston DJ (1994) Structure and mechanism of interleukin-1β converting enzyme. Nature 370:270–275

Wu D, Wallen HD, Nuñez G (1997) Interaction and regulation of subcellular localization of CED-4 by CED-9. Science 275:1126–1129

Wyllie AH (1980) Glucocorticoid-induced thymocyte apoptosis is associated with endogenous endonuclease activation. Nature 284:555–556

Wyllie AH, Kerr JFR, Currie AR (1980) Cell death: the significance of apoptosis. Int Rev Cytol 68:251–306

Xiang J, Chao DT, Korsmeyer SJ (1996) BAX-induced cell death may not require interleukin 1β-converting enzyme-like proteases. Proc Natl Acad Sci USA 93:14559–14563

Xue D, Horvitz HR (1995) Inhibition of the Caenorhabditis elegans cell-death protease CED-3 by a CED-3 cleavage site in baculovirus p35 protein. Nature 377:248–251

Yang E, Korsmeyer SJ (1996) Molecular thanatopsis: a discourse on the Bcl-2 family and cell death. Blood 88:386–401

Yang E, Zha J, Jockel J, Boise LH, Thompson CB, Korsmeyer SJ (1995) Bad, a heterodimeric partner for Bcl-xₗ and Bcl-2, displaces Bax and promotes cell death. Cell 80:285–291

Yang J, Liu X, Bhalla K, Kim CN, Ibrado AM, Cai J, Peng T-I, Jones DP, Wang X (1997) Prevention of apoptosis by Bcl-2: release of cytochrome c from mitochondria blocked. Science 275:1129–1132

Yin C, Knudson CM, Korsmeyer SJ, Van Dyke T (1997) Bax suppresses tumorigenesis and stimulates apoptosis in vivo. Nature 385:637–640

Yin X-M, Oltvai ZN, Korsmeyer SJ (1994) BH1 and BH2 domains of Bcl-2 are required for inhibition of apoptosis and heterodimerization with Bax. Nature 369:321–323

Young F, Mizoguchi E, Bhan AK, Alt FW (1997) Constitutive Bcl-2 expression during immunoglobulin heavy chain-promoted B cell differentiation expands novel precursor B cells. Immunity 6:23–33

Yuan J, Horvitz HR (1992) The *Caenorhabditis elegans* cell death gene *ced-4* encodes a novel protein and is expressed during the period of extensive programmed cell death. Development 116:309–320

Yuan J, Shaham S, Ledoux S, Ellis HM, Horvitz HR (1993) The *C. elegans* cell death gene *ced-3* encodes a protein similar to mammalian interleukin-1β-converting enzyme. Cell 75:641–652

Zha H, Aimé-Sempé C, Sato T, Reed JC (1996a) Proapoptotic protein Bax heterodimerizes with Bcl-2 and homodimerizes with Bax via a novel domain (BH3) distinct from BH1 and BH2. J Biol Chem 271:7440–7444

Zha J, Harada H, Yang E, Jockel J, Korsmeyer SJ (1996b) Serine phosphorylation of death agonist BAD in response to survival factor results in binding to 14-3-3 not BCL-X$_L$. Cell 87:619–628

Zhou Q, Snipas S, Orth K, Muzio M, Dixit VM, Salvesen GS (1997) Target protease specificity of the viral serpin CrmA. J Biol Chem 272:7797–7800

Apoptosis and the Proteasome

Lisa M. Grimm[1] and Barbara A. Osborne[1,2]

1
Introduction

When a cell is signaled to undergo apoptosis, a rapid change in the intracellular environment ensues as the cell becomes engaged in the regulated dismantlement of its structural framework. This dismantlement is mediated in part through the degradation of proteins. Protein degradation is an effective way of regulating cellular processes because it is selective, irreversible, and highly regulated. A variety of proteins have been identified that are cleaved specifically during apoptosis, and many proteases have been shown to be involved in the process. These studies have been the impetus for a more thorough analysis of the proteolytic events that take place in apoptotic cells.

The ubiquitin-proteasome pathway is a good candidate for the regulation of protein composition during apoptosis because the pathway is both ubiquitous and highly regulated. It is the major nonlysosomal proteolytic pathway in the mammalian cell and has been found to be instrumental in a wide variety of cellular processes such as cell cycle progression, transcription, and antigen presentation. Proteasomes are multicatalytic proteases that are located in the nuclei and cytosol of all eukaryotic cells. The pathway is regulated at multiple points to allow selective protein breakdown while preventing generalized proteolytic damage to the cell. This specificity is achieved through the careful selection of protein targets and by restricted access of proteins into the proteasome. In most cases, selection for degradation by the proteasome requires that the proteins be tagged covalently with multiubiquitin chains. Additional posttranslational modifications, such as phosphorylation, may also be required. In addition, entrance into the catalytic core of the eukaryotic proteasome is tightly controlled by associated subunits, preventing entry of untargeted proteins.

Any proteolytic pathway involved in the regulation of cell death pathways must be precise. The knowledge that the ubiquitin–proteasome pathway incor-

---
[1] Program in Molecular and Cellular Biology, University of Massachusetts, Amherst, Massachusetts 01003, USA

[2] Program in Molecular and Cellular Biology, Department of Veterinary and Animal Sciences, University of Massachusetts, Amherst, Massachusetts 01003, USA

porates many points of regulation while maintaining its status as the major processor of cytoplasmic proteins makes it an appealing pathway to investigate in the context of cell death. This chapter will introduce the ubiquitin-proteasome pathway by reviewing the regulation of proteasome structure, proteasome activity, and substrate degradation and will close with a summary of the recent literature that analyzes its role in apoptosis.

2
Proteasome Mechanics and Mechanisms

2.1
20S Proteasome Structure

20S proteasomes are found in all eukaryotic cells and in the archaebacteria *Thermoplasma acidophilum*. More recently, structures similar to the 20S proteasome have been discovered in eubacteria (Rohrwild et al. 1996, 1997; Yoo et al. 1996). The 20S proteasome is the catalytic core of the larger 26S proteasome, and its three-dimensional structure has been deciphered through electron microscopic and X-ray crystallographic studies. X-ray crystallographic analyses of proteasomes from *T. acidophilum* (Löwe et al. 1995) and the yeast *Saccharomyces cerevisiae* (Groll et al. 1997) provide insight into the similarities and differences between proteasomes in different organisms. The 20S proteasome is a barrel-shaped structure composed of multiple subunits ranging in size from 25 to 35 kDa. The subunits are arranged into four stacked rings with each ring containing seven subunits. There are two families of subunits, α and β. The two outer rings of the 20S proteasome are composed of α subunits and the two inner rings of β subunits. This arrangement produces an elongated cylinder that contains three large cavities. 20S proteasomes from both eukaryotes and archaebacteria exhibit this basic quaternary structure, but eukaryotic proteasomes exhibit greater diversity and complexity. Proteasomes from archaebacteria are composed of one type of α and one type of β subunit, while in eukaryotes there are at least seven α and seven β subfamilies (Coux et al. 1994). Therefore, seven distinct α-like subunits make up each outer ring and seven distinct β-like subunits make up each inner ring. This variation in subunit composition produces structural differences in the proteasomes of these two organisms. In yeast, the α ring at each end of the 20S proteasome is closed off to proteins, while in archaebacteria the entrances are accessible. It is likely that the regulatory complexes that bind at each end of the 20S complex in eukaryotes mediate the entry of substrates into the catalytic chamber.

There are several differences in the sequence and function of the α and β families. The α subunits, both within a species and between species, are more conserved than the β subunits (Coux et al. 1994). The sequence of an α subunit contains a conserved motif that is required for proteasome assembly (Zwickl et al. 1994), and some subunits contain a nuclear localization signal (Tanaka et al. 1990). The α subunits perform multiple structural roles. They are required for assembly of the β rings (Zwickl et al. 1994), for restricting access of proteins to the inner catalytic chamber (Löwe et al. 1995), and for binding the 19S (PA700) and 11S (PA28) regulatory complexes (Peters et al. 1993; Gray et al. 1994). The β subunits make up the catalytic component of the 20S proteasome. Almost all β subunits contain an amino-terminal pro-sequence that is cleaved prior to assembly to expose the amino-terminal threonine residue (Zwickl et al. 1994). This threonine is the active site residue and is required for 20S proteasome activity (Fenteany et al. 1995; Löwe et al. 1995; Seemüller et al. 1995). At least five distinct catalytic activities are performed by the mammalian proteasome. The 20S complex is able to cleave following large hydrophobic residues (chymotrypsin-like activity), basic residues (trypsin-like activity), acidic residues (postglutamyl hydrolase activity), branched-chain amino acids (BrAAP activity), and small neutral amino acids (SNAAP activity) (Rivett 1989; Orlowski et al. 1993). The structural α subunits and the catalytic β subunits assemble together to form a proteolytic complex with the capacity to cleave a protein at many different positions.

2.2
26S Proteasome Structure

A regulatory complex, called the 19S cap or the proteasome activator (PA700), associates at each end of the 20S proteasome in eukaryotes to form the 26S proteasome (Fig. 1; Peters et al. 1993; Yoshimura et al. 1993). The 19S caps are composed of at least 18 proteins that range in molecular weight from 25 to 110 kDa. Although the entire content of the 19S caps is unknown, the components that have been identified provide some insight into the regulatory functions of these complexes. ATPases are present and are required for both the assembly of the 26S proteasome and the subsequent degradation of proteins (Ganoth et al. 1988; Armon et al. 1990). Some ATPases show selectivity for particular proteins in that mutations in specific ATPases reduce the degradation of some proteins, but not others. For example, mutations in the yeast subunit, CIM3 or CIM5, result in the increased stability of specific short-lived proteins such as the CLB2 and CLB3 cyclins (Ghislain et al. 1993). ATPases may also help unfold proteins prior to their entrance into the proteolytic chamber. At least one subunit in the 19S cap is the ubiquitin receptor, binding with strong affinity to chains containing multiple ubiquitin proteins (Deveraux et al. 1994). Isopeptidases in the complex cleave the ubiquitin proteins from their chains, allowing these proteins to be recycled as free monomers (reviewed in Hershko and Ciechanover 1992). Recently, a bovine isopeptidase has been isolated that introduces another level of specificity by editing the ubiquitin chains according to their length (Lam et al. 1997). This isopeptidase rescues selectively from degradation proteins with chains containing less than four

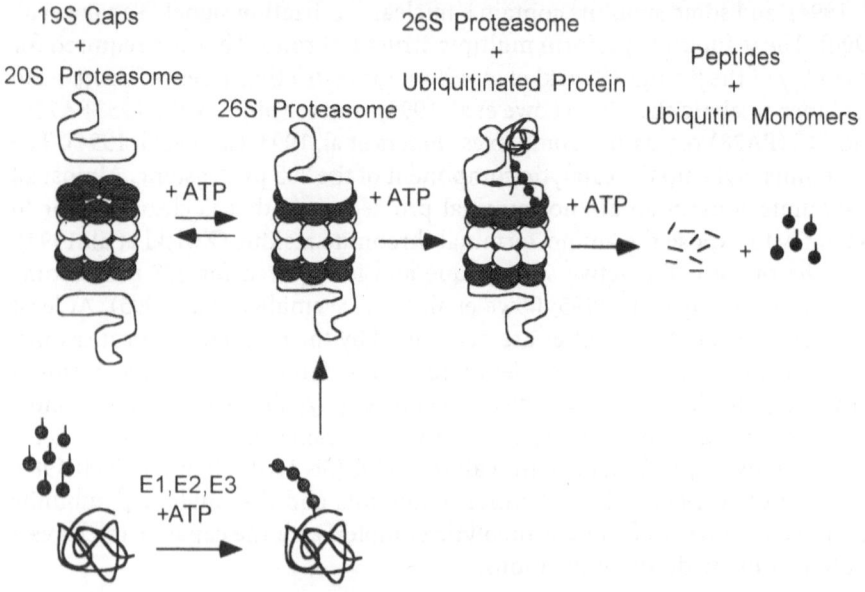

Fig. 1. Mechanism of substrate degradation by the ubiquitin–proteasome pathway

ubiquitin proteins. This means that not only is the presence of ubiquitination a factor in substrate degradation, but also the extent of ubiquitination is important as well. The assembly of these individual subunits produces a complex that preferentially binds, unfolds, and feeds highly ubiquitinated substrates into the catalytic chamber of the 20S proteasome (Fig. 1; reviewed in Coux et al. 1996).

The 26S proteasome is a dynamic structure that is able to disassemble and reassemble (Ganoth et al. 1988; Orino et al. 1991). The composition and function of this dynamic structure can vary with the physiological status of a cell. A well-described example of this variability is found in the mammalian immune system. The stimulation of antigen presentation induced by exposure to the cytokine γ-interferon is accompanied by a change in the composition of proteasomes. Three new subunits, LMP2, LMP7, and MECL1, replace three constitutive β subunits in the 20S proteasome. The incorporation of these subunits leads to an increase in the chymotrypsin-like and trypsin-like activities (Driscoll et al. 1993; Gaczynska et al. 1993; Aki et al. 1994) and a decrease in post-glutamyl activity (Gaczynska et al. 1993). As a result, proteasomes from γ-interferon-treated cells should produce more peptides with hydrophobic or basic carboxy-termini and fewer with acidic termini. It turns out that the vast majority of peptides presented by major histocompatibility complex (MHC) class I molecules contain hydrophobic or basic carboxy-termini. In addition, there occurs a γ-interferon-induced association of the PA28 complex with the 20S proteasome. This association changes the processing mechanism of the catalytic chamber, appearing to favor the production of dominant T-cell epitopes (Dick et al. 1996). As more studies analyze the structure and content of the 20S proteasome, it becomes apparent that this complex is both rigid and flexible. It tightly regulates what proteins enter and leave its catalytic chamber, yet is able to redesign its content and function to meet a particular need.

2.3
Protein Processing by the 20S Proteasome

X-ray crystallographic analysis of the eukaryotic 20S proteasome both confirms previous assessments of its basic quaternary structure and extends our knowledge of how it degrades proteins (Groll et al. 1997). Important questions about how proteins enter the complex, how proteins are cleaved, and the distance between cuts can be evaluated. There is no access to the interior of the eukaryotic 20S proteasome at each end of the cylinder without substantial rearrangement of the α rings. There are very small side openings located where the α and β rings join that may allow the passage of unfolded and extended polypeptides. However, the 20S proteasome is probably closed off to proteins in the absence of additional regulatory mechanisms. One possibility is that the association of the 19S complexes with the 20S proteasome induces the rearrangement that would be required to allow a protein in; thereby working as a gate into and out of the proteasome.

The proteasome has not been placed successfully into any one of the traditional classes of proteases because its subunits lack homology to those of other proteases (Zwickl et al. 1991; Lupas et al. 1993; Coux et al. 1994) and because its pattern of sensitivity to standard inhibitors differs from the classical families of proteases. In addition, the 20S proteasome employs a novel reaction mechanism that involves the amino-terminal threonine residue (Thr1) of some β subunits (Groll et al. 1997). This mechanism is used by other hydrolases that are known as the amino-terminal nucleophile (Ntn) family of hydrolases. The internal cavity is lined by Ntn active sites, and when a protein enters the cavity it hydrogen bonds to a Thr1 site. A water molecule is positioned appropriately at each site and acts as a nucleophile. Hydrolysis may proceed either through the deacylation of an acyl enzyme intermediate or through direct attack of the peptide by the water molecule. The crystal structure also reveals a second active site located at the carboxy-terminal end of an α helix present in all seven β subunits (Groll et al. 1997). It is likely that both active sites work together to cleave protein substrates.

The standard length of the peptides produced by proteasome processing is an important consideration when evaluating the candidacy of the proteasome in the regulation of a cellular process such as apoptosis. Many proteins that are cleaved during apoptosis are cleaved in only one or two locations to produce large and variably sized polypeptide fragments. However, some studies suggest that proteasomes generate peptides within a narrow size range (Wenzel et al.

1994). During antigen presentation, the mean length is about eight amino acids. The distance between active sites in neighboring β chains is about 28 Å, which corresponds to the length of a hepta- or octapeptide in an unfolded and extended polypeptide (Groll et al. 1997). Therefore, the distance between active sites may be an important determinant in the length of peptides generated by the 20S proteasome. There is some evidence that the proteasome may not only mediate the complete degradation of proteins but may also partially process precursors. The p50 subunit of the transcription factor, NF-κB, is processed into its active form by the cleavage of the carboxy-terminal end of the precursor protein, p105 (Fan and Maniatis 1991). Studies using proteasome inhibitors have shown that proteasome activity is required for p105 processing in vitro and in vivo (Palombella et al. 1994). These inhibitor experiments do not demonstrate that the proteasome is responsible directly for this cleavage, and it is possible that the proteasome plays an indirect role. Additional experiments in the study, however, demonstrate the ubiquitination of p105 prior to its processing. These data suggest that the proteasome cleaves directly the carboxy-terminal end of the p105 while sparing the rest of the molecule. However, the universality of partial processing and how it is performed by the proteasome complex are still unknown.

3
Substrates

3.1
Regulation of Substrate Degradation

The list of potential targets of the proteasome has expanded greatly in recent years. The proteasome has long been known to degrade damaged and abnormal proteins, but more recently this complex has been shown to degrade short-lived and long-lived proteins as well (Rock et al. 1994). The existence of so many potential substrates has made it necessary for the pathway to regulate precisely its activities. Several mechanisms are in place that allow for the selective proteolysis of proteins by the proteasome. In many cases, a substrate is targeted for degradation by the attachment of a chain of ubiquitin proteins. This attachment is mediated by a series of enzymes; E1, E2, and E3 (Fig. 1). Cooperation between ubiquitin activating (E1) enzyme, ubiquitin-conjugating (E2) enzymes and ubiquitin-ligating (E3) enzymes results in the formation of an isopeptide bond between the activated C terminal Gly of ubiquitin and an ε-NH_2 group of a Lys residue of the substrate. A polyubiquitin chain develops when additional ubiquitin molecules are transferred to Lys-48 of a previously conjugated ubiquitin. The degradation of ubiquitinated substrates is performed by the 26S proteasome (2000 kDa), which is formed by the association of the 20S catalytic core (700–750 kDa) with the 19S cap complexes in an ATP-dependent manner. The resulting 26S proteasome is able to unfold and feed the proteins into the processing chamber, where the proteins are hydrolyzed into small peptides.

There are multiple E2 and E3 enzymes, and many of them are responsible for the ubiquitination of distinct substrates (reviewed in Hochstrasser 1995). For example, the association of the E3 enzyme, E6-AP, with the human papillomavirus E6 protein triggers the degradation of p53. It is also possible that the specificity provided by these enzymes might be expanded through the combination of different E2–E3 partners. The ubiquitination of some proteins requires the presence of a molecular chaperone. In order for proteins such as actin, α-crystallin, glyceraldehyde-3-dehydrogenase, α-lactalbumin, or histone H2A to be degraded in a ubiquitin-dependent manner, Hsc70 must be present, possibly to unfold proteins to expose a ubiquitin ligase site (Bercovich et al. 1997). In addition to ubiquitination, a second post-translational modification may be required. Many proteins are phosphorylated prior to their ubiquitination and degradation by the proteasome. One well-described example is found in the degradation of IϰB. IϰB protein inhibits the activity of the transcription factor, NF-ϰB. When this inhibitory protein is degraded, NF-ϰB is released and translocates to the nucleus where it regulates the transcription of genes involved in immune and inflammatory responses (Henkel et al. 1993). Although phosphorylation of Iϰ-B is required for its degradation, this modification is not sufficient (Finco et al. 1994; Traenckner et al. 1994; Alkalay et al. 1995; Di Donato et al. 1995; Lin et al. 1995; Whiteside et al. 1995).

Particular sequences contained within a substrate may also help target the protein for destruction by the proteasome. There is a 27 amino acid domain in c-Jun, called the δ domain, that is required for the protein to be ubiquitinated (Musti et al. 1996). The deletion of this domain from c-Jun eliminates the ubiquitination of this protein, and fusion of the δ domain to β-galactosidase is sufficient for its ubiquitination (Treier et al. 1994). Another sequence that is required for some proteins to be ubiquitinated is a nine amino acid sequence known as the "destruction box". During anaphase, yeast cyclin mutants that lack this sequence fail to ubiquitinate cyclins A and B and arrest in anaphase (King et al. 1996; Klotzbucher et al. 1996). In most cases, the destruction box is necessary but not sufficient to direct ubiquitination. Sequences rich in proline, glutamic acid, serine, and threonine (PEST sequences) also contribute to the degradation of proteins such as IϰBβ and parathyroid hormone-related protein, but are neither essential nor sufficient for their proteolysis (Van Antwerp and Verma 1996; Meerovitch et al. 1997; Weil et al. 1997). Specific sequences can, therefore, provide information to direct the appropriate ubiquitination of a substrate, but by themselves are not sufficient for targeting a protein for destruction.

Not all proteins must be ubiquitinated to be targeted for degradation by the 26S proteasome. Ornithine decarboxylase (ODC) is an enzyme involved in polyamine biosynthesis that is degraded in a ubiquitin-independent manner. Because polyamines are toxic to cells, negative feedback mechanisms have evolved to rapidly degrade ODC. The buildup of polyamines induces the specific inhibitory protein antizyme (AZ). When associated with AZ, ODC is rapidly hydrolyzed by the 26S complex in an ATP-dependent manner (Murakami et al. 1992, 1993, 1996; Tokunaga et al. 1994; Elias et al. 1995). The N terminus of AZ is thought to be the region responsible for promoting degradation and, when fused to ODC, promotes its degradation. The fusion of this region to a protein not normally associated with AZ, such as cyclin B, induces cyclin B degradation in vitro (Li et al. 1996). A variety of different mechanisms are in place in the proteasome pathway, ubiquitin-dependent and ubiquitin-independent, that promote the selective degradation of proteins. The establishment of these mechanisms in the presence of a tightly sealed proteasome makes this pathway a discriminating one.

3.2
Proteolytic Targets in Apoptosis

The interleukin-1β-converting enzyme (ICE) family of cysteine proteases is the major executioner of proteolysis in an apoptotic cell. This family has come into prominence since it was discovered that a developmentally regulated cell death gene from *Caenorhabditis elegans*, ced-3, contains significant sequence homology to ICE (Yuan et al. 1993). Mammalian ICE was originally discovered to be the enzyme responsible for the processing of inactive IL-1β to active IL-1β (Thornberry et al. 1992). Since the discovery of homology between ICE and Ced-3, additional investigations into this protease have resulted in the identification of ICE as one member of a large multigene family, collectively referred to as the caspases (see Chapter by Dorstyn et al. in Vol. 24 of this series). Studies have demonstrated a role for the caspases in many cell death pathways. They are activated in dying cells, and inhibitors that inactivate these proteases inhibit apoptosis. Although many different proteases make up the family, all family members share several unique features. In order for these proteases to become active, their inactive proenzyme forms must be cleaved after specific aspartate residues. This processing releases the subunits which will make up the active enzyme tetramers (Thornberry et al. 1992). The picture that emerges from the caspase studies is that the proteases cleave and activate each other to produce a cascade of protease activity. The activated caspases then degrade other proteins located in the nucleus and cytoplasm of cells.

Evidence is accumulating to support the idea that several proteins important for cellular structure and function are cleaved by caspases during apoptosis. Many of the nuclear substrates are required to maintain the integrity of DNA and chromatin. The enzyme poly (ADP-ribose) polymerase (PARP) was first demonstrated to be cleaved when HL-60 cells were treated with topoisomerase II inhibitors (Kaufmann et al. 1993), but has since been shown to be cleaved in many apoptotic cells. At sites of DNA strand breaks, activated PARP catalyzes the transfer of ADP-ribose moieties from the substrate NAD^+ to protein acceptors involved in chromatin architecture (reviewed in de Murcia and Ménissier de Murcia 1994). Modification of proteins such as histones H1 and H2B causes a relaxation of the chromatin superstructure, facilitating the access of repair enzymes to sites of DNA strand breaks. In a dying cell, this repair mechanism is shut down through the cleavage and inactivation of PARP. The processing of a second DNA repair enzyme, DNA-dependent protein kinase (DNA-PK), results in its inactivation when the phosphatidylinositol 3-kinase (PI3-kinase) domain separates from the rest of the molecule (Song et al. 1996). Lamins A and B are degraded in dying embryonic fibroblasts (Kaufmann 1989; Oberhammer et al. 1994), and lamin B has also been shown to be degraded in TNF-treated 3T3-like fibroblasts (Voelkel-Johnson et al. 1995) and in glucocorticoid-treated thymocytes (Neamati et al. 1995). Additional nuclear proteins such as U1 small ribonucleoprotein (Casciola-Rosen et al. 1994), topoisomerases I and II (Kaufmann 1989; Voelkel-Johnson et al. 1995), and retinoblastoma (An and Dou 1996) are degraded during apoptosis. The degradation of DNA repair enzymes, lamins, and U1 small ribonucleoprotein results in the loss of integrity in DNA and chromatin and in defective processing of mRNAs. The combination of these degradative events results in the breakdown of the nucleus in a dying cell.

Proteins that are located outside the nucleus are also targeted for destruction during apoptosis. Some of these proteins such as the adenomatous polyposis coli (APC) protein (Browne et al. 1994) and teminin antigen (Hébert et al. 1994) are degraded by unidentified proteases. A variety of other proteins such as fodrin (Martin et al. 1995b), actin (Kayalar et al. 1996), Gas2 (Brancolini et al. 1995), and protein kinase C δ (Emoto et al. 1995) have been suggested to be the targets of caspases. Actin and fodrin degradation may be responsible for membrane blebbing and condensation of the cytoplasm that occur during apoptosis. Recent in vivo studies throw into question the significance of actin degradation during apoptosis. The hypothesis that actin is cleaved by caspases is supported by the presence of caspase recognition sites and in vitro data showing actin cleavage by these proteases in cellular extracts (Mashima et al. 1995; Kayalar et al. 1996). However, in many different cell types, actin is not degraded during apoptosis in vivo (Song et al. 1997). These data suggest that although a protein may contain appropriate recognition sites, factors in vivo may prevent access to these sites. Most of the work looking for caspase substrates has been performed in vitro, and may not reflect what happens inside a real cell. The study by Song et al. (1997) points out potential flaws in the in vitro experiments and reaffirms the importance of substantiating all work with in vivo studies.

4
Proteasomes and Apoptosis

4.1
Requirement for Proteasomes During Apoptosis

Investigating a role for the ubiquitin–proteasome pathway in apoptosis has become appealing because apoptosis requires proteolytic activity, and the proteasome is capable of processing a wide variety of proteins in a highly regulated way. The identification of inhibitors that block proteasome activitiy has allowed investigators to study the requirement for proteasomes in a variety of cellular processes, including apoptosis. Several peptide aldehydes are commonly used in studies; including acetyl-leu-leu-methioninal (LLM, calpain inhibitor II), acetyl-leu-leu-norleucinal (LLnL, calpain inhibitor I), carbobenzoxyl-leu-leu-norvalinal (MG115), and carbobenzoxyl-leu-leu-leucinal (MG132). LLM and LLnL inhibit most efficiently the chymotryptic-like activity of the proteasome (Vinitsky et al. 1992), while the more potent inhibitor, MG132, blocks chymotryptic, postglutamyl hydrolase, and BrAAP activities. These peptide aldehydes are reversible and protein degradation is restored after their removal. These inhibitors are not specific for the proteasome but also block the activities of calpain and some lysosomal proteases. However, there is a way to discriminate whether any effect they produce on a cellular process is due to inhibition of the proteasome or the other proteases. Because the different peptides inhibit calpain and cathepsins with equal effectiveness but inhibit proteasome activity with very different effectiveness (Rock et al. 1994), treatment of cells with a wide dose range of each of the inhibitors should distinguish which protease is responsible for any observed effects.

A much more specific proteasome inhibitor, lactacystin, was originally isolated from actinomycetes because of its ability to induce the differentiation of neuroblastoma cells (Ōmura et al. 1991). Subsequently, lactacystin was shown to be attached covalently to a specific β subunit of the mammalian proteasome. The attachment of lactacystin to some β subunits results in the modification of the active site threonine residue, thereby blocking chymotryptic-like, tryptic-like, and peptidyl glutamyl hydrolase activities. The inactivation of the chymotryptic- and tryptic-like activities is permanent, so treatment of cells with this inhibitor damages proteasomes irreversibly. Lactacystin is the most specific inhibitor of the proteasome currently available and does not inhibit the activities of a variety of other proteases, including calpain and cathepsins (Fenteany et al. 1995).

Analysis of a role for the ubiquitin–proteasome pathway in cell death is in its infancy. The most extensive studies have been performed in the intersegmental muscles (ISMs) of the hawkmoth, *Manduca sexta*. These ISMs are required for emergence of the insect from its pupal case. During a 30-h period following this emergence, the circulating titer of 20-hydroxyecdysone declines, and this decline triggers the programmed death of the ISMs. The commitment to death is accompanied by increases in polyubiquitin gene expression (Schwartz et al. 1990) and in the BrAAP and caseinolytic activities of the 20S proteasome (Jones et al. 1995). The activity changes are hypothesized to be the result of the incorporation of four new subunits into the 20S proteasome (Jones et al. 1995). Other studies have identified changes in the ATPase subunits in the 19S cap complex (Dawson et al. 1995). The mRNA of an ATPase regulatory subunit, MS73, increases more than twofold just before the ISMs commit to die. An ecdysteroid agonist that blocks programmed cell death in the ISMs also blocks the increase in levels of MS73 mRNA (Löw et al. 1997). Studies of the developmentally regulated death of ISMs were the first to suggest a role for the ubiquitin–proteasome pathway during cell death and to document changes in the composition and activity of proteasomes in response to a specific death signal. The ubiquitin–proteasome pathway may be activated to mediate the large amounts of protein degradation required for the complete destruction of these cells. Whether this pathway is involved actively in the commitment process or is activated after commitment has occurred is unclear.

The early insect studies set a precedent for thinking about the proteasome in the context of cell death, and now studies are investigating a role for proteasomes in apoptosis. One of the first studies to ask the question of whether the ubiquitin–proteasome pathway is required for apoptosis looked specifically at a requirement for ubiquitin. The introduction of antisense ubiquitin RNA into peripheral T cells inhibits significantly irradiation-induced death (Delic et al. 1993). More recently, the availability of proteasome inhibitors has allowed investigators to assess whether the proteasome is essential for apoptosis. Two different cell death systems have been explored. In one system, primary mouse thymocytes are induced to die by a variety of treatments, including exposure to the phorbol ester phorbol 12-myristate 13-acetate (PMA), the synthetic glucocorticoid dexamethasone, or γ-radiation (Grimm et al. 1996). In the second system, sympathetic neurons are triggered to die after withdrawl of nerve growth factor (NGF) (Sadoul et al. 1996). Treatment of thymocytes or neurons with peptide proteasome inhibitors or lactacystin inhibits the induction of death in both cell types (Grimm et al. 1996; Sadoul et al. 1996). The inhibitor studies suggest that in these two systems, proteasome activity is required for the progression of apoptosis.

Inhibition of proteasome activity in other cell types produces an outcome very different from the thymocyte and neuron studies. Treatment of the human monoblast line, U937, and two different human T-cell leukemia lines, MOLT-4 and HL-60, with peptide inhibitors or lactacystin results in the induction of apoptosis (Imajoh-Ohmi et al. 1995; Shinohara et al. 1996; Drexler 1997). Also, proteasome inhibitors enhance rather than inhibit TNF-induced death in U937 cells (Fujita et al. 1996). These results are initially confusing in

light of the thymocyte and neuron data. Two factors may help explain the
contradictory data. First, the proliferative status of the cells in these studies is
different. Thymocytes and differentiated neurons are mainly noncycling
populations of cells while the MOLT-4, HL-60, and U937 cells are cycling.
Second, the proteasome regulates cell cycle progression. Recently, a role for
the proteasome has been evaluated with these two factors in mind (Drexler
1997). When HL-60 cells are induced to undergo apoptosis after treatment
with proteasome inhibitors, levels of the Cdk inhibitor (p27$^{\text{Kip1}}$) accumulate
causing cell cycle arrest, CPP32-like caspases are active, and levels of c-Myc
protein remain constant (Drexler 1997). Normally, c-Myc is downregulated in
a G1-arrested cell. Therefore, the author of this study hypothesizes that in the
inhibitor-treated cells, the presence of c-Myc and p27$^{\text{Kip1}}$ produce conflicting
signals that drive the cell into apoptosis. When HL-60 cells are induced to
differentiate by treatment with PMA, proteasome inhibitors cause an increase
in p27$^{\text{Kip1}}$ protein, but c-Myc levels decline and so the conflict is avoided, and
the cells survive.

Because proteasome inhibitors are much more toxic to cycling cells than to
quiescent or differentiated cells, these inhibitors may mask in cycling cells any
underlying role the proteasome may play during apoptosis. A hypothetical
scheme for proteasome involvement in cycling and noncycling cells is depicted
in Fig. 2. In a variety of cells, an apoptotic signal results in the cleavage and

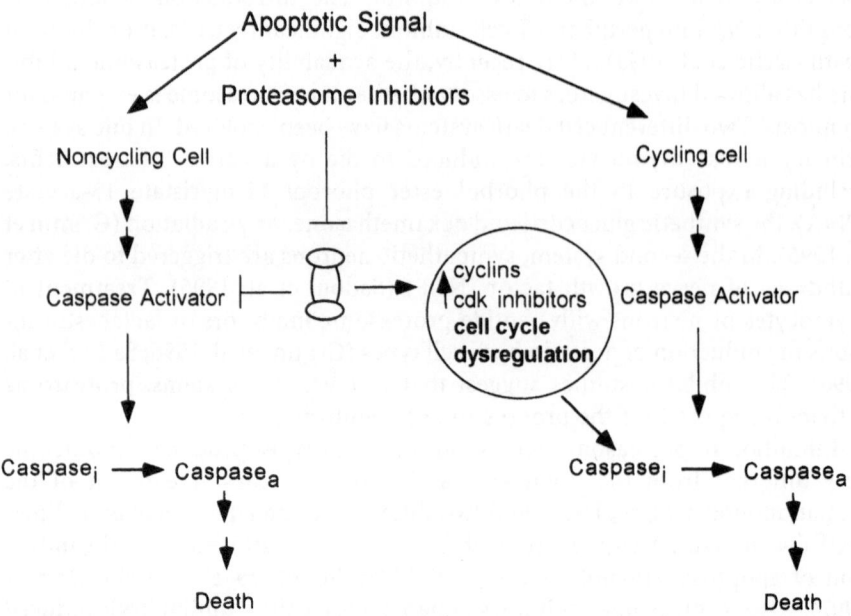

Fig. 2. Hypothetical involvement of the proteasome in apoptotic cycling and noncycling cells

activation of a caspase cascade. The protease responsible for the initial activation of the cascade is not known and is represented here as the caspase activator. The effect of proteasome inhibition on cells that have received a signal to die is cell type-dependent. In noncycling cells, the proteasome may be responsible for cleaving and activating the caspase activator or may degrade an associated inhibitory protein. Blocking proteasome activity, therefore, would inhibit the activation of the caspase cascade and the death of the cell. It is possible that the proteasome works farther upstream from the cascade, but a clarification of the proteasome's position(s) in the cell death pathway will require the identification of its targets. In cycling cells, inhibition of proteasome activity results in the accumulation of cell cycle regulators such as cyclins and cdk inhibitors. Their accumulation results in dysregulation of the cell cycle, and this dysregulation drives the cells into apoptosis, possibly by activating the caspase cascade. Therefore, inhibition of proteasome activity accelerates rather than inhibits the apoptotic pathway.

4.2
Subcellular Localization of Proteasomes

One way to understand the involvement of proteasomes in apoptosis is to compare the localization of these complexes in healthy and dying cells. Any observed translocations of proteasomes in apoptotic cells may help reveal proteolytic substrates. Immunofluorescence microscopic studies in two different apoptotic cells reveal the translocation of proteasomes. When apoptosis is induced in a lung carcinoma by treatment with a cyclin-dependent kinase inhibitor, proteasomes translocate (Machiels et al. 1996). In a nonapoptotic cell, proteasomes are dispersed rather diffusely throughout the nucleus and cytoplasm, while in an apoptotic cell proteasomes surround the condensed chromatin and reside in the apoptotic bodies. Proteasomes are retained and active in the dying cells when structural elements such as cytokeratins, lamins, and intermediate filaments are no longer detectable (Machiels et al. 1996). This study demonstrates not only that proteasomes relocate during apoptosis but also that these complexes are present and active when many other proteins have been eliminated.

More dramatic changes in translocation take place during death in rat ovarial granulosa cells. There is a major reorganization of the underlying structural network of the cells (Pitzer et al. 1996). The proteasomes are removed from the nucleus and are located entirely within the apoptotic blebs. The actin cytoskeleton has detached itself from the inner face of the plasma membrane and has reorganized itself into a ring that separates the main body from the apoptotic blebs. One hypothesis is that the proteasome triggers the cytoskeletal rearrangement by degrading proteins that connect the actin network with the plasma membrane. Although these localization studies conclude that the proteasome is active and mobile during apoptosis, questions about which proteins are targeted and the uniformity of proteasome composition and function remain unanswered.

4.3
Proteasome Substrates During Apoptosis

The requirement for proteasome activity during apoptosis in nonproliferating cells suggests that the proteasome is specifically degrading proteins which make the cells vulnerable to death. A true understanding of the role proteasomes play during apoptosis awaits the identification of these proteolytic targets. Although these targets have not been identified, some work has been done to determine where the proteasome might be working in the apoptotic pathway. Peptide inhibitors were used to determine whether the proteasome is required early or late in apoptosis (Grimm et al. 1996). The addition of peptide inhibitors to primary thymocytes up to 1 h after the induction of death results in significant inhibition of death, while addition at 3 and 5 h after induction eliminates this inhibition. This suggests that the proteasome is required sometime in the first 5 h after the induction of death, probably within 1-3 h. The more specific placement of the proteasome in the cell death pathway was attempted by asking whether the proteasome works upstream or downstream of the caspase cascade. PARP cleavage, which is an assay for caspase activity, is inhibited when thymocytes and neurons are treated with proteasome peptide inhibitors (Grimm et al. 1996; Sadoul et al. 1996). In addition, lactacystin blocks the release of active IL-1β from Lipopolysaccharide (LPS)-treated macrophages, suggesting that ICE is inactived in these cells. Lactacystin is not responsible directly for ICE inactivation, since purified ICE remains active after treatment with the inhibitor (Sadoul et al. 1996). In apoptotic thymocytes, CPP32-like activity increases significantly by 5 h, and treatment of these cells with lactacystin blocks this increase (L. M. Grimm and B. A. Osborne, unpubl. results). These data suggest that the proteasome acts, probably indirectly, upstream of the protease cascade (Fig. 2). Its precise role, however, will be incompletely understood without the identification of proteolytic targets.

5
Summary

The ubiquitin–proteasome pathway is responsible for the regular turnover of a wide variety of proteins and is a critical regulator of many cellular processes. Although this pathway is abundant and ubiquitous, it is also discriminating. This specificity is achieved because there are multiple levels of regulation at work in the pathway. X-ray crystallographic data on the eukaryotic 20S proteasome suggest that substantial rearrangement of the alpha rings, probably mediated by the association of additional regulatory complexes, is required to allow access of substrates into the inner core of the complex. The associated complexes also confer a ubiquitin-dependence on the proteasome, requiring that potential substrates be tagged with chains of ubiquitin proteins. The presence of multiple ubiquitinating enzymes that favor distinct substrates provides a way for a cell to regulate what proteins are to be ubiquitinated. In some cases ubiquitination is not required, but we now know that other modifications, such as phosphorylation and protein–protein interactions, are also important for targeting proteins for degradation.

Even with the existence of so many regulatory controls, it is difficult to imagine how one complex can perform so many tasks. As more information is gathered about the proteasome, we begin to understand that all proteasomes are not exactly the same. For example, there is strong evidence that proteasomes involved in antigen presentation differ in both composition and function from proteasomes involved in other processes. The past image of the proteasome as a static structure is being shed, and a new image is emerging that portrays the complex as dynamic and flexible, able to tailor its composition and function to meet a particular need.

With this new image of the proteasome in mind, investigators are looking at the potential involvement of the proteasome in cell death. Inhibitor studies have demonstrated a requirement for proteasomes during apoptosis in noncycling and differentiated cells. Similar studies in cycling cells suggest that the proteasome may regulate a cell's decision to proliferate, differentiate, or die. It will be necessary in the future to supplement the peptide and lactacystin studies with work that is not inhibitor-driven since the specificity of an inhibitor for a particular protease is always in question. In addition, a real understanding of how proteasomes may regulate this process awaits the identification of its substrates. With cell death investigators showing increased interest in proteasomes, it may be possible in the next few years to determine the precise role of the proteasome in the pathways that lead to the death of a cell.

Acknowledgments. We thank Dr. Larry Schwartz for critical review of the manuscript and Steve Charron for assistance in the preparation of the manuscript.

References

Aki M, Shimbara N, Takashina M, Akiyama K, Kagawa S, Tamura T, Tanahashi N, Yoshimura T, Tanaka K, Ichihara A (1994) Interferon-gamma induces different subunit organizations and functional diversity of proteasomes. J Biochem (Tokyo) 115(2):257–269

Alkalay I, Yaron A, Hatzubai A, Jung S, Avraham A, Gerlitz O, Pashut-Lavon I, Ben-Neriah Y (1995) In vivo stimulation of I kappa B phosphorylation is not sufficient to activate NF-kappa B. Mol Cell Biol 15(3):1294–1301

An B, Dou QP (1996) Cleavage of retinoblastoma protein during apoptosis: an interleukin 1 beta-converting enzyme-like protease as candidate. Cancer Res 56(3):438–442

Armon T, Ganoth D, Hershko A (1990) Assembly of the 26S complex that degrades proteins ligated to ubiquitin is accompanied by the formation of ATPase activity. J Biol Chem 265(34):20723–20726

Bercovich B, Stancovski I, Mayer A, Blumenfeld N, Lazlo A, Schwartz AL, Ciechanover A (1997) Ubiquitin-dependent degradation of certain protein substrates in vitro requires the molecular chaperone Hsc 70. J Biol Chem 272(14):9002–9010

Brancolini C, Benedetti M, Schneider C (1995) Microfilament reorganization during apoptosis: the role of Gas2, a possible substrate for ICE-like proteases. EMBO J 14(21):5179–5190

Browne SJ, Williams AC, Hague A, Butt AJ, Paraskeva C (1994) Loss of APC protein expressed by human colonic epithelial cells and the appearance of a specific low-molecular-weight form is associated with apoptosis in vitro. Int J Cancer 59(1):56–64

Casciola-Rosen LA, Miller DK, Anhalt GJ, Rosen A (1994) Specific cleavage of the 70-kDa protein component of the U1 small nuclear ribonucleoprotein is a characteristic biochemical feature of apoptotic cell death. J Biol Chem 269(49):30757–30760

Coux O, Nothwang HG, Silva Pereira I, Recillas Targa F, Bey F, Scherrer K (1994) Phylogenic relationships of the amino acid sequencess of prosome (proteasome, MCP) subunits. Mol Gen Genet 245(6):769–780

Coux O, Tanaka K, Goldberg AL (1996) Structure and functions of the 20S and 26S proteasomes. Annu Rev Biochem 65:801–847

Dawson SP, Arnold JE, Mayer NJ, Reynolds SE, Billett MA, Gordon C, Colleaux L, Kloetzel PM, Tanaka K, Mayer RJ (1995) Developmental changes of the 26S proteasome in abdominal intersegmental muscles of *Manduca sexta* during programmed cell death. J Biol Chem 270(4):1850–1858

Delic J, Morange M, Magdelenat H (1993) Ubiquitin pathway involvement in human lymphocyte gamma-irradiation-induced apoptosis. Mol Cell Biol 13(8):4875–4883

de Murcia G, Mènissier de Murcia J (1994) Poly (ADP-ribose) polymerase: a molecular nick-sensor. Trends Biochem Sci 19(4):172–176

Deveraux Q, Ustrell V, Pickart C, Rechsteiner M (1994) A 26S protease subunit that binds ubiquitin conjugates. J Biol Chem 269(10):7059–7061

Dick TP, Ruppert T, Groettrup M, Kloetzel PM, Kuehn L, Koszinowski UH, Stevanovic S, Schild H, Rammensee HG (1996) Coordinated dual cleavages induced by the proteasome regulator PA28 lead to dominant MHC ligands. Cell 86(2):253–262

DiDonato JA, Mercurio F, Karin M (1995) Phosphorylation of I kappa B alpha precedes but is not sufficient for its dissociation from NF-kappa B. Mol Cell Biol 15(3):1302–1311

Drexler HC (1997) Activation of the cell death program by inhibition of proteasome function. Proc Natl Acad Sci USA 94(3):855–860

Driscoll J, Brown MG, Finley D, Monaco JJ (1993) MHC-linked LMP gene products specifically alter peptidase activities of the proteasome. Nature 365(6443):262–264

Elias S, Bercovich B, Kahana C, Coffino P, Fischer M, Hilt W, Wolf DH, Ciechanover A (1995) Degradation of ornithine decarboxylase by the mammalian and yeast 26S proteasome complexes requires all the components of the protease. Eur J Biochem 229(1):276–283

Emoto Y, Manome Y, Meinhardt G, Kisaki H, Kharbanda S, Robertson M, Ghayur T, Wong WW, Kamen R, Weichselbaum R, Kufe D (1995) Proteolytic activation of protein kinase C delta by an ICE-like protease in apoptotic cells. EMBO J 14(24):6148–6156

Fan CM, Maniatis T (1991) Generation of p50 subunit of NF-kappa B by processing of p105 through an ATP-dependent pathway. Nature 354(6352):395–398

Fenteany G, Standaert RF, Lane WS, Choi S, Corey EJ, Schreiber SL (1995) Inhibition of proteasome activities and subunit-specific amino-terminal threonine modification by lactacystin. Science 268(5211):726–731

Finco TS, Beg AA, Baldwin AS Jr (1994) Inducible phosphorylation of I kappa B alpha is not sufficient for its dissociation from NF-kappa B and is inhibited by protease inhibitors. Proc Natl Acad Sci USA 91(25):11884–11888

Fujita E, Mukasa T, Tsukahara T, Arahata K, Ōmura S, Momoi T (1996) Enhancement of CPP32-like activity in the TNF-treated U937 cells by the proteasome inhibitors. Biochem Biophys Res Commun 224(1):74–79

Gaczynska M, Rock KL, Goldberg AL (1993) Gamma-interferon and expression of MHC genes regulate peptide hydroysis by proteasomes. Nature 365(6443):264–267

Ganoth D, Levinsky E, Eytan E, Hershko A (1988) A multicomponent system that degrades proteins conjugated to ubiquitin. Resolution of factors and evidence for ATP-dependent complex formation. J Biol Chem 263(25):12412–12419

Ghislain M, Udvardy A, Mann C (1993) S. cerevisiae 26S protease mutants arrest cell division in G2/metaphase. Nature 366(6453):358–362

Gray CW, Slaughter CA, DeMartino GN (1994) PA28 activator protein forms regulatory caps on proteasome stacked rings. J Mol Biol 236(1):7–15

Grimm LM, Goldberg AL, Poirier GG, Schwartz LM, Osborne BA (1996) Proteasomes play an essential role in thymocyte apoptosis. EMBO J 15(15):3835–3844

Groll M, Ditzel L, Löwe J, Stock D, Bochtler M, Bartunik HD, Huber R (1997) Structure of 20S proteasome from yeast at 2.4 Å resolution. Nature 386(6624):463–471

Hébert L, Pandey S, Wang E (1994) Commitment to cell death is signaled by the appearance of a terminin protein of 30 kDa. Exp Cell Res 210(1):10–18

Henkel T, Machleidt T, Alkalay I, Kronke M, Ben-Neriah Y, Baeuerle PA (1993) Rapid proteolysis of I kappa B-alpha is necessary for activation of transcription factor NF-kappa B. Nature 365(6442):182–185

Hershko A, Ciechanover A (1992) The ubiquitin system for protein degradation. Annu Rev Biochem 61:761–807

Hochstrasser M (1995) Ubiquitin, proteasomes, and the regulation of intracellular protein degradation. Curr Opin Cell Biol 7:215–223

Imajoh-Ohmi S, Kawaguchi T, Shinji S, Tanaka K, Ōmura S, Kikuchi H (1995) Lactacystin, a specific inhibitor of the proteasome, induces apoptosis in human monoblast U937 cells. Biochem Biophys Res Commun 217(3):1070–1077

Jones ME, Haire MF, Kloetzel P-M, Mykles DL, Schwartz LM (1995) Changes in the structure and function of the multicatalytic proteinase (proteasome) during programmed cell death in the intersegmental muscles of the hawkmoth, Manduca sexta. Dev Biol 169(2):436–447

Kaufmann SH (1989) Induction of endonucleolytic DNA cleavage in human acute myelogenous leukemia cells by etoposide, camptothesin, and other cytotoxic anticancer drugs: a cautionary note. Cancer Res 49(21):5870–5878

Kaufmann SH, Desnoyers S, Ottaviano Y, Davidson NE, Poirier GG (1993) Specific proteolytic cleavage of poly (ADP-ribose) polymerase: an early marker of chemotherapy-induced apoptosis. Cancer Res 53(17):3976–3985

Kayalar C, Ord T, Testa MP, Zhong LT, Bredesen DE (1996) Cleavage of actin by interleukin 1 beta-converting enzyme to reverse Dnase I inhibition. Proc Natl Acad Sci USA 93(5):2234–2238

King RW, Glotzer M, Kirschner MW (1996) Mutagenic analysis of the destruction signal of mitotic cyclins and structural characterization of ubiquitinated intermediates. Mol Biol Cell 7(9):1343–1357

Klotzbucher A, Stewart E, Harrison D, Hunt T (1996) The "destruction box" of cyclin A allows B-type cyclins to be ubiquitinated, but not efficiently destroyed. EMBO J 15(12):3053–3064

Lam YA, Xu W, DeMartino GN, Cohen RE (1997) Editing of ubiquitin conjugates by an isopeptidase in the 26S proteasome. Nature 385(6618):737–740

Li X, Stebbins B, Hoffman L, Pratt G, Rechsteiner M, Coffino P (1996) The N terminus of antizyme promotes degradation of heterologous proteins. J Biol Chem 271(8):4441–4446

Lin YC, Brown K, Siebenlist U (1995) Activation of NF-kappa B requires proteolysis of the inhibitor I kappa B-alpha: signal-induced phosphorylation of I kappa B-alpha alone does not release active NF-kappa B. Proc Natl Acad Sci USA 92(2):552–556

Löw P, Bussell K, Dawson SP, Billett MA, Mayer RJ, Reynolds SE (1997) Expression of a 26S proteasome ATPase subunit, MS73, in muscles that undergo developmentally programmed cell death, and its control by ecdysteroid hormones in the insect *Manduca sexta*. FEBS Lett 400(3):345–349

Löwe J, Stock D, Jap B, Zwickl P, Baumeister W, Huber R (1995) Crystal structure of the 20S proteasome from the archaeon *T. acidophilum* at 3.4 Å resolution. Science 268(5210):533–539

Lupas A, Koster AJ, Baumeister W (1993) Structural features of 26S and 20S proteasomes. Enzyme Protein 47(4–6):252–273

Machiels BM, Henfling MER, Schutte B, van Engeland M, Broers JLV, Ramaekers FCS (1996) Subcellular localization of proteasomes in apoptotic lung tumor cells and persistence as compared to intermediate filaments. Eur J Cell Biol 70(3):250–259

Mashima T, Naito M, Fujita N, Noguchi K, Tsuruo T (1995) Identification of actin as a substrate of ICE and an ICE-like protease and involvement of an ICE-like protease but not ICE in VP-16-induced U937 apoptosis. Biochem Biophys Res Commun 217(3):1185–1192

Martin SJ, O'Brien GA, Nishioka WK, McGahon AJ, Mahboubi A, Saido TC, Green DR (1995b) Proteolysis of fodrin (non-erythroid spectrin) during apoptosis. J Biol Chem 270:6425–6428

Meerovitch K, Wing S, Goltzman D (1997) Preproparathyroid hormone-related protein, a secreted peptide, is a substrate for the ubiquitin proteolytic system. J Biol Chem 272(10):6706–6713

Murakami Y, Matsufuji S, Kameji T, Hayashi S, Igarashi K, Tamura T, Tanaka K, Ichihara A (1992) Ornithine decarboxylase is degraded by the 26S proteasome without ubiquitination. Nature 360(6404):597–599

Murakami Y, Matsufuji S, Tanaka K, Ichihara A, Hayashi S (1993) Involvement of the proteasome and antizyme in ornithine decarboxylase degradation by a reticulocyte lysate. Biochem J 295(Pt1):305–308

Murakami Y, Tanahashi N, Tanaka K, Ōmura S, Hayashi S (1996) Proteasome pathway operates for the degradation of ornithine decarboxylase in intact cells. Biochem J 317(Pt1):77–80

Musti AM, Treier M, Peverali FA, Bohmann D (1996) Differential regulation of c-Jun and Jun D by ubiquitin-dependent protein degradation. Biol Chem 377(10):619–624

Neamati N, Fernandez A, Wright S, Kiefer J, McConkey DJ (1995) Degradation of lamin B1 precedes oligonucleosomal DNA fragmentation in apoptotic thymocytes and isolated thymocyte nuclei. J Immunol 154(8):3788–3795

Oberhammer FA, Hochegger K, Fröscl G, Tiefenbacher R, Pavelka M (1994) Chromatin condensation during apoptosis is accompanied by degradation of lamin A + B, without enhanced activation of cdc 2 kinase. J Cell Biol 126(4):827–837

Ōmura S, Matsuzaki K, Fujimoto T, Kosuge K, Furuya T, Fujita S, Nakagawa A (1991) Structure of lactacystin, a new microbial metabolite which induces differentiation of neuroblastoma cells. J Antibiot (Tokyo) 44(1):117–118

Orino E, Tanuka K, Tamura T, Stone S, Ogura T, Ichihara A (1991) ATP-dependent reversible association of proteasomes with multiple protein components to form 26S complexes that degrade ubiquitinated proteins in human HL-60 cells. FEBS Lett 284(2):206–210

Orlowski M, Cardozo C, Michaud C (1993) Evidence for the presence of five distinct proteolytic components in the pituitary multicatalytic proteinase complex. Properties of two components cleaving bonds on the carboxyl side of branched chain and small neutral amino acids. Biochemistry 32(6):1563–1572

Palombella VJ, Rando OJ, Goldberg AL, Maniatis T (1994) The ubiquitin–proteasome pathway is required for processing the NF-kappa B1 precursor protein and activation of NF-kappa B. Cell 78(5):773–785

Peters JM, Cejka Z, Harris JR, Kleinschmidt JA, Baumeister W (1993) Structural features of the 26S proteasome complex. J Mol Biol 234(4):932–937

Pitzer F, Dantes A, Fuchs T, Baumeister W, Amsterdam A (1996) Removal of proteasomes from the nucleus and their accumulation in apoptotic blebs during programmed cell death. FEBS Lett 394(1):47–50

Rivett AJ (1989) The multicatalytic proteinase. Multiple proteolytic activities. J Biol Chem 264(21):12215–12219

Rock KL, Gramm C, Rothstein L, Clark K, Stein R, Dick L, Hwang D, Goldberg AL (1994) Inhibitors of the proteasome block the degradation of most cell proteins and the generation of peptides presented on MHC class I molecules. Cell 78(5):761–771

Rohrwild M, Coux O, Huang HC, Moerschell RP, Yoo SJ, Seol JH, Chung CH, Goldberg AL (1996) HslV-HslU: a novel ATP-dependent protease complex in Escherichia coli related to the eukaryotic proteasome. Proc Natl Acad Sci USA 93(12):5808–5813

Rohrwild M, Pfeifer G, Santarius U, Muller SA, Huang HC, Engel A, Baumeister W, Goldberg AL (1997) The ATP-dependent HslVU protease from Escherichia coli is a four-ring structure resembling the proteasome. Nat Struct Biol 4(2):133–139

Sadoul R, Fernandez PA, Wuiquerez AL, Martinou I, Maki M, Schroter M, Becherer JD, Irmler M, Tschopp J, Martinou JC (1996) Involvement of the proteasome in the programmed cell death of NGF-deprived sympathetic neurons. EMBO J 15(15):3845–3852

Schwartz LM, Myer A, Kosz L, Engelstein M, Maier C (1990) Activation of polyubiquitin gene expression during developmentally programmed cell death. Neuron 5(4):411–419

Seemüller E, Lupas A, Stock D, Löwe J, Huber R, Baumeister W (1995) Proteasome from Thermoplasma acidophilum: a threonine protease. Science 268(5210):579–582

Shinohara K, Tomioka M, Nakano H, Toné S, Ito H, Kawashima S (1996) Apoptosis induction resulting from proteasome inhibition. Biochem J 317(Pt2):385–388

Song Q, Lees-Miller SP, Kumar S, Zhang Z, Chan DW, Smith GC, Jackson SP, Alnemri ES, Litwack G, Channa KK, Lavin MF (1996) DNA-dependent protein kinase catalytic subunit: a target for an ICE-like protease in apoptosis. EMBO J 15(13):3238–3246

Song Q, Wei T, Lees-Miller S, Alnemri E, Watters D, Lavin MF (1997) Resistance of actin to cleavage during apoptosis. Proc Natl Acad Sci USA 94(1):157–162

Tanaka K, Yoshimura T, Tamura T, Fujiwara T, Kumatori A, Ichihara A (1990) Possible mechanism of nuclear translocation of proteasomes. FEBS Lett 271(1–2):41–46

Thornberry NA, Bull HG, Calaycay JR, Chapman KT, Howard AD, Kostura MJ, Miller DK, Molineaux SM, Weidner JR, Aunins J, Elliston KO, Ayala JM, Casano FJ, Chin J, Ding GJ-F, Egger LA, Gaffney EP, Limjuco G, Palyha OC, Raju SM, Rolando AM, Salley JP, Yamin T-T, Lee TD, Shively JE, MacCross M, Mumford RA, Schmidt JA, Tocci MJ (1992) A novel heterodimeric cysteine protease is required for interleukin-1β processing in monocytes. Nature 356(6372):768–774

Tokunaga F, Goto T, Koide T, Murakami Y, Hayashi S, Tamura T, Tanaka K, Ichihara A (1994) ATP- and antizyme-dependent endoproteolysis of ornithine decarboxylase to oligopeptides by the 26S proteasome. J Biol Chem 269(26):17382–17385

Traenckner EB, Wilk S, Baeuerle PA (1994) A proteasome inhibitor prevents activation of NF-kappa B and stabilizes a newly phosphorylated form of I kappa B-alpha that is still bound to NF-kappa B. EMBO J 13(22):5433–5441

Treier M, Staszewski LM, Bohmann D (1994) Ubiquitin-dependent c-Jun degradation in vivo is mediated by the delta domain. Cell 78(5):787–798

Van Antwerp DJ, Verma IM (1996) Signal-induced degradation of I(kappa)B(alpha): association with NF-kappa B and the PEST sequence in I(kappa)B(alpha) are not required. Mol Cell Biol 16(11):6037–6045

Vinitsky A, Michaud C, Powers JC, Orlowski M (1992) Inhibition of the chymotrypsin-like activity of the pituitary multicatalytic proteinase complex. Biochemistry 31(39):9421–9428

Voelkel-Johnson C, Entingh AJ, Wold WSM, Gooding L-R, Laster SM (1995) Activation of intracellular proteases is an early event in TNF-induced apoptosis. J Immunol 154(4):1707–1716

Weil R, Laurent-Winter C, Israel A (1997) Regulation of I kappa B beta degradation. Similarities to and differences from I kappa B alpha. J Biol Chem 272(15):9942–9949

Wenzel T, Eckerskorn C, Lottspeich F, Baumeister W (1994) Existence of a molecular ruler in proteasomes suggested by analysis of degradation products. FEBS Lett 349(2):205–209

Whiteside ST, Ernst MK, LeBail O, Laurent-Winter C, Rice N, Israel A (1995) N- and C-terminal sequences control degradation of MAD3/I kappa B alpha in response to inducers of NF-kappa B activity. Mol Cell Biol 15(10):5339–5345

Yoo SJ, Seol JH, Shin DH, Rohrwild M, Kang MS, Tanaka K, Goldberg AL, Chung CH (1996) Purification and characterization of the heat shock proteins HslV and HslU that form a new ATP-dependent protease in *Escherichia coli*. J Biol Chem 271(24):14035–14040

Yoshimura T, Kameyama K, Takagi T, Ikai A, Tokunaga F, Koide T, Tanahashi N, Tamura T, Cejka Z, Baumeister W, Tanaka K, Ichihara A (1993) Molecular characterization of the "26S" proteasome complex from rat liver. J Struct Biol 111(3):200–211

Yuan J, Shaham S, Ledoux S, Ellis H, Horvitz HR (1993) The *C. elegans* cell death gene *ced*-3 encodes a protein similar to mammalian interleukin-1β-converting enzyme. Cell 75(4):641–652

Zwickl P, Lottspeich F, Dahlmann B, Baumeister W (1991) Cloning and sequencing of the gene encoding the large (alpha-) subunit of the proteasome from *Thermoplasma acidophilum*. FEBS Lett 278(2):217–221

Zwickl P, Kleinz J, Baumeister W (1994) Critical elements in proteasome assembly. Nat Struct Biol 1(11):765–770

Subject Index

Springer
and the
environment

At Springer we firmly believe that an
international science publisher has a
special obligation to the environment,
and our corporate policies consistently
reflect this conviction.
We also expect our business partners –
paper mills, printers, packaging
manufacturers, etc. – to commit
themselves to using materials and
production processes that do not harm
the environment. The paper in this
book is made from low- or no-chlorine
pulp and is acid free, in conformance
with international standards for paper
permanency.

 Springer